Fracture Mechanics: Fundamentals and Applications

Fracture Mechanics: Fundamentals and Applications

Editor: Jeff Campbell

MURPHY & MOORE
www.murphy-moorepublishing.com

Published by Murphy & Moore Publishing
1 Rockefeller Plaza,
New York City, NY 10020, USA
www.murphy-moorepublishing.com

Fracture Mechanics: Fundamentals and Applications
Edited by Jeff Campbell

International Standard Book Number: 978-1-63987-240-4 (Hardback)

Cataloging-in-Publication Data

Fracture mechanics : fundamentals and applications / edited by Jeff Campbell.
p. cm.
Includes bibliographical references and index.
ISBN 978-1-63987-240-4
1. Fracture mechanics. 2. Deformations (Mechanics). 3. Strength of materials. I. Campbell, Jeff.
TA409 .F73 2022
620.112 6--dc23

Contents

Preface VII

Chapter 1 **Laboratory Investigations of Hydraulic Fractures in Naturally Fractured Formations** 1
Guangqing Zhang and Tiegang Fan

Chapter 2 **Fracture Mechanics in Adhesive Dentistry** 29
Mügem Aslı Gürel, Bağdagül Helvacıoğlu Kıvanç and Hacer Deniz Arısu

Chapter 3 **Toughness Assessment and Fracture Mechanism of Brittle Thin Films Under Nano-Indentation** 53
Kunkun Fu, Youhong Tang and Li Chang

Chapter 4 **Fracture Toughness of Ferritic Steels in the Ductile-to-Brittle Transition Region** 77
Carlos Berejnoi and Juan E. Perez Ipiña

Chapter 5 **The Strain Energy Release Rate for Stack of Coated Conductors with Interface Crack in Perpendicular Magnetic Field** 96
Meng Zhao, Huadong Yong and Youhe Zhou

Chapter 6 **Investigation of Mechanical Behaviour of a Bioceramic** 115
Achouak Elghazel, Rym Taktak and Jamel Bouaziz

Chapter 7 **Fatigue Crack Propagation Rates Prediction Using Probabilistic Strain-Based Models** 135
José António Fonseca De Oliveira Correia, Abílio M.P. De Jesus, Pedro M.G.P. Moreira, Rui A.B. Calçada and Alfonso Fernández-Canteli

Chapter 8 **Toughening of Low-Alloy Steel by Ultrafine-Grained Structure (Development of Fracture Control from Microstructure Design)** 166
Tadanobu Inoue

Chapter 9 **On the Fractography of Impact-Tested Samples of Al-Si Alloys for Automotive Alloys** 184
Zheyuan Ma, Agnes M. Samuel, Herbert W. Doty and Fawzy H. Samuel

Chapter 10 **Fractography on Rigid Ceramics with Ultra-High-Molecular-Weight Polyethylene Fabric after Ballistic Impacts** 214
Giovanni Barrera Torres, Juan Manuel Vélez and Carlos Mario Gutierrez

Permissions

List of Contributors

Index

Preface

The field of mechanics which focuses on the study of the propagation of cracks in materials is referred as fracture mechanics. It comprises the analysis of flaws that is used to discover which structures are safe and which ones are liable to propagate as cracks and thereby, resulting in the failure of the flawed structure. Safe operation of a structure can be achieved through damage tolerance analysis. The two kinds of fracture mechanics are linear elastic fracture mechanics and elastic-plastic fracture mechanics. Linear elastic fracture mechanics covers the concepts of Griffith's criterion, Irwin's modification, stress intensity factor, strain energy release and crack tip plastic zone. Elastic-plastic fracture mechanics focuses upon the concepts of crack tip opening displacement (CTOD), R-curve, J-integral, cohesive zone model and transitions flaw size. This book will also provide interesting topics for research which interested readers can take up. It studies, analyses and upholds the pillars of fracture mechanics and its utmost significance in modern times. This book includes contributions of experts and scientists which will provide innovative insights into this field.

This book has been the outcome of endless efforts put in by authors and researchers on various issues and topics within the field. The book is a comprehensive collection of significant researches that are addressed in a variety of chapters. It will surely enhance the knowledge of the field among readers across the globe.

It gives us an immense pleasure to thank our researchers and authors for their efforts to submit their piece of writing before the deadlines. Finally in the end, I would like to thank my family and colleagues who have been a great source of inspiration and support.

Editor

1

Laboratory Investigations of Hydraulic Fractures in Naturally Fractured Formations

Guangqing Zhang and Tiegang Fan

Abstract

Researchers have recently realized that hydraulic fracture networks are significant for the exploitation of unconventional reservoirs (tight gas, shale gas, coalbed methane, etc.). Laboratory experiments are proposed to study the evolution of fracture networks in natural fractures. The density of natural fractures and injection rate were selected as the primary dominating factors. It was concluded that the interaction between reopening and connecting of natural fractures led to hydraulic fracture networks, which can be detected by pressure fluctuations. The fracture network is composed of multiple fractures, resembling an ellipsoid with the major axis different from the maximum horizontal stress direction. The treatment pressure will be substantially raised by both the great natural fracture density and high injection rates.

Keywords: hydraulic fractures, naturally fractured formation, fracture initiation, interferences

1. Introduction

Natural fractures have substantial effects on hydraulic fracture propagation, since hydraulic fractures will interact with the natural fractures during propagation, leading to complex growth patterns of fracture systems.

Researchers employed both experimental [1–5] and numerical [6–12] means to investigate the interaction between hydraulic and natural fractures. It is found that the controlling factors include horizontal stress difference, the angle of approaching, and treatment pressure [2–4]. The interaction between hydraulic fracture and natural fractures can be categorized into four types. Hydraulic fracture can cross natural fractures, as if natural fractures do not exist. The hydraulic fracture may be completely hindered or arrested at the natural fractures, either by opening natural fractures or causing both faces slipping. Or, hydraulic fractures may start new fractures to form fracture branches. And hydraulic fracture might follow natural fractures for a short distance and initiate a new fracture on natural fractures along the maximum stress direction [1, 2, 13–15]. Multiple hydraulic fractures were also observed in field experiments [16–19], inducing abnormally high treating pressure and implying reduced fracture width. The reduced width of each fracture is caused by competition among multiple fractures [9].

The effect of natural fractures on hydraulic fracture was realized and investigated very early [3, 20, 21], with the emphasis on a single natural fracture. De Pater and Beugelsdijk [22] investigated the effect of the injection rate on fracture propagation, with Portland cement model blocks containing heating-induced prefractures. The tests showed that a high flow rate or viscosity was necessary for generating new fractures. Zhou and Xue [23] conducted similar tests, and they found three types of fracture geometries: a vertical main fracture with various branches; radial fractures around a wellbore; and one wing vertical fracture with branches. However, heating-induced fractures are not easy to control to simulate natural fractures, thus causing much uncertainty in pinpointing the factors for hydraulic fractures. Olson et al. [24] used cast hydrostone blocks embedded with planar glass sheet to study the effects of cemented natural fractures on hydraulic fracture propagation. They found that oblique natural fractures will divert a hydraulic fracture more easily than the orthogonal natural fractures.

Laboratory experiments are considered as the direct way to investigate fracture propagation, because numerical simulation needs to be further improved in terms of constitutive modeling, boundary conditions. Meanwhile, it is very costly and hard to retrieve cores for research purposes, and the size of cores is usually too small to meet the needs of laboratory tests for hydraulic fracturing. Therefore, researchers tend to select outcrops and artificial specimens for laboratory tests of hydraulic fracturing. And homogeneous and continuous blocks are often used, in which no natural fractures are included. However, natural fractures are present frequently in formations, and in laboratory tests it seems necessary to consider the effects of natural fractures on hydraulic fracturing. Many researchers [4, 22–24] proposed different methods to take natural fractures into consideration.

Natural fractures seldom are found alone, however, several groups of natural fractures exist in reservoir formations. La Pointe and Hudson [25] presented methods for describing parameters of natural fractures, such as spacing, density, and trace length. As per Lachenbruch [26], there are two types of fracture intersection: orthogonal and nonorthogonal, which can be divided into three sets: all continuous; part continuous and part discontinuous; and all discontinuous. In this study, laboratory experiments were conducted with natural and artificial specimens, which contain natural fractures or prefractures to investigate the influence of natural fracture on the hydraulic fracture propagation behaviors.

In this chapter, laboratory work is conducted on specimens of coal, shale, and artificial hydrostone, for simulating hydraulic fracturing in naturally fractured reservoirs. The equipment used in the laboratory work is capable of applying independent stresses along three directions, while injecting pressurized fluid into the specimens. Hydraulic fractures induced during the experiments can be observed directly either by slicing the specimens or just splitting along the fractures.

2. Experimental setup and procedures

2.1. Experimental equipment

The hydraulic fracturing experiments reported here were conducted using a true tri-axial hydraulic fracturing test system (**Figure 1**).

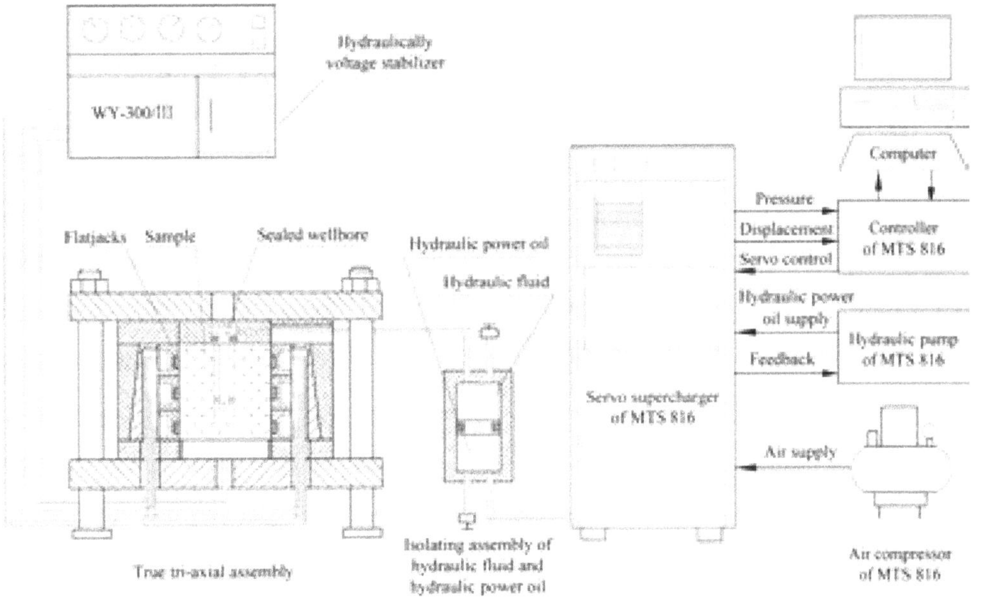

Figure 1. Schematic of the true tri-axial hydraulic fracturing test system.

All of the external stresses are applied using a set of hydraulic stabilizer, and the injection pressure is provided by a servo-controlled hydraulic pump of MTS 816. Up to 28 MPa of external stresses and injection pressures of up to 140 MPa can be supplied to cubic samples whose side is 300 or 400 mm long. Various injection scheme of pumping fracturing fluids can be supported. The experimental implementation and data collection are conducted using software on a PC computer. Various testing materials and fracturing fluids were used to initiate and propagate hydraulic fractures.

2.2. Experimental arrangements

2.2.1. Hydraulic fracturing experiment of coal

Fractures occur in nearly all coal seams. Coal fractures have been investigated since the early days of coal mining, and published descriptions and speculation on the origin of fractures date from early in the nineteenth century [27]. Over the years, many different terms were used for natural fractures in coal, people prefer to the ancient mining term: cleats [28].

In coal-seam, water and gas flow in the cleat system, which is composed of two or more sets of parallel or subparallel fractures. However, natural fractures are irregular discontinuities.

The coal specimens for the hydraulic fracturing were cut from the outcrop, Shenmu, China. Usually, the blocks from outcrops should be much larger in size than the required, for there are many weak discontinuities inside. Therefore, the coal blocks were cut and finished with a saw-cutter into cubes of 300 mm on each side.

After being cut, further treatments of polishing the surface are needed to fit the loading frame dimensions, and a borehole of 20 mm in diameter, 160 mm in depth is prepared, before a metal casing of 10 mm in diameter is inserted and cemented. An open borehole of 200 mm interval at the end of the block should be left for easy fracture initiation.

During testing, the true tri-axial equipment was used to load the cubic specimens by three independent flat jacks to simulate three orthogonal stresses. And thin Teflon sheets were also inserted between the specimen and the flat jacks, for reducing shearing stress [22]. A normal-faulting stress regime was adopted in our experiments, and the three orthogonal stresses were applied hydrostatically to the minimum stress level firstly, and the other two stresses were raised to the designed levels respectively (see **Table 1**). About 30 min were allowed to hold at the predefined stresses before the hydraulic fracturing tests begin [29].

Test	β^a degree	Vertical stress σ_v, MPa	Maximum horizontal stress σ_H, MPa	Minimum horizontal stress σ_h, MPa	Injection rate of the fracturing fluid Q, ml/min	Fracturing fluid[b]
1	5	18	7	5	20	Guar-based gel
2	7	18	7	5	20	Guar-based gel
3	10	18	7	5	20	Guar-based gel
4	11	18	7	5	20	Guar-based gel
5	20	18	7	5	20	Guar-based gel
6	39	18	7	5	20	Guar-based gel

[a]β is the angle between the face cleat and the maximum horizontal stress, degree.

[b]Fracturing fluid is mixed with luminous yellow fluorescent dye.

Table 1. Parameters used in the experiment for the Shenmu coal.

2.2.2. *Hydraulic fracturing experiment of shale*

The shale specimens for the tests were cored from outcrops of Longmaxi formation of shale in Sichuan, China. The outcrops obtained in the field were irregular in shape and could not be used directly in the experiments. Therefore, in the experiments cubic specimens of 400 mm in lateral were used, and broken parts along the boundary were repaired with cements. A hole of 18 mm in diameter was made up to 210 mm deep, and then a metal casing was put and cemented in the hole, leaving about 20 mm open interval for fracture initiation. **Figure 2** shows a picture of a specimen.

Figure 2. The shale specimen for hydraulic fracturing test.

For each experiment, fracturing fluid with luminous yellow fluorescent dye was injected at a constant flow rate (see **Table 2**). The fracturing fluid used was a guar-based gel mixed at 5 g gel powder/1000 ml water. This gel has an apparent viscosity of about 24.75 cP. The fluorescent dye is nonpenetrating and therefore shows the created fracture surface.

Test	β[a] degree	Vertical stress σ_v, MPa	Maximum horizontal stress σ_H, MPa	Minimum horizontal stress σ_h, MPa	Injection rate of the fracturing fluid Q, ml/min	Fracturing fluid[b]
1	–	15	12	10	10	Guar-based gel
2	–	20	15	13	10	Guar-based gel
3	–	20	15	13	20	Guar-based gel

[a]β is the angle between the face cleat and the maximum horizontal stress, degree.

[b]Fracturing fluid is mixed with luminous yellow fluorescent dye.

Table 2. Parameters used in the experiment for the Sichuan shale.

In the experiments, the friction between the specimen and flat jacks can be great, leading to substantial shear force on the boundary. In order to minimize the shear forces [22], a thin Teflon sheet covered on both sides with Vaseline was inserted between the block and flat jacks. After each test, observation of friction scratch on the block surface was necessary for evaluation of shear force along the boundary. Before testing, the confining stresses along the three orthogonal directions were kept for about 30 min, allowing stress equilibrium inside the specimen [29].

After a test, the specimen was opened with a hammer and chisel along the expected orientation. Under ultraviolet light, the luminous yellow fluorescent dye can clearly show the traces of hydraulic fracture.

2.2.3. *Hydraulic fracturing experiment of artificial specimens*

In this part, we used paper as the separating agent embedding the blocks for simulating a single natural fracture. There were types of paper: type one is rice paper of 0.06 mm thick, type two printing paper of 0.11 mm thick, and type 3 wrapping paper of 1.2 mm thick.

When the hydraulic fracture was approaching a natural fracture, the interaction between the hydraulic fracture and the natural fracture was determined by several factors, including the approaching angle, stress difference, and net pressure inside the hydraulic fracture. All the experiments were conducted with vertical wellbores. And horizontal stress difference along with stress regime was designed from the beginning of the tests. For each test conditions, three tests were performed to ensure the repeatability of the results.

In fracturing tests, red-dyed water was injected at a constant flow rate (see **Table 3**). The dye is nonpenetrating and therefore highlights the fracture surface generated by the experiment.

Test	Fracture spacing (mm)[a]	Vertical stress σ_v, MPa	Maximum horizontal stress σ_H, MPa	Minimum horizontal stress σ_h, MPa	Injection rate of the fracturing fluid Q, ml/min	Fracturing fluid
1	30	15	8	5	6	Red-dyed water
2	35	15	8	5	6	Red-dyed water
3	40	15	8	5	9.6	Red-dyed water
4	45	15	8	5	9.6	Red-dyed water
5	50	15	8	5	9.6	Red-dyed water
6	40	15	8	5	19.8	Red-dyed water

[a]The distance between two adjacent cemented fractures in the specimens, mm.

Table 3. Parameters used in the experiment for the artificial cement paste.

Six cubic specimens of 300 mm in lateral were used in the experiments, made out of a mixture of cement and 20–40 mesh siliceous sand. Small cuboid blocks of various sizes were prepared to assemble into a final specimen. And the scheme is shown in **Table 4**.

Test	Size (length × width × height)[a]	Amount/set[b]	Sets[c]	Amount[d]
1	30 mm × 30 mm × 200 mm	49	1	49
2	35 mm × 35 mm × 200 mm	49	1	49
3	40 mm × 40 mm × 200 mm	25	2	50
4	45 mm × 45 mm × 200 mm	25	1	25
5	50 mm × 50 mm × 200 mm	25	1	25

[a]The three dimensions of the cuboid blocks.

[b]The amount of cuboid blocks in one set.

[c]The amount of sets which composed by a certain number of cuboid blocks.

[d]The amount of cuboid blocks which prepared for tests with a certain size.

Table 4. Experimental parameters and results of the artificial specimens with continuous orthogonal prefractures.

When simulating closed natural fractures, cement paste was used to glue neighboring small blocks to form a big cuboid block. And mechanical properties were obtained through standard test procedures. For the open natural fractures, the small blocks were left intact, and they were simply arranged according to the designed scheme. Then the big cuboid block was put in the central part of the metal mould, with the internal dimensions of 300 mm × 300 mm × 300 mm. The cement paste between the small blocks was weaker than the small blocks, which can be seen as a group of continuous cemented fractures, perpendicular to each other. **Figure 3** shows a schematic of the facture system.

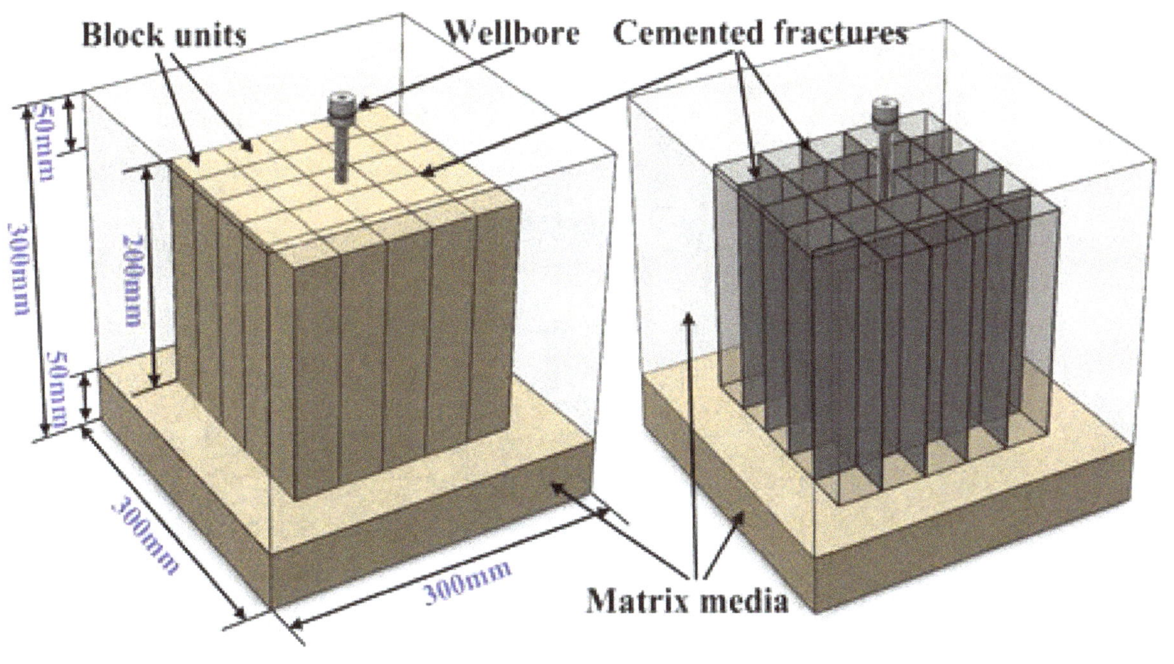

Figure 3. Schematics of the fracture system.

3. Results

3.1. Experiment results of coal

The geometry of the hydraulic fractures was observed and analyzed by cutting the specimens after each test. The results showed a similar feature compared to that in naturally fractured rocks, however, hydraulic fractures in coal seam may initiate from opening cleats intersected by the wellbore.

3.1.1. Initiation from cleats

A main vertical fracture was usually observed along the maximum horizontal stress, as shown by specimen C1. And the fracture was also rather straight and planar overall. The fracture surface was quite rough (**Figure 4(a)**), caused by a large number of cleats and microfracture.

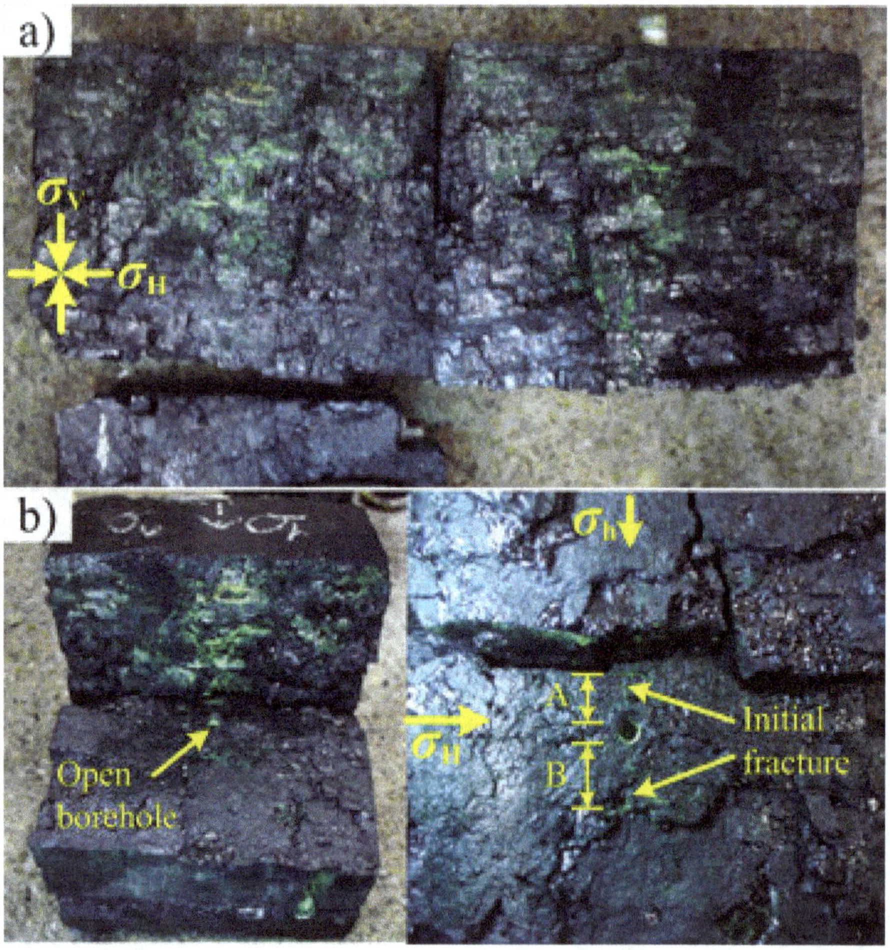

Figure 4. The main fracture (a) and cross-section (b) of specimen C1.

Figure 4(b) shows the photo of a specimen after testing. The specimen was split apart along the horizontal plane. And the vertical plane of the hydraulic fracture did not cross the borehole

straight. The hydraulic fractures did not start along the maximum horizontal stress, instead the minimum horizontal stress direction was followed. Fracture section A met a face cleat when it was 30 mm away from the one side, then it coincided the face cleat. Fracture section B just propagated along its initial direction and did not meet any face cleats and the propagation halted after fracture section A connected the face cleat.

3.1.2. Hydraulic fracture initiation in coal matrix

In specimen C2, no cleats and cracks around the borehole were found, thus the hydraulic fracture started in the coal matrix along the maximum horizontal stress.

In **Figure 5**, the specimen was split for direct observation with the vertical main fracture of elliptic area along the maximum horizontal stress around the borehole. To the left, the hydraulic fracture met a face cleat about 60 mm away from the wellbore and then it extended along the face cleat surface until the boundary of the specimen was reached. To the right, the hydraulic fracture propagated approximately 90 mm.

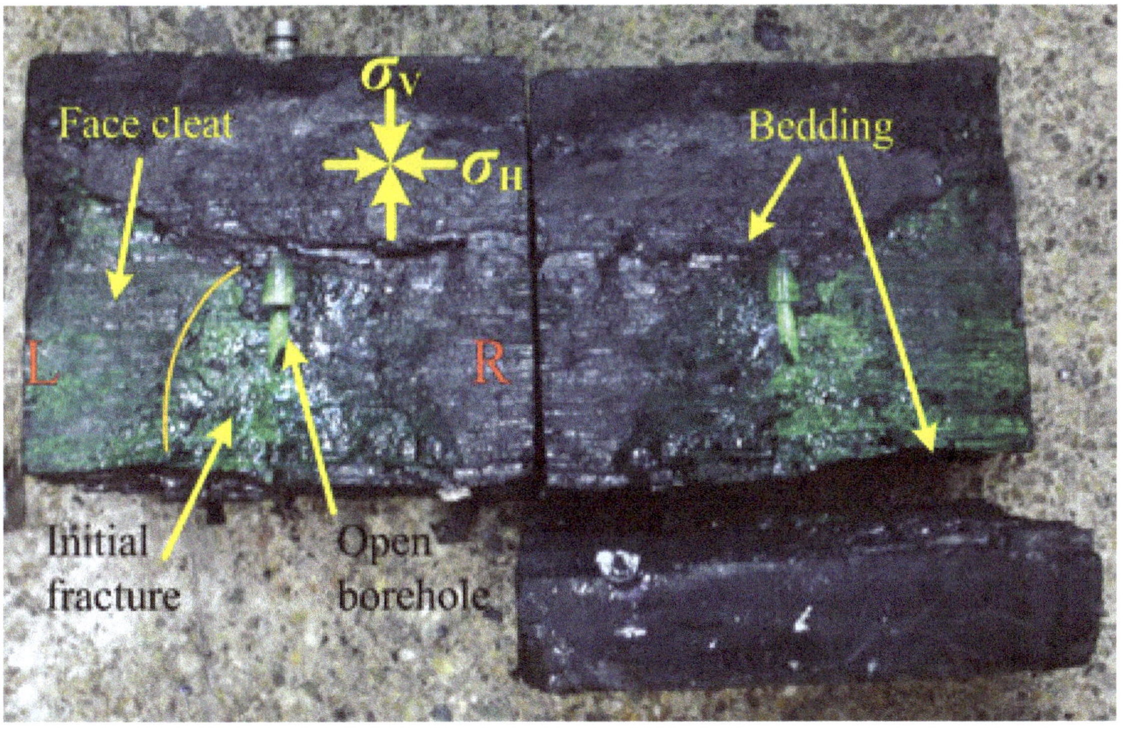

Figure 5. The controlled fracture of specimen C2.

The hydraulic fracture height was confined by bedding during propagation. The fractures were mainly distributed in the range of 60–200 mm away from the bottom of the specimen.

The fracture geometry characteristics of specimen C5 are almost the same as specimen C2, except that both wings of the fracture encountered the external surface and multiple fractures initiated at the borehole (see **Figure 6**).

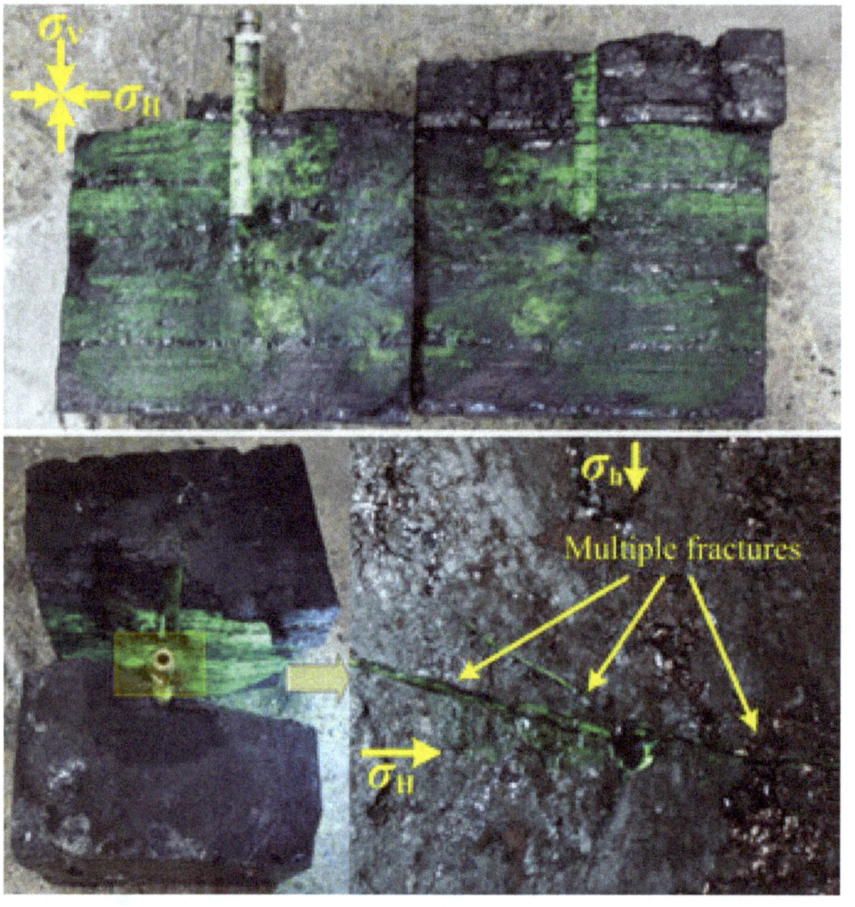

Figure 6. Multiple fractures around the borehole of specimen C5.

3.1.3. Reorientation of hydraulic fractures

When a vertical fracture initiates along the direction other than the maximum horizontal stress, the fracture will change its direction gradually to the maximum horizontal stress, due to strong control by the stress difference. Such phenomenon also was overserved in cases of hydraulic fracturing of coal seam tests.

The hydraulic fractures are shown in **Figure 7** for specimen C3, after being opened along the horizontal directions. It was obverses that there was a hydraulic fracture of S shape.

At the upper part, there were two fractures which propagated initially subparallel and subsequently coincided into one fracture. Both fractures connected the cleats during reorientation. The hydraulic fracture ultimately turned to the maximum horizontal stress and reached the specimen boundary. At last, the width of the fracture network increased to 45 mm.

At the lower part, there are three hydraulic fractures at the wellbore, one of which propagated 50 mm, and the other two intersect a cleat and extend straight along the cleat to the specimen boundary. There is 59° apart between the opened cleat and the maximum horizontal stress.

Figure 7. View of cross-section of specimen C3.

3.1.4. *The influence of cleats on hydraulic fractures*

The hydraulic fracturing of coal specimens shows complicated results, compared with the conventional single fracture. Usually, the specimens will be cut and split for visual examination

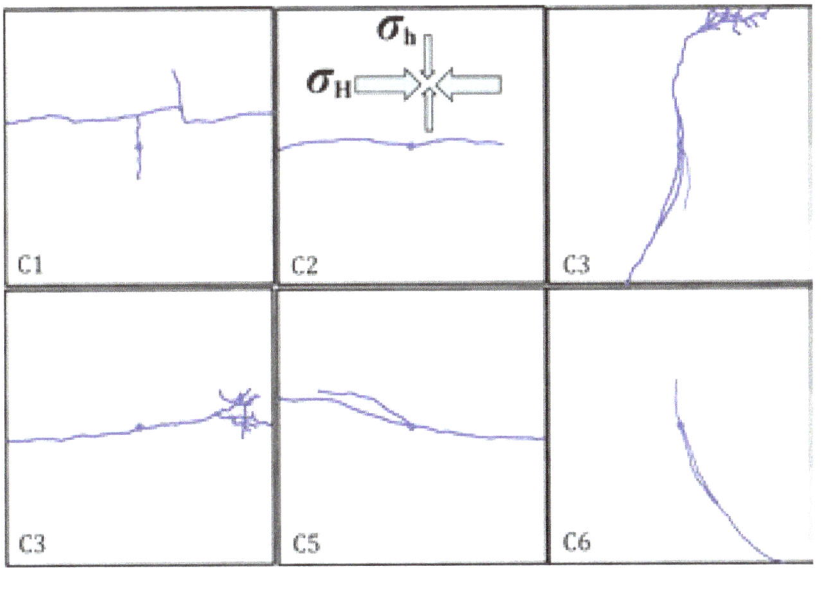

Figure 8. The distribution of the hydraulic fractures in the central horizontal plane for the tested specimens.

of the hydraulic fractures. **Figure 8** shows the hydraulic fracture patterns are very diverse, and they are not just a single planar fracture, but multiple fracture system connected to cleats.

The results of tests, in **Figure 8**, show that hydraulic fractures in coal seam can be started along directions other than the maximum horizontal stress. For example, hydraulic fractures could initiate from face or butt cleats, under the combined influences of stress state, coal mechanical properties, and borehole positions. There are three types of hydraulic fractures in coal seam.

Type 1. The hydraulic fracture extended along a face cleat (**Figure 9**). For example, type 1 includes specimen C4, where there is a long face cleat, intersecting the borehole. As in **Table 1**, the angle β is 11° and the hydraulic fracture started along an open cleat. Here, much low fluid pressure is required to open the cleat (Eq. (1)).

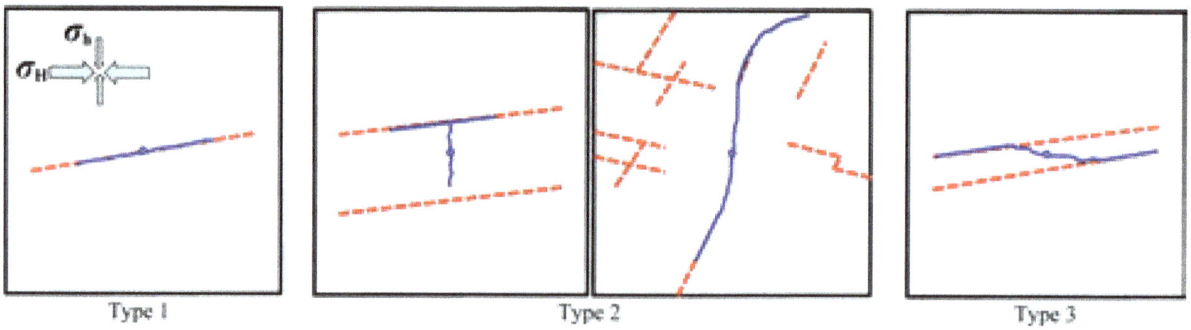

Figure 9. The observed three possible types of hydraulic fracture initiation and propagation patterns.

$$p_{b1} < p_{bE} = 3\sigma_h - \sigma_H + \sigma_t - p \tag{1}$$

where p_{b1} is the tested breakdown pressure for case 1 type systems; p_{bE} is the breakdown pressure predicted by the elastic model [30]; σ_h the minimum horizontal stress; σ_H the maximum horizontal stress; σ_t the tensile strength of coal; and p the pore pressure. The breakdown pressure for specimen C4 is 8.8 MPa.

Type 2. The hydraulic fracture started inside a butt cleat, almost parallel to the minimum horizontal stress, then propagated along a face cleat (**Figure 9**, type 2), including specimens C1, C3, and C6. The fluid pressure in the wellbore is higher than the breakdown pressure of type 1 (Eq. (2)), and not sufficient to create a new fracture.

$$p_{b1} < p_{b2} < p_{bE} = 3\sigma_h - \sigma_H + \sigma_t - p \tag{2}$$

where p_{b2} is the breakdown pressure for type 2.

For type 2, a new fracture can form along σ_H much easier than those along σ_h to propagate, if $3\sigma_h - \sigma_H + \sigma_t < 3\sigma_H - \sigma_h$.

Type 3. Figure 9 shows type 3 of fracture propagation (specimens C2 and C5), in which the hydraulic fracture initiated along the maximum stress, as depicted by hydraulic fracturing theory. If follows from **Figure 9** that the hydraulic fracture tends to extend the cleat if the propagation direction is close to the maximum horizontal stress (<20°). Hubbert and Willis [30] proposed a breakdown pressure model for impermeable rocks (Eq. (3))

$$p_{bE} = 3\sigma_h - \sigma_H + \sigma_t - p \tag{3}$$

For type 3, it is known that:

$$p_{b3} = p_{bE} \tag{4}$$

where p_{b3} is the measured breakdown pressure for case of type 3.

The pressure records during tests of specimens C2 and C5 give the breakdown pressure of 24.2 and 32.8 MPa. In our tests, pore pressure is not applied inside the specimens, and tensile strengths of 16.2 and 24.8 MPa can be determined for coal.

The tensile strengths determined above are much greater than those reported (0.2–3.7 MPa) for coal [31]. The reported tensile strengths were possibly obtained through rock mechanical tests (like Brazilian test or uniaxial tension test), with the specimens be limited in size, not many cleats included. As has been depicted for type 3, the tensile fracture initiated from the wellbore does not meet any cleats, implying initiation of hydraulic fracture on an intact surface. Furthermore, it has been recognized that coal shows significant scale effect, therefore the small-scale borehole (10 mm in diameter and 20 mm high) may also affect the tensile strength. Additional factors may contribute to the abnormal high breakdown pressures. However, there is currently insufficient research to address this question and it remains a topic for future investigation. Although the reason that our experiments derived such high tensile strengths is undetermined, it is evident that the breakdown pressure and derived tensile strength cannot be used directly at a field scale.

All of the above three types of hydraulic fracture initiation and propagation patterns indicate that the cleats near the borehole have a significant influence on the initiation of hydraulic fractures. When the fracturing fluid is continuously injected into the borehole and the fluid pressure is greater than the minimum fracture pressure of coal, cleats in coal open and probably leading to multiple hydraulic fractures near the borehole simultaneously. There may be uncertainty for the initial fracture orientation, and the final orientation is determined by cleat distribution and stress state. The initial fracture orientation is uncertain and determined by the distribution of the cleats on the borehole wall and the stress condition around the borehole. As the hydraulic fractures extend further, the number and appearance of the hydraulic fractures evolve, leading to multiple hydraulic fractures to the end.

There are many cleats in coal seam to be hydraulically fractured, which affect the propagation of hydraulic fractures. Following conventional hydraulic fracturing, the hydraulic fractures always extend along the maximum stress, where the least pressure is required. If cleats, in other directions, are connected by the hydraulic fractures, additional pressure is needed to prop those cleats, causing more cleats to be connected. At last, multiple hydraulic fractures will be induced, much like a network of hydraulic fractures in coal seam.

Since stress states are more responsible for the initiation and growth of the hydraulic fractures than cleats, the hydraulic fractures concentrate on a dominant direction, along with fracture branches of various complexities. If the approximate outline of such complex fractures in coal seam is made, an ellipse can be drawn on the horizontal plane, with the major axis following the maximum stress.

3.2. Experiment results of shale

3.2.1. Combination of vertical and horizontal fracture

Hydraulic fracturing experiments of shale specimens in laboratory were made to consider stress states, bedding, and natural fractures, and it is found that the hydraulic fractures of shales are usually combined by vertical, horizontal, and oblique fractures.

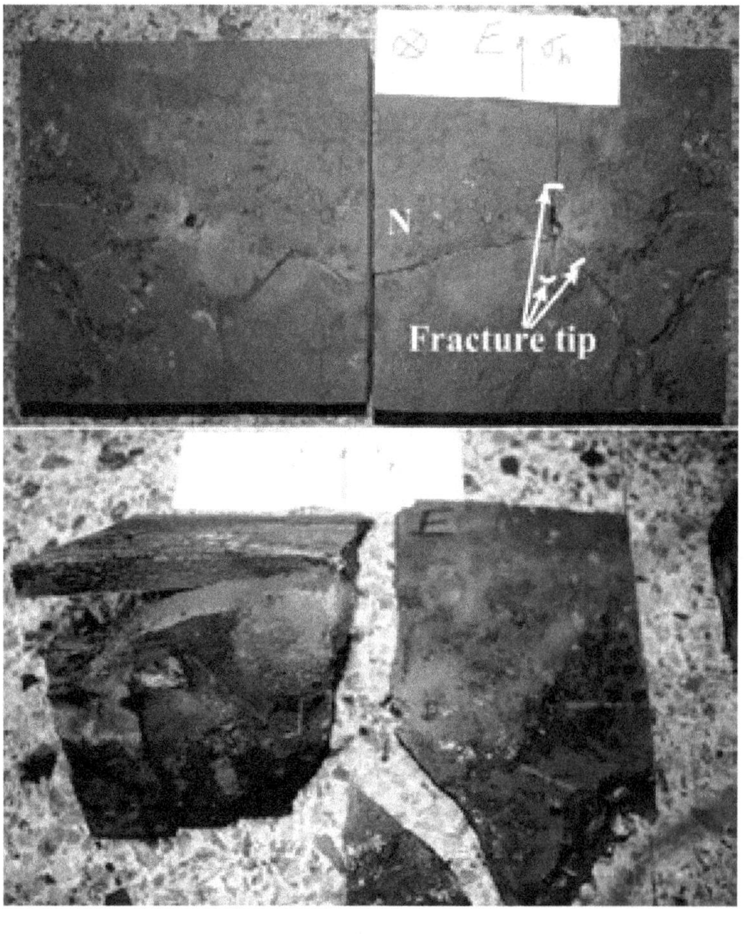

Figure 10. Hydraulic fractures of specimen S1.

Specimen S1 of shale after hydraulic fracturing is outlined in **Figure 10**. There are two sets of hydraulic fractures as shown in **Figure 10**, one set being multiple vertical fractures of various shapes and orientations. It is conceivable that one long fracture extends along the maximum horizontal stress, and three other small fractures along the minimum horizontal stress. The other set of hydraulic fractures in shale is following the beddings, or horizontal hydraulic fractures. It is not clear to tell how such fracture system developed in time only by visual checking the final fractures.

It is reasonable to postulate the possibilities of initiation and propagations of the multiple vertical and horizontal fractures. The hydraulic fractures are vertical on the wellbore at the beginning and extend to reach the beddings, where hydraulic fractures deviate from planar vertical to horizontal, and the beddings' effect becomes obvious.

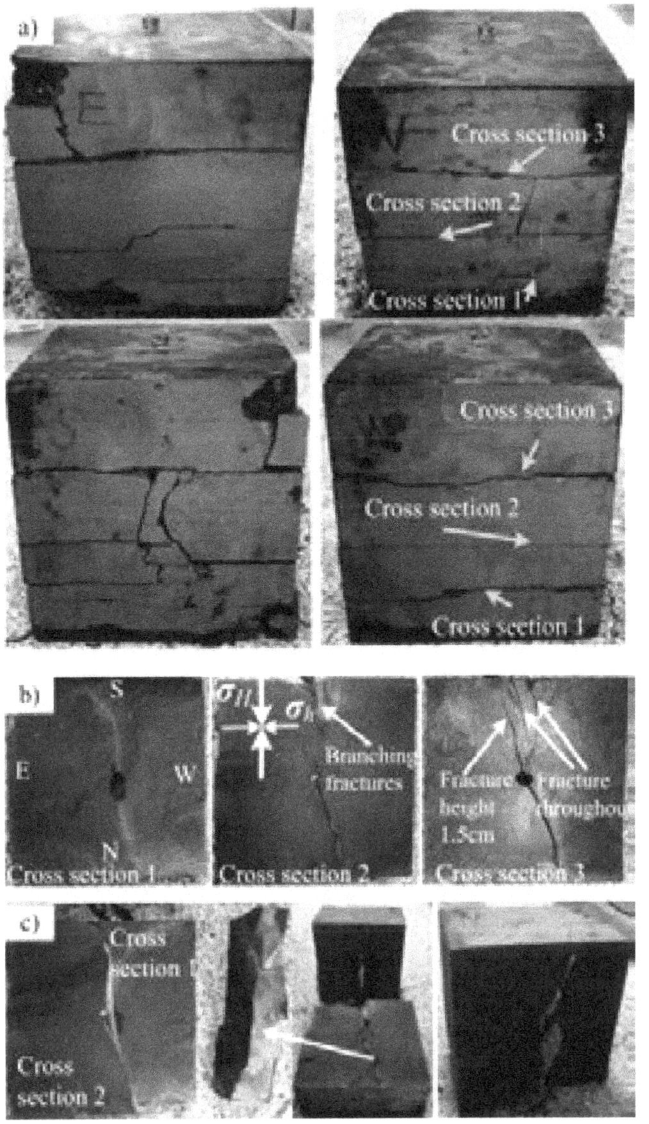

Figure 11. (a) The outside surfaces of specimen S2; (b) view of each cross-section; (c) the multifracture between cross-section 1 and cross-section 2.

3.2.2. *Multiple hydraulic fractures*

Multiple hydraulic fractures are also observed, shown in **Figure 11**. Along the beddings and horizontal fractures, the specimens are disintegrated for specifying hydraulic fractures. As depicted in **Figure 11**, fracture extension tends to extend in the maximum horizontal stress, in accordance with the least resistance. However, in south and north directions, there are multiple fractures, with much closely-spaced multiple fractures in the north direction. And the hydraulic fractures in the south are radiating from the wellbore and become subparallel later up to the side of the specimen, about 30 mm apart.

The hydraulic fractures in the north are sub-parallel with the maximum horizontal stress, with 20° of angle. Currently, stimulated reservoir volume is becoming popular in industry, especially in shale and tight reservoir stimulations, and we believe that the hydraulic fractures in shale specimens can be an indicator for evidence.

3.2.3. *Oblique hydraulic fractures*

In the fracturing experiments of shales, other than vertical, horizontal, and multiple fractures, twisted oblique fractures are also present, shown in specimen S3.

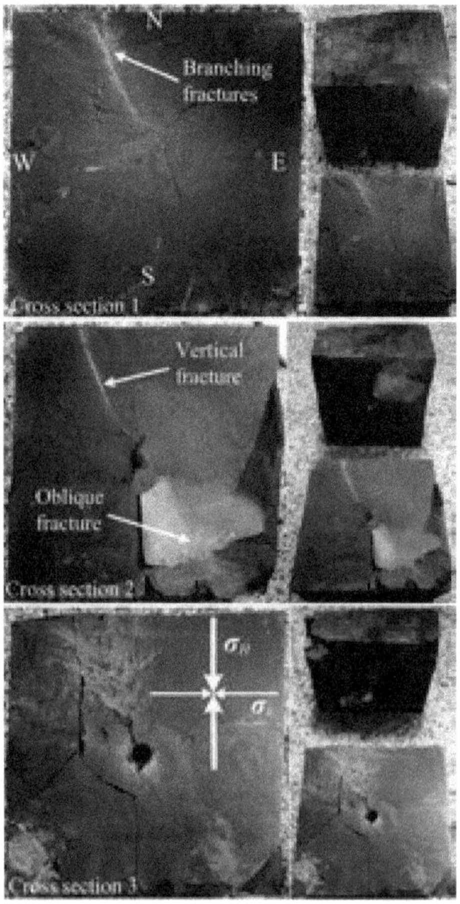

Figure 12. View of each cross-section of specimen S3.

In **Figures 12** and **13**, along the fractured faces, specimen S3 is pull apart for observation. On cross-section 1, branched fractures appear along N-W directions; on cross-section 2, subvertical fracture and oblique fracture occur simultaneously. And cross-section 3 shows a horizontal fracture along the bedding. The most distinct result is found from the south direction, characterized by a curved surface of hydraulic fractures, originated from the well bore, extended outward in S shape.

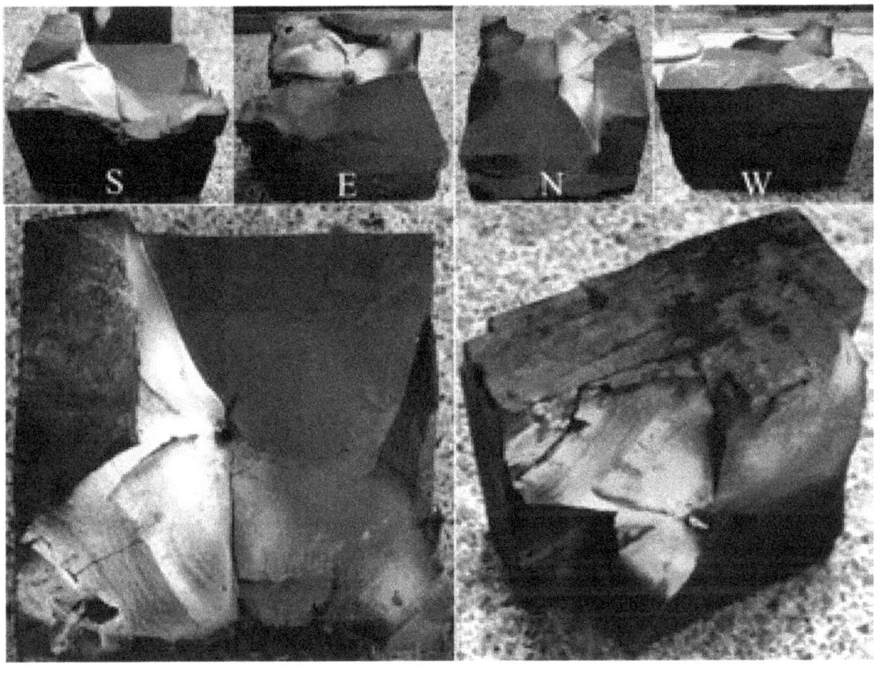

Figure 13. The oblique fracture between cross-section 2 and cross-section 3.

3.3. Experiment results of artificial specimens

In this part, we used cementation specimens in the experiment, for apparent reasons of easy preparation of specimens of various natural fractures. Emphasis is put on natural fracture density and the injection rate.

3.3.1. Hydraulic fractures of artificial specimens

The direct observation of the hydraulic fractures can provide useful information on the influences of testing conditions. As in **Figure 14**, the specimen can be split easily along the natural fracture for the cases of orthogonal natural fractures.

1. Spatial configuration of hydraulic fractures

The projections of the hydraulic fractures in the horizontal plane are shown in **Figure 15**. **Figure 15** shows that the hydraulic fractures are interconnected by the natural fractures, and their dimensions varied with locations. The overall affected zone by hydraulic fractures appears to be a horizontal ellipse.

Figure 14. Dismantled specimen.

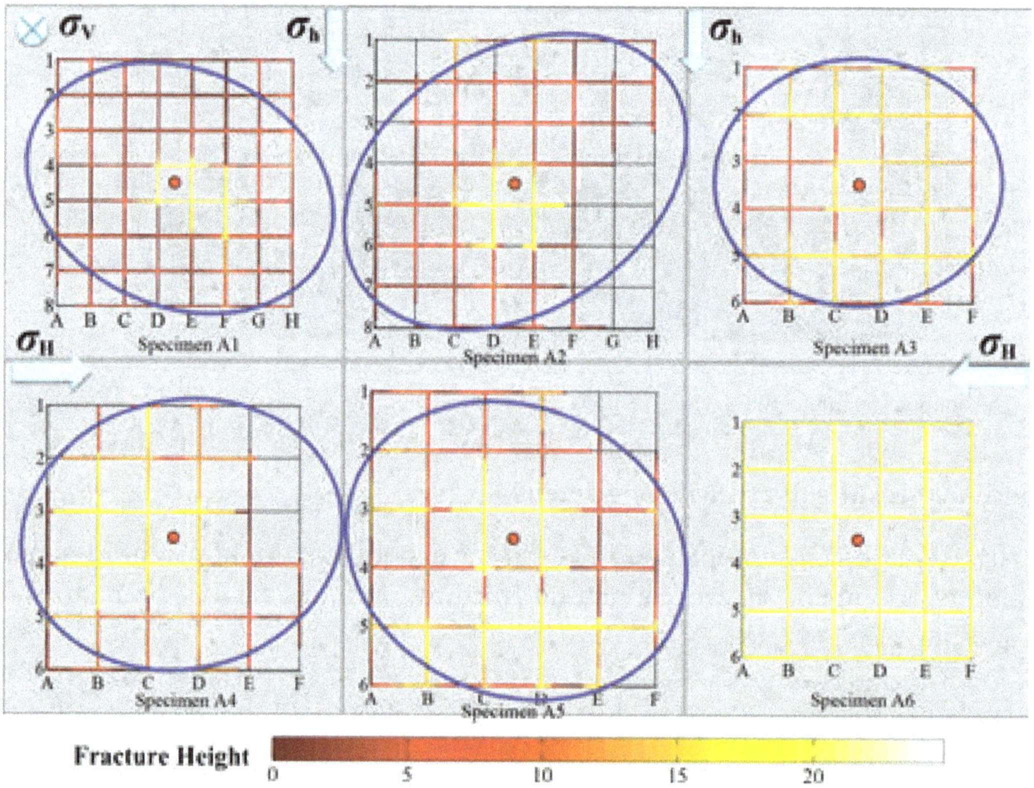

Figure 15. Height variation by projection onto the horizontal plane.

The major axis of the fracture swarm ellipsis is not the same as the maximum horizontal stress and the fracture height decreases gradually away from the wellbore, as shown in specimens A1–A5. In accordance with the conventional fracturing, in a homogeneous formation the hydraulic fracture is a planar along the maximum horizontal stress. And the major axis is not along the maximum stress in our experiments, implying that the natural fractures have apparent effects on the hydraulic fractures, through stress redistribution.

2. *Area of hydraulic fractures*

The hydraulic fractures are composed of a system of complicated branched fractures, and it is not easy to have a simple calculation for area of hydraulic fractures, each of which varies with locations in terms of length and height. In the calculation of areas, a process of scanning and digitization is employed to get an entire spatial geometry of the fracture network, as in **Figure 16**. The fracture set 1 (fractures 1, 2, 3 ...) and fracture set 2 are parallel to the maximum and minimum stresses, respectively.

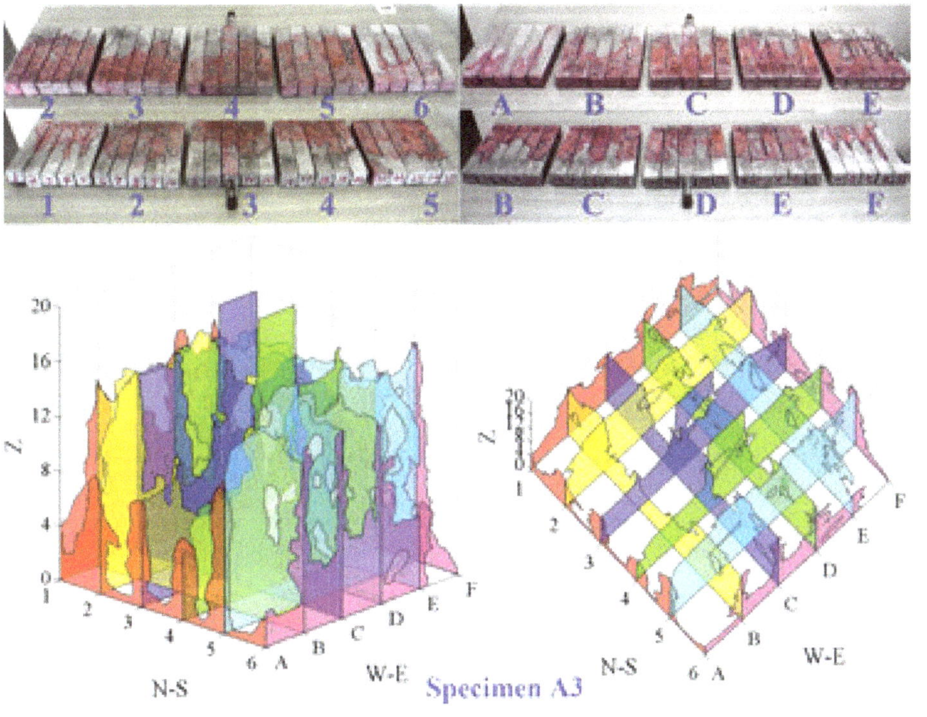

Figure 16. The geometry of the hydraulic fracture network in 3D space.

The final shape of multiple fracture system created in tests of natural fractured blocks, show an approximate ellipsoid envelope, with three orthogonal axes, in which the major one corresponding to fracture length along the maximum horizontal stress, the minor one corresponding to fracture width along the minimum horizontal stress, and the third one along the vertical direction. Such an ellipsoid embraces two groups of hydraulically induced fractures, most of which are along the natural fracture. And the three orthogonal axes are not exactly along the stress directions, i.e., there are some deviations between stress directions and the three orthogonal axes. This means that both in-situ stress state and multiple natural fractures determine the hydraulic fracture network at the same time.

Due to the weak cementation of the natural fractures, the hydraulic fractures mainly extend in the natural fractures, and at the beginning the hydraulic fracture started in the central block (no. 25 in specimens A1 and A2 and as no. 13 in specimens A3–A6, in **Figure 17**). Different from single hydraulic fracture, a complicated hydraulic fracture network results.

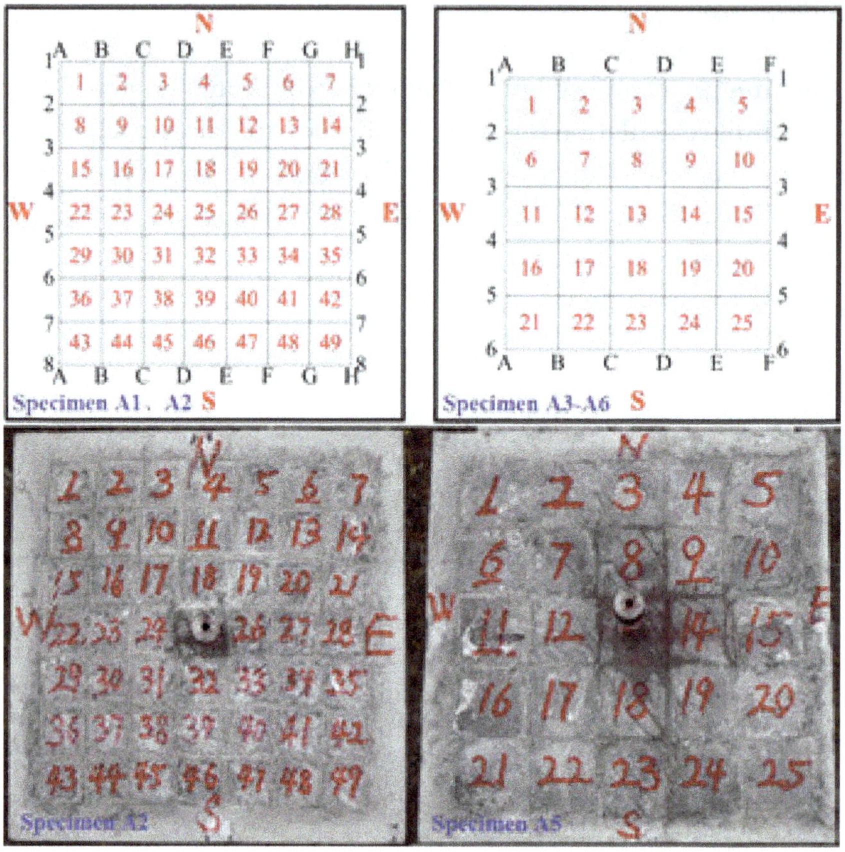

Figure 17. Block units and cemented fractures with numbers on the top of the specimens. (Notes: numbers in red are block units, letters in red are the orientation of the specimens, and those in black are cemented fractures).

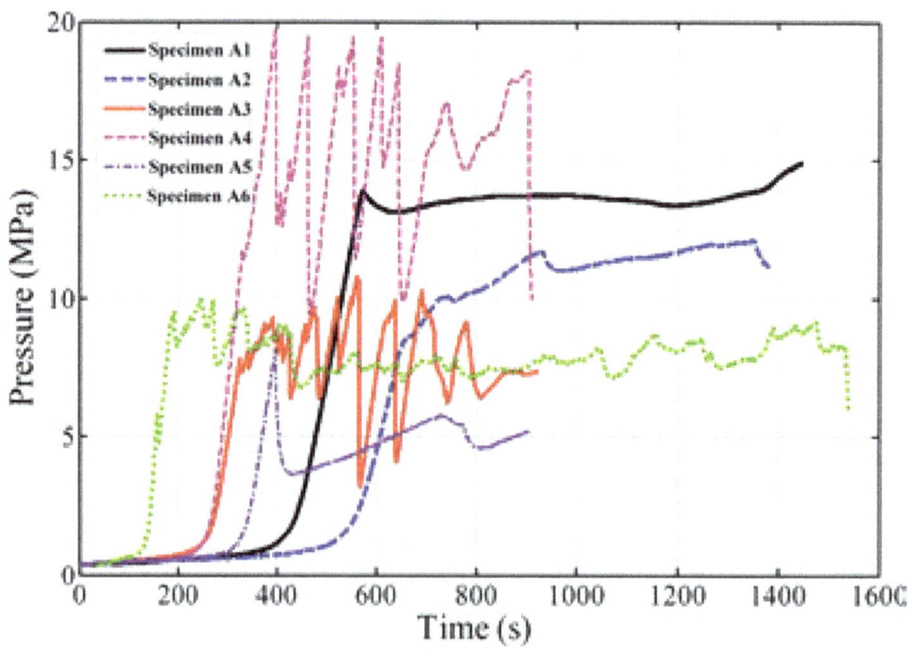

Figure 18. Injection pressure versus time for the specimens.

Here, we present the physical explanation on the possible fracture extension process in the natural fractured blocks, because it is difficult to conduct an accurate measurement of the initiation of fracture growth in a routine test. And in the next paragraph, will discuss to characterize fractures by acoustic emission. In testing of specimen A3 as no. 13 in the upper right quadrant of **Figure 17**, a vertical fracture initiated in the central block, and extended along E-W direction until the natural fractures C and D were encountered, with the fracture tips touching the adjacent bounders of blocks 12 and 14, and no obvious penetration. Such observation can imply the hydraulic fractures became blunt when intersecting with natural fracture C or D, and then fracture propagation paused for 10 s. The rapid drop of fluid pressure (**Figure 18**) also confirmed the assumed behavior of hydraulic fracture, indicating the initiation of a hydraulic fracture or shear slipping between both fracture faces.

Acoustic emission was used to locate breaking events, caused during rock failures in hydraulic fracturing. It was applied in fields to provide real-time fracture propagation in terms of length and height [32–35]. There are also many different arguments for or against the mechanism of acoustic emission for locating failure events. Rock failures can be roughly categories into two types, tensile, and shear failures. There is no consensus that acoustic receivers collect tensile or shear failures, however, tensile and shear failures both emit acoustic signals. Shear slippage caused by injection of water pressure is a process of a local stress redistribution, which could accommodate fracture branching or dendritic. Many fracture branches at different locations could lead to a volume of reservoir stimulation, with each local failure indicating an acoustic emission.

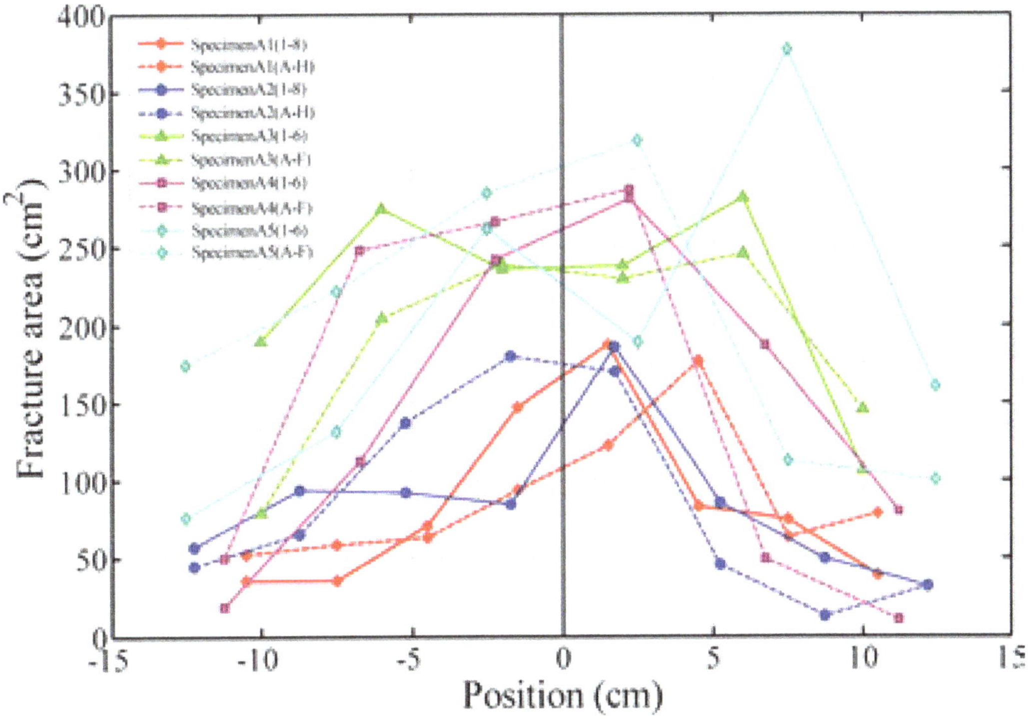

Figure 19. Variations of the hydraulic fracture areas with their positions. (The origin is located on the borehole axis, north to south and west to east is positive.)

In order to calculate the area and the geometry of the fractures, the area of each hydraulic fracture swept by dyed fracturing fluid. In **Figure 19**, the fracture areas for specimens A1–A5 after tests are shown. In our tests, all hydraulically induced fractures are along the natural fracture, due to weak mechanical properties of natural fractures.

Next, the calculated total fracture area was compared with the total volumes of injected fluid. **Figure 20** depicted the relationship between the total fracture area and the total volumes, and it was shown that the total fracture areas of specimens A1 and A2 are less than others, even though the total volumes of injected fluid were quite similar. In our tests, the permeability of all specimens are very slight, then the specimens A1 and A2 are much wider than those of specimens A3, A4, and A5, based on material balance.

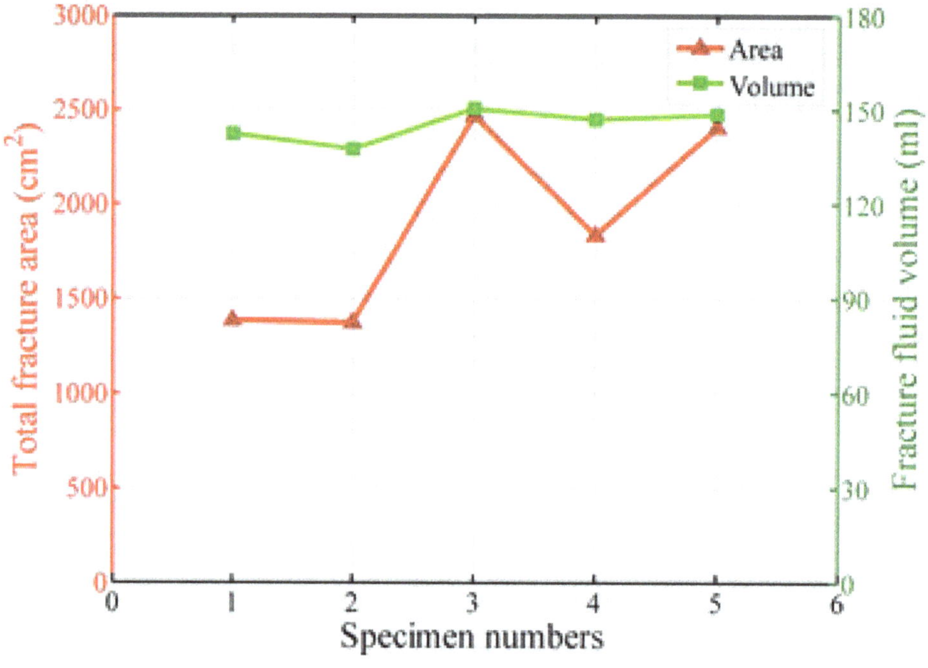

Figure 20. Hydraulic fracture area and injection fluid volume of each specimen.

3.3.2. *Influences of natural fracture and injection rate*

Figure 18 shows the pressure records of all the six specimens. These records can be classified into two types according to the fluctuation features: (1) The pressure records changed very slowly and smoothly (specimens A1 and A2) after initiation. (2) The injection pressure changed abruptly in the entire process (specimens A3, A4, A5, and A6). It can be attributed to the influences of both the natural fracture and the injection rate.

(1) Natural fracture

Pressure records can be seen as a measure of the influences of fracture density on fracture propagation. From **Figure 18**, it seems that the extension pressure of specimens A1 and A2 are greater than A3 and A5, less than specimen A4.

Firstly, for rate of 0.1 ml/s, the total fracture areas and injection fluid volumes are quite close for specimens A1 and A2, however, the extension pressure of specimen A1 (13.63 MPa) is much higher than that of specimen A2 (11.28 MPa) (**Figure 21a**). Such difference can reflect that fracture spacing affects the interaction between multiple fractures. When fracture opens, it will exert additional stresses on the neighboring fractures, so less fracture spacing can improve fracture interference. **Figure 21b** for specimens A3 and A5 show the same results.

In **Figure 21**, the extension pressure of specimen A4 was much higher than expected. The cementation between failed fracture faces is still strong, in specimen A4. Stronger cementation applies more restriction on fracture extension.

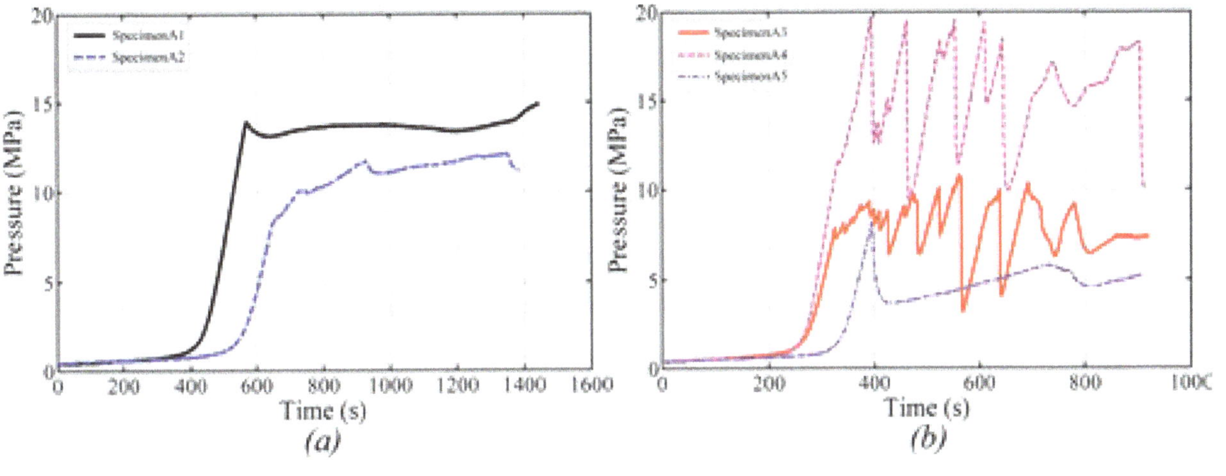

Figure 21. Injection pressure records: (a) injection rate 0.10 ml/s, (b) injection rate 0.16 ml/s.

(2) Influences of injection rates

The pressure records can also as the indication of changes in injection rate, and then some useful information can be obtained from response pressure caused by injection rate.

If the injection rate is less than 0.1 ml/s, in **Figure 18**, changes in flow rate do not cause great change in extension pressure.

However, if the injection rate is high (up to 0.16 ml/s), changes in extension pressure are frequent and great.

If the injection rate is higher than 0.33 ml/s, the changes in extension pressure is more frequent, however is more mild.

The laboratory tests show that injection rate can influence the extension pressure to a great degree. The extension pressure can become more frequent.

3.4. Result comparisons

Laboratory experiments are conducted with three types of rocks: coal seam, shale, and artificial rock, which include natural fractures or simulated natural fractures. It is found that natural fractures play an important role in hydraulic fracturing. Hydraulic fractures always extend and follow the natural fractures, since natural fractures in rocks usually are weaker in mechanical properties than rocks. However, it seems the difference between natural fractures and rocks also affect the behavior of hydraulic fracturing. The mechanical properties of coal seam, artificial rock, and shale become stronger, so it is much easier for hydraulic fractures in shale to follow the natural fractures, and in coal seams, hydraulic fractures not only follow the natural fractures, but also extend in coal matrix.

4. Conclusions

Natural fractures are complicated, composed of multiple cracks of different origins and mechanical properties. Usually, the cracks or natural fractures are weak in mechanical properties. And natural fractures present themselves in various configurations. In coal seams, two orthogonal groups of natural fractures, butt and face cleats, determine the overall properties of coals. In shales, natural fractures are distributed more or less at random, filled with minerals or open, mostly controlled by geological processes. For artificial rocks, natural fractures can be designed in terms of geometrical and mechanical properties.

In this chapter, laboratory tests were conducted, with three rocks with natural fractures. Even though the rocks are different, shales, coals, and artificial rocks, experiments suggested a strong influence of natural fractures and injection rates.

The process of hydraulic fracture networks is definitely complex, however, in our laboratory tests we can infer from our testing data that hydraulic fracture networks are caused by interaction between hydraulic fracturing and natural fractures.

(1) Cleats, the natural fractures in coal seam, have significant influence on hydraulic fracture initiation. Because of low strength of cleats, hydraulic fracture in most cases takes the extension path as the cleats. So, stress state underground and cleat parameters can affect the interactive process of natural fractures and hydraulic fractures.

(2) Hydraulic fractures in shale in laboratory can be vertical, horizontal, or combinations of both cases. And we can infer that the stimulated reservoir volume in shale can be seen as a group of different hydraulic fractures, and much of which are augmented by the natural fractures.

(3) Based on influences of stress difference and natural fracture directions on hydraulic fractures, the conditions for hydraulic fracture crossing the natural fractures can be obtained experimentally as $\Delta\theta = 90°/\Delta\sigma \geq 2$ MPa are $\Delta\theta = 60°/\Delta\sigma \geq 4$ MPa.

(4) Hydraulic fractures, or stimulated reservoir volumes, is the combined effects of reopening of natural fractures and interaction between hydraulic fracture and natural fractures. In naturally fractured formations, fracture network can be characterized by a three dimensional ellipsoid, with its three axes being aligned the orthogonal stresses approximately, but the deviation can be substantial.

Author details

Guangqing Zhang* and Tiegang Fan

*Address all correspondence to: zhangguangqing@cup.edu.cn

Department of Engineering Mechanics, China University of Petroleum in Beijing, Beijing, China

References

[1] Lamont N, Jessen FW. The effects of existing fractures in rocks on the extension of hydraulic fractures. Journal of Petroleum Technology. 1963;5(2):203–209. DOI: 10.2118/419-PA.

[2] Blanton TL. An experimental study of interaction between hydraulically induced and pre-existing fractures. In: SPE Unconventional Gas Recovery Symposium; 16–18 May 1982; Pittsburgh, Pennsylvania. Society of Petroleum Engineers; 1982. p. 559–571. DOI: 10.2118/10847-MS.

[3] Warpinski NR, Teufel LW. Influence of geologic discontinuities on hydraulic fracture propagation (includes associated papers 17011 and 17074). Journal of Petroleum Technology. 1987;39(2):209–220. DOI: 10.2118/13224-PA.

[4] Zhou J, Chen M, Jin Y, Zhang GQ. Analysis of fracture propagation behavior and fracture geometry using a tri-axial fracturing system in naturally fractured reservoirs. International Journal of Rock Mechanics and Mining Sciences. 2008;45(7):1143–1152. DOI: 10.1016/j.ijrmms.2008.01.001.

[5] Athavale AS, Miskimins JL. Laboratory hydraulic fracturing tests on small homogeneous and laminated blocks. In: 42nd US Rock Mechanics Symposium and 2nd US-Canada Rock Mechanics Symposium; 29 June–2 July; San Francisco, California. American Rock Mechanics Association; 2008.

[6] Jeffrey RG, Zhang X, Thiercelin MJ. Hydraulic fracture offsetting in naturally fractured reservoirs: quantifying a long-recognized process. In: SPE Hydraulic Fracturing Technology Conference; 19–21 January 2009; The Woodlands, Texas. Society of Petroleum Engineers; 2009. p. 1–15. DOI: 10.2118/119351-MS.

[7] Rahman MM, Aghighi MA, Shaik AR. Numerical modelling of fully coupled hydraulic fracture propagation in naturally fractured poro-elastic reservoirs. In: EUROPEC/EAGE Conference and Exhibition; 8–11 June 2009; Amsterdam, The Netherlands. Society of Petroleum Engineers; 2009. p. 1–7. DOI: 10.2118/121903-MS.

[8] Dahi Taleghani A. Analysis of hydraulic fracture propagation in fractured reservoirs: an improved model for the interaction between induced and natural fractures [Ph.D. dissertation]. Texas, United States: The University of Texas at Austin; 2009.

[9] Chuprakov DA, Akulich A, Siebrits E, Thiercelin MJ. Hydraulic fracture propagation in a naturally fractured reservoir. In: SPE Oil and Gas India Conference and Exhibition; 2010; Mumbai, India. Society of Petroleum Engineers; 2010. p. 1–12. DOI: 10.2118/128715-PA.

[10] McLennan JD, Tran DT, Zhao N, Thakur SV, Deo MD, Gil IR, et al. Modeling fluid invasion and hydraulic fracture propagation in naturally fractured formations: a three-dimensional approach. In: SPE International Symposium and Exhibition on Formation Damage Control; 10–12 February 2010; Lafayette, Louisiana, USA. Society of Petroleum Engineers; 2010. p. 1–13. DOI: 10.2118/127888-MS.

[11] Min KS, Zhang Z, Ghassemi A. Numerical analysis of multiple fracture propagation in heterogeneous rock. In: 44th US Rock Mechanics Symposium and 5th US-Canada Rock Mechanics Symposium; 27–30 June 2010; Salt Lake City, Utah. American Rock Mechanics Association; 2010. p. 1–10. DOI: 10.2118/127888-MS.

[12] Akulich AV, Zvyagin AV. Interaction between hydraulic and natural fractures. Fluid Dynamics 2008;43(3):428–435. DOI: 10.1134/S0015462808030101.

[13] Potluri NK, Zhu D, Hill AD. Effect of natural fractures on hydraulic fracture propagation. In: SPE European Formation Damage Conference; 25–27 May 2005; Sheveningen, The Netherlands. Society of Petroleum Engineers; 2005. p. 1–6. DOI: 10.2118/94568-MS.

[14] Zhang X, Jeffrey RG. Thiercelin M. Deflection and propagation of fluid-driven fractures at frictional bedding interfaces: a numerical investigation. Journal of Structural Geology. 2007;29(3):396–410. DOI: 10.1016/j.jsg.2006.09.013.

[15] Daneshy AA. Hydraulic fracture propagation in the presence of planes of weakness. In: SPE European Spring Meeting; 29–30 May 1974; Amsterdam, Netherlands. Society of Petroleum Engineers; 1974. p. 1–8. DOI: 10.2118/4852-MS.

[16] Stadulis JM. Development of a completion design to control screenouts caused by multiple near-wellbore fractures. In: Low Permeability Reservoirs Symposium; 19–22 March 1995; Denver, Colorado; 1995. Copyright 1995, Society of Petroleum Engineers, Inc.; 1995. p. 69–84. DOI: 10.2118/29549-MS.

[17] Britt LK, Hager CJ, Thompson JW. Hydraulic fracturing in a naturally fractured reservoir. In: International Petroleum Conference and Exhibition of Mexico; 10–13 October 1994; Veracruz, Mexico. Society of Petroleum Engineers; 1994. p. 483–492. DOI: 10.2118/28717-MS.

[18] Rodgerson JL. Impact of natural fractures in hydraulic fracturing of tight gas sands. In: SPE Permian Basin Oil and Gas Recovery Conference; 21–23 March 2000; Midland, Texas. Society of Petroleum Engineers; 2000. p. 1–9. DOI: 10.2118/59540-MS.

[19] Jeffrey RG, Zhang X, Bunger AP. Hydraulic fracturing of naturally fractured reservoirs. In: Thirty-fifth Workshop on Geothermal Reservoir Engineering; February 1–3, 2010; Stanford University, Stanford, California.

[20] Blanton TL. Propagation of hydraulically and dynamically induced fractures in naturally fractured reservoirs. In: SPE Unconventional Gas Technology Symposium; 18–21 May 1986; SPE Unconventional Gas Technology Symposium. Society of Petroleum Engineers; 1986. p. 613–627. DOI: 10.2118/15261-MS.

[21] Renshaw CE, Pollard DD. An experimentally verified criterion for propagation across unbonded frictional interfaces in brittle, linear elastic materials. International Journal of Rock Mechanics and Mining Science & Geomechanics. 1995;32(2):237–249. DOI: 10.1016/0148-9062(94)00037-4.

[22] De Pater CJ, Beugelsdijk LJL. Experiments and numerical simulation of hydraulic fracturing in naturally fractured rock. In: Alaska Rocks 2005, The 40th U.S. Symposium on Rock Mechanics (USRMS); 25–29 June 2005; Anchorage, Alaska. American Rock Mechanics Association; 2005.

[23] Zhou J, Xue C. Experimental investigation of fracture interaction between natural fractures and hydraulic fracture in naturally fractured reservoirs. In: SPE EUROPEC/EAGE Annual Conference and Exhibition; 23–26 May 2011; Vienna, Austria. Society of Petroleum Engineers; 2011. p. 1–12. DOI: 10.2118/142890-MS.

[24] Olson JE, Bahorich B, Holder J. Examining hydraulic fracture: natural fracture interaction in hydrostone block experiments. In: SPE Hydraulic Fracturing Technology Conference; 6–8 February 2012; The Woodlands, Texas. Society of Petroleum Engineers; 2012. p. 1–10. DOI: 10.2118/152618-MS.

[25] La Pointe PR, Hudson JA. Characterization and interpretation of rock mass joint patterns. Geological Society of America Special Papers. 1985;199:1–38. DOI: 10.1130/SPE199-p1.

[26] Lachenbruch AH. Depth and spacing of tension cracks. Journal of Geophysical Research. 1961;66(12):4273–4292. DOI: 10.1029/JZ066i012p04273.

[27] Laubach SE, Marrett RA, Olson JE, et al. Characteristics and origins of coal cleat: a review. International Journal of Coal Geology. 1998;35(1–4):175–207. DOI: 10.1016/S0166-5162(97)00012-8.

[28] Dron RW. Notes on cleat in the Scottish coalfield. Transactions of the Institute of Mining Engineers. 1925;70:115–117. DOI: 10.1016/S0166-5162(97)00012-8.

[29] Kim CM and Abass HH. Hydraulic fracture initiation from horizontal wellbores: laboratory experiments. In: The 32nd U.S. Symposium on Rock Mechanics (USRMS), 10–12 July, Norman, Oklahoma; 10–12 July 1991; Norman, Oklahoma. American Rock Mechanics Association; 1991. p. 231–240.

[30] Hubbert MK and Willis DG. Mechanics of hydraulic fracturing. Transactions of Society of Petroleum Engineers of AIME. 1957;210:153–168.

[31] Shen W, Zhang B. Testing study on mechanical parameters of coal. Chinese Journal of Rock Mechanics and Engineering. 2000;19(S1):860–862 (in Chinese).

[32] Barree RD, Fisher MK, Woodroof RA. A practical guide to hydraulic fracture diagnostic technologies. In: SPE Annual Technical Conference and Exhibition; 29 September–2 October 2002; San Antonio, Texas. Society of Petroleum Engineers; 2002. p. 1–12. DOI: 10.2118/77442-MS.

[33] Fisher MK, Heinze JR, Harris CD, Davidson BM, Wright CA, Dunn KP. Optimizing horizontal completion techniques in the Barnett shale using microseismic fracture mapping. In: SPE Annual Technical Conference and Exhibition; 26–29 September 2004; Houston, Texas. Society of Petroleum Engineers; 2004. p. 1–11. DOI: 10.2118/90051-MS.

[34] Warpinski N, Kramm RC, Heinze JR, Waltman CK. Comparison of single- and dual-array microseismic mapping techniques in the Barnett shale. In: SPE Annual Technical Conference and Exhibition; 9–12 October 2005; Dallas, Texas. Society of Petroleum Engineers; 2005. p. 1–10. DOI: 10.2118/95568-MS.

[35] Tezuka K, Kamitsuji R, Tamagawa T. Fractured reservoir characterization incorporating microseismic monitoring and pressure analysis during massive hydraulic injection. In: International Petroleum Technology Conference; 3–5 December 2008; Kuala Lumpur, Malaysia. International Petroleum Technology Conference; 2008. p. 1–7. DOI: 10.2523/IPTC-12391-MS.

2

Fracture Mechanics in Adhesive Dentistry

Mügem Aslı Gürel,

Bağdagül Helvacıoğlu Kıvanç and Hacer Deniz Arısu

Abstract

Bond strength between adhesive systems and the tooth structure is influenced by a large number of variables. The bond strength tests are important tools for improving resin-tooth adhesion to increase the service life of dental resin–based composite restorations. This chapter discusses the materials used in adhesive dentistry and test methods applied for evaluating bond strength between tooth structure and adhesive materials.

Keywords: adhesion, adhesive dentistry, bond strength, fracture strength, tooth

1. Introduction

Adhesion or bonding is the process of forming an adhesive joint, which consists of two substrates joined together [1]. Adhesives join materials together to resist separation and transmit loads across the bonds [2]. In dentistry, the adherend is the substrate to which the adhesive—enamel and dentin, rarely cementum—is applied [1]. The mechanisms of adhesion to the inorganic and organic components of teeth have relied primarily on the evaluation of morphologic relations of materials with tooth substrates [2]. In 1952, Kramer and McLean published an article about adhesion, in which altered impregnation of dentin surface, resulting from the interaction with glyrecophosphate dimethacrylate, was observed [3]. After few years, the pioneering work of Michael Buonocore marked the beginning of successful "adhesive dentistry". Buonocore's discovery of the acid-etch technique has led to major changes in recent dental practice. He showed that the treatment of enamel with phosphoric acid resulted in a porous surface, which could be infiltrated by resin material, to produce a micromechanical bond [4]. On the other hand, chemical bonding was developed by Smith and resulted in the introduction of polycarboxylate cement into the dentistry. The basic bonding mechanism was an ionic attraction between cement and enamel or dentin [5].

2. Adhesion to tooth structure

Tooth is composed of enamel, the pulp-dentin complex, and cementum (**Figure 1**). Enamel covers the anatomic crown of the tooth and varies in thickness in different areas. Enamel is composed of hydroxyapatite, organic matrix proteins, and water. Enamel has high elastic modulus, high compressive strength, low tensile strength, and supports dentin to withstand masticatory forces. Dentin and pulp tissues are specialized connective tissues. Dentin forms the largest portion of the tooth structure, extending almost the full length of the tooth. Human dentin is composed of approximately 50% inorganic material and 30% organic material by volume [6]. Dentin is considerably more complex and consists of solid and porous phases. The porous phase consists of numerous fluid-filled tubules emanating from the pulp. They transverse the dentin to the dentino-enamel junction, making dentin a highly permeable tissue. Each dentinal tubule is surrounded by a collar of hypermineralized, peritubular dentin [7]. The structure surrounding the peritubular dentin is intertubular dentin, which is approximately 9% less mineralized than peritubular dentin. The intertubular dentin is a biphasic biologic composite that contains mineral and organic components [8]. The humidity and organic nature of dentin makes bonding to this hard tissue extremely difficult [1]. Dentin has a low elastic modulus, high compressive strength, and high tensile strength and increases the fracture toughness of enamel. Externally, dentin is covered by enamel on the anatomic crown and cementum on the anatomic root. Cementum is a thin layer of hard dental tissue covering the anatomic roots of teeth. Cementum is slightly softer than dentin and consists of about 45–50% inorganic material (hydroxyapatite) and 50–55% organic matter and water by weight. Internally, dentin forms the walls of the pulp cavity (pulp chamber and pulp canals). The dental

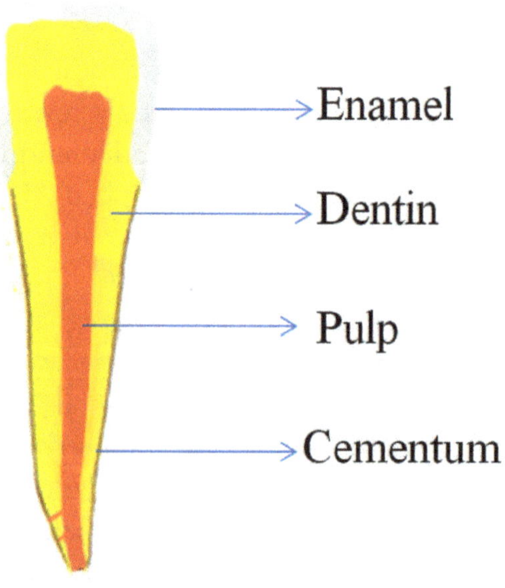

Figure 1. Structures of tooth.

pulp occupies the pulp cavity in the tooth and is a unique, specialized organ of the human body [6]. The pulp contains nerves, arterioles, venules, capillaries, lymph channels, connective tissue cells, intercellular substance, odontoblasts, fibroblasts, macrophages, collagen, and fine fibers [9].

During the past decades, restorative concepts have been continually changing and adhesive technology has become more important. Adhesion to enamel was followed by adhesion to dentin [10]. Adhesion to tooth structure primarily depends on the displacement of inorganic tooth substrate with the resinous materials [11]. In this circumstance, there are two steps. In the first step, inorganic part of tooth (calcium, phosphate) is removed from enamel and dentin surfaces to create microporosities. In the second step, when resin materials are applied to these surfaces, they infiltrate into these microporosities and subsequently polymerized. This process is named as 'hybridization'. This is a mechanical interlocking between the tooth structure and resin material and depends primarily on the diffusion mechanism of the resin material into the microporosities of tooth structure. Although it is believed that micromechanical interlocking has a primary role on achieving a clinically good bonding, there are chemical interactions between functional monomers and tooth components that have a potential benefit on additional bond strength [12].

3. Classification of adhesive materials

3.1. Bonding agents

One of the greatest challenges in restorative dentistry is to obtain an effective seal of the tooth-restoration interface. Adhesive restoration (composite resin, etc.) rely on bonding systems that form a micromechanical bond with the tooth structure [13]. During the tooth preparation with burs or other instruments, smear layer (composed of hydroxyapatite and altered collagen) forms a coating on enamel and dentin, reducing the permeability of dentin [14]. The smear layer acts as a physical barrier against the penetration of adhesive monomers into the dentin tubules. Therefore, for adhesive penetration into dentin surface, it must be dissolved or made permeable [1]. Contemporary adhesives interact with the dental substrates using one of the two different bonding strategies in order to remove or alter smear layer: (1) the etch and rinse technique, which requires smear layer removal before the use of dentin adhesive, and (2) the self-etch technique, in which the smear layer is maintained as a substrate for bonding. In addition, these two bonding strategies may be classified according to the number of application steps (step 1 to step 3) required to couple the resin composites to dental substrates. Etch and rinse adhesives require a separate etching step. Thus, an inorganic acid (mostly 30–40% phosphoric acid) is applied to dental substrates and then rinsed off. This step is followed by a priming treatment, wherein amphiphilic functional resin monomers are applied to dental substrates to make them prone to receiving a mixture of relatively more hydrophobic resin monomers that will complete the bonding procedure. This sequence of events exemplifies a three-step application procedure. Simplified two-step etch and rinse adhesives combine the primer and adhesive resin into a single application step. On the other hand, self-etch adhesives no longer require a separate etching step. This approach requires the use of nonrinse acidic monomers that simultaneously etch and prime dentin. The bonding procedure with self-etch adhesives can be achieved using either two- or one-step systems, depending on whether the etching/primer agent is separated from the adhesive resin or combined with it to allow a single application step [15].

3.2. Glass-ionomer cements

Glass-ionomer cements were introduced to the profession 20 years ago and have been shown to be a very useful adjunct to restorative dentistry [16]. Glass-ionomer cements remain as the only materials that are self-adhesive to tooth tissue, without any surface pretreatment. Glass ionomers are composed of organic acid and glass component and set with an acid-base reaction [17]. The major advantages of glass-ionomer cements are a continuing fluoride release throughout life of the restoration and the ion exchange adhesion to both enamel and dentin. However, apparent lack of physical strength and translucency are disadvantages of glass-ionomer cements. Small additions of resin increase the physical properties to a degree and allow for a light-initiated setting mechanism [16].

3.3. Compomers

This material is a polyacrylic/polycarboxylic acid–modified composite. Polyacid-modified resin-based composites or compomers combine the characteristics of both composites and glass ionomers into a single component. It is an attempt to take advantage of the desirable qualities of both materials: the fluoride release and ease of use of the glass ionomer cement and the superior material qualities and aesthetics of the composites [18]. They contain 72% (by weight) strontium fluorosilicate glass and the average particle size is 1.5 μm [19]. The mechanical properties are superior to glass-ionomer cements as clinically demonstrated by lower fracture rates. Furthermore, wear does not seem to be critical after short-term [20].

3.4. Composite resin materials (CRMs)

CRMs are the universally used tooth-colored direct/indirect restorative materials in dentistry. CRMs were developed by combining dimethacrylates (epoxy resin and methacrylic acid) with silanized quartz powder in 1962 [21]. They have taken over some restorative materials depending on their properties and advantages. CRMs are composed of resin matrix (organic content), fillers (inorganic part), and coupling agents. Bis-GMA (bisphenol-A glycidyldimethacrylate) and TEGDMA (triethylenglycol-dimethacrylate) constitute resin matrix of CRMs. The fillers are made of quartz, ceramic, or silica. With increasing filler content, the polymerization shrinkage, the linear coefficient, and water absorption are reduced and the compressive and tensile strength, and the modulus of elasticity and wear resistance are improved [22]. The coupling agent is a molecule that connects resin and inorganic matrix of composites [23].

Choosing a suitable composite resin material for a restoration requires functional properties, including excellent mechanical properties such as high strength, fracture toughness, surface hardness, optimized modulus of elasticity, low wear, low water sorption and solubility, low polymerization shrinkage, low fatigue and degradation, and high radiopacity [24].

3.5. Root canal sealers

A successful root canal treatment depends on proper diagnosis, adequate cleaning, shaping, and finally, the three-dimensional (3D) obturation of the root canal system. The complete obturation of the root canal system and the creation of a fluid tight seal have been proposed as a goal for successful endodontic treatment [25]. Gutta-percha and root canal sealers are usually used for obturation of root canals. Improvements in adhesive technology directed their attention toward the characteristics of the filling material, for increasing the adhesion of the endodontic sealers to root canal walls. The adhesion between root canal walls and resin-based sealers is the result of a physicochemical interaction across the interface, enabling the union between the filling material and root canal walls [26]. Several resin-based sealers, such as AH 26, AH Plus, EndoREZ, Real Seal, and Epiphany, etc., have become available in dental markets. The aim of these resin bonding systems is to allow for the adhesion of the obturation material and to form a hermetic seal [27].

3.6. Postcore restorations and luting cements

Endodontically treated teeth are more susceptible to fracture possibly due to extensive loss of tooth structure which may be arising from extensive decays, previous large restorations, broad access cavities, aggressively flared, and overinstrumented canals [28, 29]. When a significant coronal tooth structure has been lost, a full crown may be restoration of choice. More frequently, the cementation of a post inside the root canal is necessary to provide attention for the core material and the crown. The core is anchored to the tooth by extension by the root canal through the post and replaces missing coronal structures. The crown covers the core and restores the esthetics and function of the tooth. The post's ability to anchor the core is also an important factor for successful reconstruction. Finally, luting material used to cement the post, the core, and the crown to the tooth will also influence the longevity of the restoration. The post, the core, and adhesive materials together form a foundation restoration to support the future crown [30] (**Figure 2**).

In recent years, nonmetalic posts, such as zirconia, fiber-reinforced posts, and polyethylene-woven fiber-reinforced posts, have become quite popular. Among these posts, glass fiber-reinforced posts are most popular. These posts could be made from glass or silica fibers (white or translucent) but the most commonly used fibers are silica based. The matrix for this post is an epoxy resin [31]. The clinical success of fiber-reinforced post systems is due to their presenting good retention properties under mechanical strain as a result of their low elastic modulus (17.5–21.6 GPa), which is similar to that of dentin (14.0–18.6 GPa) [32]. Therefore, by using a fiber post with a modulus of elasticity very similar to that of dentin, tooth-postcore monoblock can be achieved instead of an assembly of heterogeneous materials. This can help to distribute masticatory loads to homogenously reduce stress during function [33].

Retention is important for the use of intracanal posts in endodontically treated teeth [34]. The major factors affecting post retention are their dimensions (length, diameter), shape (conical, cylindrical), type of surface (serrated, screw, and smooth), intracanal shape preparation, type of cement, and operator skills [35].

Adhesive resin-based cements contain 4-methyacrylate-ethyl-trimethyl-anhidride that reacts chemically with the oxide metallic layer increasing post retention compared to nonadhesive resin cements [36]. The capacity of different cements to retain posts is related to mechanical properties, adhesion capacity to dentin, and durability [37]. The elastic modulus of resin cements is close to dentin and fiber post, and its stress concentration in dentin is low. Therefore, roots reinforced with posts that are cemented with dentin adhesives and resin cements are more fracture resistant than those cemented with other cements [33].

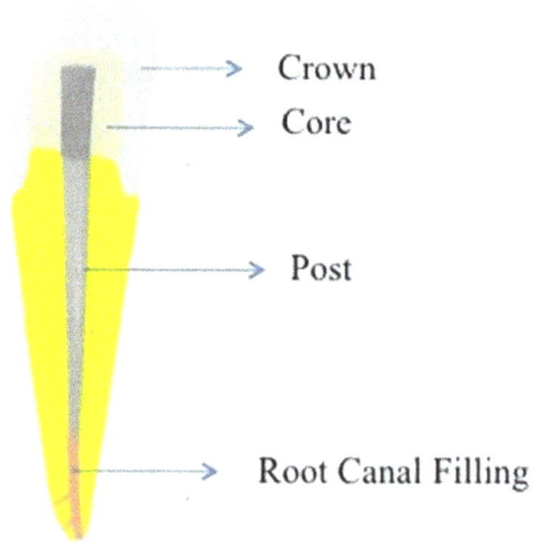

Figure 2. Schematic presentation of a postcore and crown restoration.

3.7. Ceramic crown restorations

Dental crowns are used to replace the natural crowns of teeth when the enamel and dentin are lost through dental caries, tooth wear, root canal treatment, or trauma. The aims of a dental crown are to restore function, occlusion, and contact points with adjacent teeth and aesthetics. In addition, a full coverage crown may be used to prevent further controlled loss of tooth substance by catastrophic fracture of weakened tooth cusps [38]. All ceramic restorations have gained more popularity due to their high esthetic, high improvement in their fracture strength, and good biocompatibility properties [39]. These restorations are luted in position with a resin cement to tooth structure after using bonding agent [40]. Micromechanical retention of the luting cement to the restoration is obtained by etching the fitting surface of the crown with hydrofluoric acid. In addition, the ceramic surface is treated by a silane bond enhancing agent. These agents have been suggested to use for improving the bond strength to resin-based cements [41]. This luting procedure is used to provide bonding at both the dentin-luting cement and luting cement-ceramic crown interfaces [42].

4. Test methods in adhesive dentistry

Laboratory tests are useful for testing new operative techniques and materials before they are clinically implemented. The methods employed, however, should meet the following requirements (FDA 1978): The results must be reproducible, the parameters which influence the test results must be known, the variability of the measured values must be low and within an acceptable range and if devices are employed for the test itself and/or to measure parameters and posttesting conditions of the specimens, then these devices must be suitable for the given purpose, that is, they must be qualified. These requirements were described for medical devices and compiled under the name "Good Laboratory Practice" [43].

Flexural strength, composite resin	699
Flexural strength, fiber post	50
Flexural strength, ceramic	491
Elastic modulus, composite resin	570
Elastic modulus, fiber post	96
Elastic modulus, ceramic	518
Shear bond strength, composite resin	2240
Shear bond strength, fiber post	30
Shear bond strength, ceramic	853
Tensile bond strength, composite resin	2222
Tensile bond strength, fiber post	95
Tensile bond strength, ceramic	497
Dynamic fatigue, composite resin	24
Dynamic fatigue, fiber post	5
Dynamic fatigue, ceramic	51
Push-out, composite resin	191
Push-out, fiber post	217
Push-out, ceramic	58
Pull-out, composite resin	52
Pull-out, fiber post	40
Pull-out, ceramic	54
Fracture toughness, composite resin	261
Fracture toughness, fiber post	9
Fracture toughness, ceramic	422
FEA, composite resin	345
FEA, fiber post	96
FEA, ceramic	562

Table 1. Searching PubMed with chosen keywords (March, 2016).

Nowadays, numerous new adhesive systems are put on dental markets. *In vitro* test methods are necessary to evaluate test materials' suitability for the clinical and physical properties within reasonable conditions. **Table 1** shows the PubMed analysis of adhesive materials and test methods used in dentistry with chosen key words.

Adhesion quality of adhesive materials on enamel and dentin may be quantified using several methodologically distinct approaches [44], roughly divided into macro- and microsetups, depending on the size of adhesion area. The macrobond strength can be evaluated with shear or tensile mode, a push- or pull-out methods at a bonded area larger than 3 mm² [45]. The ability of microbond strength tests is to evaluate the effect of local tooth structure on bond strength [46, 47] and to allow depth profiling of different substrates [45].

4.1. Flexural strength and elastic modulus

The flexural or bending strength is a measure of the fracture resistance of a material. For restorative materials in occlusion bearing areas, the ISO standard demands a flexural strength of at least 80 MPa [48]. For this test, bar-shaped specimens (25 × 2 × 2 mm) are made, stored in water for 24 h and at 37°C, and loaded until failure in a universal testing machine (crosshead speed 0.75 mm/min [±0.25]). The flexural strength in three-point bending test is calculated with the following formula:

$$\mathrm{BF} = 3Fd / 2wh^2$$

where F is the maximum force, d is the distance between the two anchors, w is the width of the specimen, and h is the height of the specimen.

Because the flexural strength changes after water storage, the value at 24 h only provides limited information. Reliable data on the behavior of the material are obtained when the value after 1-day storage is compared to that after 1 month of water storage [49].

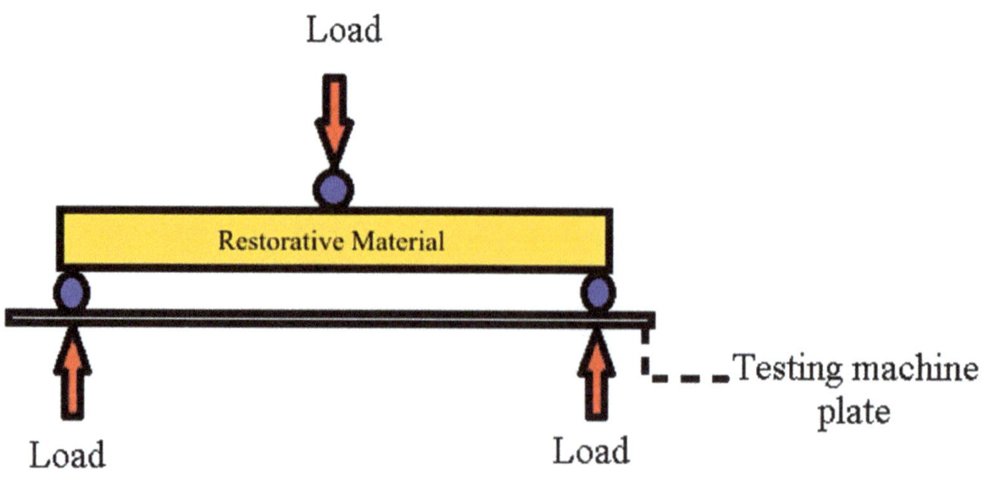

Figure 3. Schematic presentation of a three-point bending test.

The elastic modulus of the dental restorative materials should be close to that of the enamel and dentin to allow better stress distribution. The existence of large modulus gradient between restorative materials and dental hard tissues may lead to fracture [50]. Elastic modulus represents the stiffness of a material within the elastic range when tensile or compressive forces are applied [51]. A bar-shaped sample fixed at three points is loaded to failure in a testing machine at a crosshead speed of 0.5 mm/min (**Figure 3**). The elastic modulus can be read from the stress-strain diagram [49].

4.2. Shear bond strength measurement

In macroshear strength tests, composite cylinders with a diameter of about 3 or 4 mm are adhered to flat ground tooth tissue (dentin or enamel) surfaces after application of the adhesive systems and then sheared off with a special testing machine (**Figure 4**).

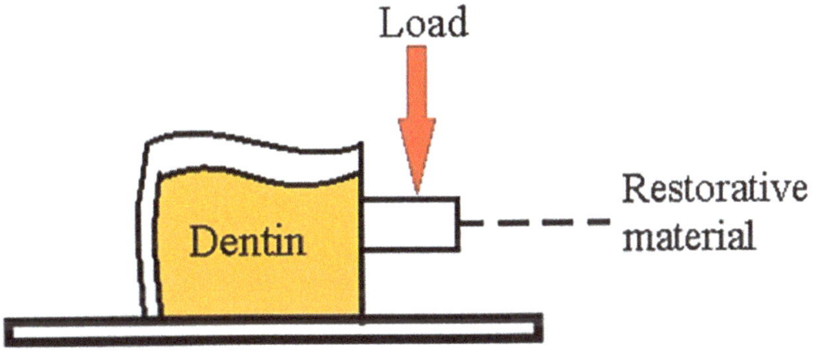

Figure 4. Schematic presentation of a shear bond strength test.

The force in Newtons required to debond the composite cylinder from the substrate is measured. Finally, this force is applied parallel to the area of the bonding surface to yield the bond strength in megapascals (MPa = 1 N/mm^2) [49]. The stress distribution is seen as a disadvantage in shear strength testing. For instance, enormous forces are exerted on the site at which the shearing blade contacts the specimen [52]. The absolute bond strength also depends on whether the shearing blade is flat or if it bears a notch which surrounds half of the specimen [53]. After preparing samples, test is applied immediately, after 24 h or after up to several months of water storage at 37°C [45].

Special jigs have been prepared which have different configurations such as wire loops, notched chisels, and knife edges are used to apply shear force. A knife-edge chisel causes severe stress concentration at the force application area; however, wire loop and the notched chisel render a more even stress distribution at the edge of the bonding area [52]. Peak stress during application of force by blunt knife edges may explain the frequently observed cohesive composite failure close to loading point [54].

Microshear bond strength test is applied on bonded cross-sectional areas of 1 mm^2 or less [55]. This test permits efficient screening of adhesive systems, regional and depth profiling of a variety of substrates and conservation of teeth. Aqueous storage durability studies are also possible microshear bond strength due to the relatively short diffusional distances (0.02–0.05 mm) from the cavosurface [56]. In the microshear bond strength, specimen is prestressed prior to testing only mold removal. The use of the mold for restorative materials placement can lead to introduction of flaws and different stress concentrations upon shear loading [57]. Microshear bond strength tests are influenced by a number of factors including the thickness of adhesive layer, size of bond area, cross-head speed, the mechanical properties of adherent and adhesive and debonding procedure [55, 58]. The failure of this test method is attributed to crack initiation by tensile forces as well as shear forces that imply that a fracture mechanics approach may also be important [59].

4.3. Tensile bond strength measurement

In a tensile test, composite cylinders are adhered flat dentin or enamel surfaces and tensile force may be exerted on composite cylinders using chucks or conical composite plugs employed instead of cylinders [60]. The macrotensile bond strength test was used as frequently as the macroshear bond strength test, in 1991 to 2001 [61]. The specimen can be held by active or passive gripping technique. Active gripping technique involves mechanical locking of specimen to gripping device, such as glue or clamps, whereas in passive gripping technique, specimen is placed in a testing machine without the aid of glue or clamps or mechanical gripping [62]. Currently, the macrotensile bond strength test is not common, but it is still important to measure bond strength to restorative materials such as ceramics and metal alloys [63, 64].

It was observed that tensile bond strength is inversely related to bonded surface areas and that although much higher bond strengths were measured, most failures still occurred at the interface between tooth substrate and adhesive [55, 65].

For specimen preparation, the adhesive is applied to prepared dentin and a large restorative material plug incrementally polymerized onto adhesive. After storage duration, this assembly is sectioned using precision circular saws into 1-mm thick slices, which are then cut into sticks of about 1 × 1 mm [45]. Some approaches trim such sticks to dumbbell or hourglass shapes [66, 67]. After specimen preparation, specimens are mounted on a variety of jig designs using fast setting glues to standardize sample fixation to the universal testing machine for loading [68, 69] (**Figure 5**).

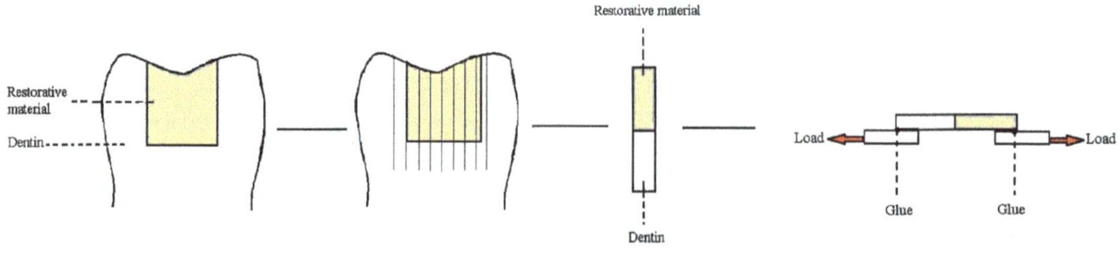

Figure 5. Schematic presentation of a microtensile bond strength test.

The advantages of this method are that, only a few extracted teeth are required, regional dentin differences can be examined and it is easier to distinguish between different materials, the better stress distribution at the true interface, ability to test irregular surfaces and very small areas and facilitates microscopic examinations of the failed bonds due to smaller areas [70, 71]. The major disadvantage of microtensile bond strength test is the rather labor-intensive, technically demanding and relatively fragile sample preparation technique. Sawing and trimming of the test samples appear to be the most technique sensitive part of microtensile bond strength test [70]. Interfacial stress during sample preparation is reflected by the number of pretest failures, as often occurs with lower-performance adhesives or on brittle substrates [45]. Special care should be taken to avoid/reduce the production of microfractures at the interface during specimen preparation. They may weaken the bond and, thus, reduce the actual bond strength [72].

4.4. Dynamic fatigue measurement

Fatigue can be defined as the failure of mechanical properties after repeated applications of stresses, at a level well below the ultimate fracture strength of the material interface [73]. During the restoration survived in mouth, a restoration is sustained to cyclic loading, in the long term this loading may possibly lead to marginal deterioration and loss of the restoration [12]. Therefore, fatigue testing of dental adhesives is expected to better determine their performance. But also, there is no standard fatigue test for dental adhesives. Possible methods are a cyclic shear test [74], a cyclic tensile test [75], or a cyclic push-out test [76]. Another possibility is loading not only the interface but the whole tooth until the tooth-restoration complex fails [77]. A microrotary fatigue device has been developed for dynamic tests of tooth-composite interfaces [78]. In this method, bar-type samples prepared with a rounded, constricted interface were clamped in a pin-chuck and connected to a stepping motor with the free end loaded with a certain weight. By rotating the specimen, each spot at the outer surface of the interface undergo compressive and tensile loading [79].

4.5. Push-out and pull-out test methods

Push-out method was first described by Roydhouse in 1970 [80]. Conical preparations are made in teeth and filled with restorative material. Beginning at the pulpal axial wall, dentin is removed up to the level of restoration, which is pushed out with the testing machine's plunger. Specimen preparation is also simplified by cutting dentin into disks and making conical preparation in it. Then the disk is placed on a glass plate and the restorative material is inserted into the cavity [49] (**Figure 6**).

The advantages of this method are the simultaneous testing of marginal seal and adhesive bond on the same specimen [81], and taking into account the effect of polymerization stress might have in the clinical situation [74]. The disadvantage of this method is that minor degrees of composite swelling upon water storage can induce a significant amount of friction, independent from the adhesive performance [45].

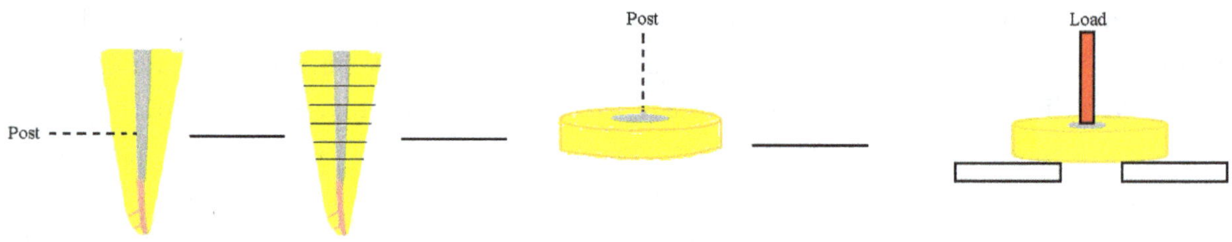

Figure 6. Schematic presentation of a push-out test.

In this method, the plunger must provide near complete coverage of the testing material without touching the root canal walls. This method is very commonly used in the analyses of post and root canal sealers adhesion in root canals. The bond strengths of cements for ceramic restorations are also measured with this method [49].

Micro push-out test is a modification of push-out test where the specimen thickness is less than or equal to 1 mm^2. Micro push-out is more dependable than microtensile bond strength test while evaluating the bond strength of luted fiber posts [82]. Castellan et al. showed that a modified push-out and microtensile bond strength test revealed higher values than traditional conventional push-out and pull-out tests [83].

A novel approach to the pull-out test was designed to eliminate the region of weakness in the postcement system, which was elsewhere reported to be sometimes the predominant failure modes. The pull-out method allows comprising the whole length of the root canal because the focus was not on regional differences in bond strength. The postcemented roots were not

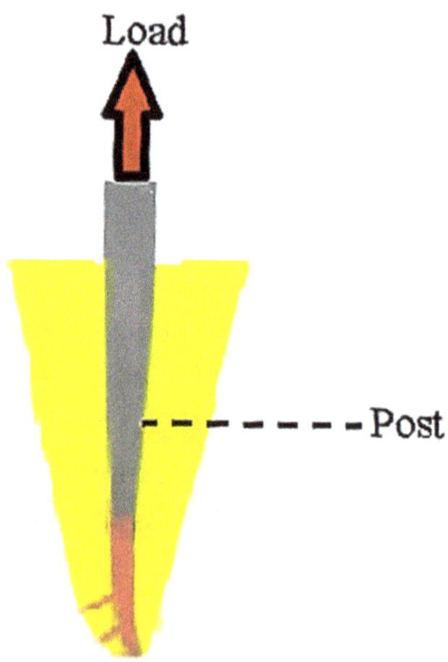

Figure 7. Schematic presentation of a pull-out test.

sectioned, so potential artifacts or premature failures caused by preparation of the specimens could be avoided [84] (**Figure 7**).

4.6. Fracture toughness

Interfacial material properties, such as crack growth resistance, are described by linear elastic fracture mechanics. Fracture mechanics is well established in the engineering community to describe the properties of monolithic materials and is becoming increasingly common to investigate the interface of dissimilar materials [85, 86]. Fracture toughness can be defined as a measure of the material's resistance to crack propagation [73]. Fracture toughness relates a loaded sample's ability to resist flaw propagation to its mechanical durability. Flaws appear as cracks, voids, metallic inclusions, weld defects, design discontinuities, or a superimposition of the above, and are considered an unavoidable consequence of specimen processing, fabrication, or service of a material/component. Fracture toughness of materials is described by the stress intensity factor (*K*) that combines load, crack size, and structural geometry. A Roman numeral subscript is used to indicate fracture mode. Mode I fracture is the most common mode and attributed to a crack plane orthogonal to the highest tensile load. The stress intensity factor is represented by the following equation:

$$K_I = \sigma\sqrt{\pi\alpha\beta}$$

where σ is the applied stress (MPa), α is the crack length (m), and β is the individual geometry factor.

According to the differences in mechanical properties, the stress distribution between the joint components follows complex patterns. *In vitro* analysis of dental adhesives is further complicated by residual polymerization stress, in homogeneity of the tooth substrate and interface (adhesive layer, hybrid layer, prism, tubule orientation, etc). Even tensile stress distribution occurs only if none of the involved components deforms laterally under tension, or if the same deformation results from tensile stress applied orthogonally to the adhesive [45].

The concept is to initiate and propagate in a stable manner a crack through the bonded interface using either the chevron notch short rod or bar design [87–89] or a modification of the chevron notched short rod known as the notchless triangular prism [90, 91] or the single-edge notched beam [89].

Differently sized specimens, even when of identical proportions, give different values for K_I stress intensity factor, since stress acting adjacent to a flaw of a given size is influenced by specimen dimensions up to a critical dimension. Beyond this critical dimension, the value K_I becomes a true material property called the plane strain fracture toughness (K_{IC}). The stress intensity K_I represents the level of stress at the tip of the crack, and the fracture toughness K_{IC} denominates the highest stress intensity a material under plane strain conditions can withstand without fracture. Fracture toughness or the strain energy release rates are tests that are considered more meaningful to measure the energy or work to separate the adhesive resin from its bond to dentin [92].

Specimen preparation for fracture toughness test is difficult and no standard procedure for dental adhesives is available [73]. Also, it is not possible to prepare multiple specimens from the same tooth. One point of concern is presence of resin flashes extending out of the chevron, if the required notch is prepared with Teflon tape. These flashes may increase the values measured [89, 93] also can be avoided if groove cuts are prepared afterward [94]. On the other hand, the prepared grooves are more apt to microcracks, which may act as crack initiators and so lower the interfacial fracture toughness [73].

In all studies on fracture toughness, failure analyses revealed primarily interfacial failures, in accordance with observed clinical failure modes [95].

4.7. Finite-element analysis

The finite-element analysis (FEA) is an upcoming and significant research tool for biomechanical analysis in biological research. It is an ultimate method for modeling complex structures and analyzing their mechanical properties. FEA has now become widely accepted as a noninvasive and excellent tool for studying the biomechanics and the influence of mechanical forces on the biological systems. It enables the visualization of superimposed structures, and the stipulation of the material properties of anatomic craniofacial structures [96]. It also allows to establish the location, magnitude, and direction of an applied force, as it may also assign stress points that can be theoretically measured. Overall stress distribution within the tooth/restoration complex is determined by not only geometry and material arrangement, but also material properties, fixation, and loading conditions determine stress distributions [97].

FEA is an analyzing method for stresses and deformations in the structures of any given geometry. In the FEA method, the computational model is developed based on the modular principle and is made from many finite size elements, thus, it is well adapted to the real structures. This procedure is called discretization [98]. The steps followed are generally constructing a finite-element model, followed by specifying appropriate material properties, loading and boundary conditions so that the desired settings can be accurately simulated. Various engineering software packages are available to model and simulate the structure of interest [99].

Most models consider isotropic behavior, since it is not possible to quantify the whole anisotropic structure of a bone, organ with current techniques [100]. The load is applied either to the tooth or to the bone as required. Although, the muscle activity and craniofacial morphology affect the occlusal load in actual clinical situation, it is presently difficult to simulate individual muscle forces to FEA modeling. So, usually vertical or oblique load on the teeth or materials is used as an input load in FEA [101, 102].

FEA has been useful to predict stress distributions within teeth and at the interface of adhesives and dentin. This modeling requires knowledge of the strength of materials vs. the strength of mineralized dental tissues and the differences in elastic moduli of materials versus dental tissues. Then 3D stress distribution within these structure can be calculated during varies types of loading [70].

The results of an FEA are expressed as stresses distributed in the structures under investigation. Using FEA during adhesive testing, applied stresses may be shear, microshear, tensile, microtensile, compressive, etc., or a combination known as von Misses stresses. von Misses stresses depend on entire stress field and are a widely used indicator of the possibility of damage occurrence [103] (**Figure 8**).

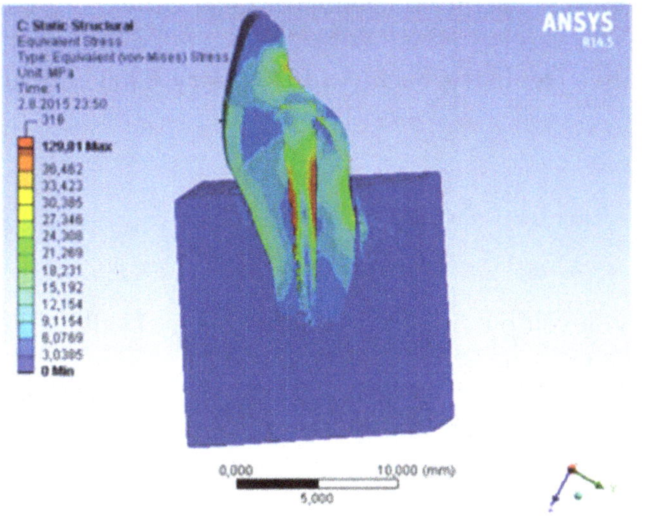

Figure 8. Distribution of von Misses stresses (MPa) in the post and core systems. Blue to red colors represent stress values from lower to higher, respectively. (Dr. N. Güven, Dr. Ö. Topuz).

5. Conclusion

During the past decades, restorative concepts have been continually changing and adhesive technology has become more important. In adhesive dentistry, many kinds of adhesive materials have been using in order to restore damaged tooth structures. Although the materials used in adhesive dentistry approach the properties of enamel and dentin, it is not always possible to replicate their mechanical behaviors. For evaluating adhesive materials' suitability for the clinical and physical properties within reasonable conditions, *in vitro* test methods are necessary. The similarity of elastic modulus and flexural strength between restorative materials and tooth structures is important. The elastic modulus of materials can be tested by a three-point bending test. The macrobond strength can be evaluated with shear or tensile mode, a push- or pull-out methods at a bonded area larger than 3 mm^2. The ability of microbond strength tests is to evaluate the effect of local tooth structure on bond strength and to allow depth profiling of different substrates. Finite-element analysis is useful to predict stress distributions within teeth and interface of adhesives and dentin. This modeling requires knowledge of the strength of materials vs. the strength of mineralized dental tissues and the differences in elastic moduli of materials versus dental tissues. These test methods used in adhesive dentistry are useful for testing new operative techniques and materials before they are clinically implemented.

Author details

Mügem Aslı Gürel[1], Bağdagül Helvacıoğlu Kıvanç[1*] and Hacer Deniz Arısu[2]

*Address all correspondence to: bagdagulkivanc@gmail.com; bagdagul@gazi.edu.tr

1 Department of Endodontics, Faculty of Dentistry, Gazi University, Ankara, Turkey

2 Department of Operative Dentistry, Faculty of Dentistry, Gazi University, Ankara, Turkey

References

[1] Perdigão J. New developments in dental adhesion. Dental Clinics of North America 2007; 51: 333–357. DOI: 10.1016/j.cden.2007.01.001.

[2] Stagel I, Ellis TH, Sacher E. Adhesion to tooth structure mediated by contemporary bonding systems. Dental Clinics of North America 2007; 51: 677–694. DOI: 10.1016/j.cden.2007.04.007.

[3] Kramer IRH, McLean JW. Alterations in the staining reactions of dentine resulting from a constituent of a new self-polymerizing resin. British Dental Journal 1952; 93: 150–153.

[4] Buonocore MG. A simple method of increasing the adhesion of acrylic filling materials to enamel surfaces. Journal of Dental Research 1955; 34: 849–853. DOI: 10.1177/00220345550340060801.

[5] Smith DC. A new dental cement. Brazilian Dental Journal 1968; 124: 381–384.

[6] Boushell LW, Sturdevant JR. Clinical significance of dental anatomy, histology, physiology and occlusion. In: Heymann HO, Swift EJ Jr, Ritter AV, editors. Art and science of operative dentistry. 6th ed. St. Louis: Elsevier/Mosby 2013. p. 2–7. ISBN: 978-0-323-08333-1. ch1.

[7] Breschi L, Gobbi P, Mazzotti G, Ellis TH, Sacher E, Stangel I. Field emission in-lens SEM study of enamel and dentin. Journal of Biomedical Materials Research 1999; 46: 315–323. DOI: 10.1002/(SICI)1097-4636(19990905)46:3<315::AID-JBM2>3.0.CO;2-6.

[8] Di Renzo M, Ellis TH, Sacher E, Stangel I. A photoacoustic FTIRS study of the chemical modifications of human dentin surfaces: I. Demineralization. Biomaterials. 2001; 22: 787–792. DOI: http://dx.doi.org/10.1016/S0142-9612(00)00240-4.

[9] Digka A, Lyroudia K, Jirasek T, Kasampalidis IN, Karayannopoulou G, Kubinova L. Visualisation of human dental pulp vasculature by immunohistochemical and immunofluorescent detection of CD34: a comparative study. Australian Endodontic Journal 2006; 32: 101–106. DOI: 10.1111/j.1747-4477.2006.00028.x.

[10] Van Meerbeek B, Perdigão J, Lambrechts P, Vanherle G. The clinical performance of adhesives. Journal of Dentistry 1998; 26: 1–20. DOI: 10.101603005712(96)00070-x.

[11] Van Meerbeek B, Vargas S, Inoue S, Yoshida Y, Peumans M, Lambrechts P, Vanherle G. Adhesives and cements to promote preservation dentistry. Operative Dentistry 2001; Suppl 6: 119–144.

[12] Van Meerbeek B, De Munck J, Yoshida Y, Inoue S, Vargas M, Vijay P, Van Landuyt K, Lambrechts P, Vanherle G. Buonocore memorial lecture. Adhesion to enamel and dentin: current status and future challenges. Operative Dentistry 2003; 28: 215–235.

[13] Manuja N, Nagpal R, Pandit IK. Dental adhesion: mechanism, techniques and durability. Journal of Clinical Pediatric Dentistry 2012; 36: 223–234. DOI: http://dx.doi.org/10.17796/jcpd.36.3.68805rl1r037m063.

[14] Bowen RL, Eick JD, Henderson DA, Anderson DW. Smear layer: removal and bonding considerations. Operative Dentistry Supplement 1984; 3: 30–34.

[15] Reis A, Carrilho M, Breschi L, Loguercio AD. Overview of clinical alternatives to minimize the degradation of the resin-dentin bonds. Operative Dentistry 2013; 38: E1–E25. DOI: 10.2341/12-258-LIT.

[16] Mount GJ. Clinical performance of glass-ionomers. Biomaterials 1998; 19: 573–579. DOI: 10.1016/S0142-9612(97)00139-7.

[17] Sidhu SK. Glass-ionomer cement restorative materials: a sticky subject? Australian Dental Journal 2011; 56, Suppl 1: 23–30. DOI: 10.1111/j.1834-7819.2010.01293.x.

[18] Zimmerli B, Strub M, Jeger F, Stadler O, Lussi A. Composite materials: composition, properties and clinical applications. A literature review. Schweizer Monatsschrift Zahnmedizin 2010; 120: 972–986.

[19] Nicholson JW. Polyacid-modified composite resins ("compomers") and their use in clinical dentistry. Dental Materials 2007; 23: 615–622 DOI: 10.1016/j.dental.2006.05.002.

[20] Krämer N, Frankenberger R. Compomers in restorative therapy of children: a literature review. International Journal of Pediatric Dentistry 2007; 17: 2–9. DOI: 10.1111/j.1365-263X.2006.00803.x.

[21] Bowen RL. Properties of a silica-reinforced polymer for dental restorations. Journal of the American Dental Association 1963; 66: 57–64.

[22] Kim KH, Ong JL, Okuno O. The effect of filler loading and morphology on the mechanical properties of contemporary composites. The Journal of Prosthetic Dentistry 2002; 87: 642–649. DOI: 10.1067/mpr.2002.125179.

[23] Ikejima I, Nomoto R, McCabe J F. Shear punch strength and flexural strength of model composites with varying filler volume fraction, filler size and silanization. Dental Materials 2003; 19: 206–211. DOI: 10.1016/S0109-5641(02)00031-3.

[24] Hickel R, Roulet JF, Bayne S, Heintze SD, Mjör IA, Peters M, Rousson V, Randall R, Schmalz G, Tyas M, Vanherle G. Recommendations for conducting controlled clinical studies of dental restorative materials. Science committee project 2/98—FDI world federation study design (part I) and criteria for evaluation (part II) of direct and indirect restorations including onlays and partial crowns. The Journal of Adhesive Dentistry 2007; 9: 121–147. DOI: 10.3290/j.jad.a11976.

[25] Brosco VH, Bernardineli N, Moraes IG."In vitro" evaluation of the apical sealing of root canals obturated with different techniques. Journal of Applied Oral Science 2003; 11: 181–185. DOI: 10.1590/S1678-77572003000300005

[26] Kouvas V, Liolios E, Vassiliadis L, Parissis-Messimeris S, Boutsioukis A. Influence of smear layer on depth of penetration of three endodontic sealers: an SEM study. Endodontics & Dental Traumatology 1998; 14: 191–195.

[27] Teixeira FB, Teixeira EC, Thompson JY, Trope M. Fracture resistance of roots endodontically treated with a new resin filling material. The Journal of American Dental Association 2004; 135: 646–652. DOI: 10.14219/jada.archive.2004.0255.

[28] Assif D, Gorfil C. Biomechanical considerations in restoring endodontically treated teeth. The Journal of Prosthetic Dentistry 1994; 71: 565–567.DOI: 10.1016/0022-3913(94)90438-3

[29] Reeh ES, Messer HH, Douglas WH. Reduction in tooth stiffness as a result of endodontic and restorative procedures. Journal of Endodontics 1989; 15: 512–516. DOI: 10.1016/S0099-2399(89)80191-8.

[30] Dietschi D, Bouillaguet S, Sadan A. Restoration of the Endodontically Treated Tooth. In: Hargreaves KM, Cohen S, editors. Cohen's Pathways of the Pulp. 10th ed. Mosby: Elsevier; 2011. p. 777–807. ch. 22.

[31] Nadim Z. Baba andCharles J. Goodacre. Restoration of endodontically treated teeth: contemporary concepts and future perspectives. Endodontic Topics 2014; 31: 68–83. DOI: 10.1111/etp.12066.

[32] Lassila LV, Tanner J, Le Bell AM, Narva K, Vallittu PK. Flexural properties of fiber reinforced root canal posts. Dental Materials 2004; 20: 29–36. DOI: 10.1016/S0109-5641(03)00065-4.

[33] Tay FR, Pashley DH. Monoblocks in root canals: a hypothetical or a tangible goal. Journal of Endodontics 2007; 33: 391–398. DOI: 10.1016/j.joen.2006.10.009.

[34] Robbins JW. Guidelines for the restoration of endodontically treated teeth. The Journal of the American Dental Association 1990; 120: 558, 560, 562, 564, 566. DOI: http://dx.doi.org/10.14219/jada.archive.1990.0087.

[35] Standlee JP, Caputo AA, Hanson EC. Retention of endodontic dowels: effects on cement, dowel length, diameter, design. The Journal of Prosthetic Dentistry 1978; 39: 401–405. DOI: http://dx.doi.org/10.1016/S0022-3913(78)80156-5.

[36] Alfredo E, de Souza ES, Marchesan MA, Paulino SM, Gariba-Silva R, Sousa-Neto MD. Effect of eugenol-based endodontic cement on the adhesion of intraradicular posts. Brazilian Dental Journal 2006; 17: 130–133. DOI: http://dx.doi.org/10.1590/S0103-64402006000200009.

[37] Chan FW, Harcourt JK, Brockhurst PJ. The effect of post adaptation in the root canal on retention of posts cemented with various cements. Australian Dental Journal 1993; 38: 39–45. DOI: 10.1111/j.1834-7819.1993.tb05449.x.

[38] Mitchell CA, Orr JF. Engineering properties and performance of dental crowns. The Journal of Engineering in Medicine, Proceedings of the Institution of Mechanical Engineers Part H 2005; 219: 245–255. DOI: 10.1243/095441105X34248.

[39] Urapepon S, Taenguthai P. The effect of zirconia framework design on the failure of all-ceramic crown under static loading. The Journal of Advanced Prosthodontics 2015; 7: 146–150. DOI: 10.4047/jap.2015.7.2.146.

[40] Burke FJ. The effect of variations in bonding procedure on fracture resistance of dentin-bonded all-ceramic crowns. Quintessence International 1995; 26: 293–300.

[41] Simonsen RJ. Calamia JR. Tensile hond strength of etched porcelain. Journal of Dental Research 1983; 62: 297. DOI: 10.1177/00220345830620S201.

[42] Crothers AJR, Wassell RW, Allen R. The resin-bonded porcelain crown: a rationale for use on anterior teeth. Dental Update 1993; 20: 388–395.

[43] FDA: Good Laboratory Practice (GLP). (PART 58 52 FR 33780, 1978, last revision 2004).

[44] Burke FJ, Hussain A, Nolan L, Fleming GJ. Methods used in dentine bonding tests: an analysis of 102 investigations on bond strength. European Journal Prosthodontics and Restorative Dentistry 2008; 16: 158–165.

[45] Salz U, Bock T. Testing adhesion of direct restoratives to dental hard tissue – a review. Journal of Adhesive Dentistry 2010; 12: 343–371. DOI: 10.3290/j.jad.a19741.

[46] Proença JP, Polido M, Osorio E, Erhardt MC, Aguilera FS, García-Godoy F, Osorio R, Toledano M. Dentin regional bond strength of self-etch and total-etch adhesive systems. Dental Materials 2007; 23: 1542–8. DOI: 10.1016/j.dental.2007.02.001.

[47] Shimada Y, Tagami J. Effects of regional enamel and prism orientation on resin bonding. Operative Dentistry 2003; 28: 20–27.

[48] ISO: No. 4049 Dentistry – Polymer-based filling, restorative and luting materials (2009).

[49] Heintze SD, Zimmerli B. Relevance of in vitro tests of adhesive and composite dental materials. A review in 3 parts. Part 3: in vitro tests of adhesive systems. Schweizer Monatsschrift Zahnmedizin 2011; 121: 1024–1040.

[50] Lambrechts P, Braem M, Vanherle G. Buonocore memorial lecture. Evaluation of clinical performance for posterior composite resins and dentin adhesives. Operative Dentistry 1987; 12: 53–78.

[51] Suansuwan N, Swain MV. Determination of elastic properties of metal alloy and dental porcelains. Journal of Oral Rehabilitation 2001; 28: 133–139. DOI: 10.1046/j.1365-2842.2001.00642.x.

[52] DeHoff PH, Anusavice KJ, Wang Z. Three-dimensional finite element analysis of the shear bond test. Dental Materials 1995; 11: 126–131. DOI: 10.1016/0109-5641(95)80047-6.

[53] Pecora N, Yaman P, Dennison J, Herrero A. Comparison of shear bond strength relative to two testing devices. Journal of Prosthetic Dentistry 2002; 88: 511–515. DOI: 10.1067/mpr.2002.129063.

[54] Versluis A, Tantbirojn D, Douglas WH. Why do shear bond tests pull out dentin? Journal of Dental Research 1997; 76: 1298–1307. DOI: 10.1177/00220345970760061001.

[55] Phrukkanon S, Burrow MF, Tyas MJ. Effect of cross-sectional surface area on bond strengths between resin and dentin. Dental Materials 1998; 14: 120–128. DOI: 10.1016/S0109-5641(98)00018-9.

[56] Sadr A, Ghasemi A, Shimada Y, Tagami J. Effects of storage time and temperature on the properties of two self-etching systems. Journal of Dentistry 2007; 35: 218–225. DOI: 10.1016/j.jdent.2006.08.004.

[57] Van Noort R, Cardew GE, Howard IC, Noroozi S. The effect of local interfacial geometry on the measurement of the tensile bond strength to dentin. Journal of Dental Research 1991; 70: 889–893. DOI: 10.1177/00220345910700050501.

[58] Tamura Y, Tsubota K, Otsuka E, Endo H, Takubo C, Miyazaki M, Latta MA. Dentin bonding: influence of bonded surface area and crosshead speed on bond strength. Dental Materials Journal 2011; 30: 206–211. DOI: 10.4012/dmj.2010-157

[59] Bucknall CB. New criterion for craze initiation. Polymer 2007; 48: 1030–1041. DOI: 10.1016/j.polymer.2006.12.033.

[60] Mota CS, Demarco FF, Camacho GB, Powers JM. Tensile bond strength of four resin luting agents bonded to bovine enamel and dentin. Journal of Prosthetic Dentistry 2003; 89: 558–564. DOI: 10.1016/S0022-3913(03)00177-x.

[61] Braga RR, Meira JB, Boaro LC, Xavier TA. Adhesion to tooth structure: a critical review of "macro" test methods. Dental Materials 2010; 26: e38–e49. DOI: 10.1016/j.dental.2009.11.150.

[62] Armstrong S, Geraldeli S, Maia R, Raposo LH, Soares CJ, Yamagawa J. Adhesion to tooth structure: a critical review of "micro" bond strength test methods. Dental Materials 2010; 26: e50–e62. DOI: 10.1016/j.dental.2009.11.155.

[63] Abreu A, Loza MA, Elias A, Mukhopadhyay S, Looney S, Rueggeberg FA. Tensile bond strength of an adhesive resin cement to different alloys having various surface treat-

ments. Journal of Prosthetic Dentistry 2009; 101: 107–118. DOI: 10.1016/S0022-3913(09)60004-4.

[64] Kern M, Barloi A, Yang B. Surface conditioning influences zirconia ceramic bonding. Journal of Dental Research 2009; 88: 817–822. DOI: 10.1177/0022034509340881.

[65] Shono Y, Terashita M, Shimada J, Kozono Y, Carvalho RM, Russell CM, Pashley DH. Durability of resin-dentin bonds. Journal of Adhesive Dentistry 1999; 1: 211–218.

[66] Betamar N, Cardew G, Van Noort R. Influence of specimen designs on the microtensile bond strength to dentin. Journal of Adhesive Dentistry 2007; 9: 159–168.

[67] Poitevin A, De Munck J, Van Landuyt K, Coutinho E, Peumans M, Lambrechts P, Van Meerbeek B. Critical analysis of the influence of different parameters on the microtensile bond strength of adhesives to dentin. Journal Adhesive Dentistry 2008; 10: 7–16.

[68] Perdigão J, Geraldeli S, Carmo AR, Dutra HR. In vivo influence of residual moisture on microtensile bond strengths of one-bottle adhesives. Journal of Esthetic and Restorative Dentistry 2002; 14: 31–38. DOI: 10.1111/j.1708-8240.2002.tb00145.x.

[69] Poitevin A, De Munck J, Van Landuyt K, Coutinho E, Peumans M, Lambrechts P, Van Meerbeek B. Influence of three specimen fixation modes on the micro-tensile bond strength of adhesives to dentin. Dental Materials Journal 2007; 26: 694–699. DOI: http://doi.org/10.4012/dmj.26.694.

[70] Pashley DH, Sano H, Ciucchi B, Yoshiyama M, Carvalho RM. Adhesion testing of dentin bonding agents: a review. Dental Materials 1995; 11: 117–125. DOI: 10.1016/0109-5641(95)80046-8.

[71] Sirisha K, Rambabu T, Ravishankar Y, Ravikumar P. Validity of bond strength tests: a critical review-Part II. Journal of Conservative Dentistry 2014; 17: 420–426. DOI: 10.4103/0972-0707.139823.

[72] Blunck U, Roulet JF. Effect of one-year water storage on the effectiveness of dentin adhesives in Class V composite resin restorations. Journal of Dental Research M 2002; 81: A139. DOI:10.1177/0022034502081S101.

[73] De Munck J, Van Landuyt K, Peumans M, Poitevin A, Lambrechts P, Braem M, Van Meerbeek B. A critical review of the durability of adhesion to tooth tissue: methods and results. Journal of Dental Research 2005; 84: 118–132. DOI: 10.1177/154405910508400204.

[74] Dewji HR, Drummond JL, Fadavi S, Punwani I. Bond strength of Bis-GMA and glass ionomer pit and fissure sealants using cyclic fatigue. European Journal of Oral Sciences 1998; 106: 594–599. DOI: 10.1046/j.0909-8836.1998.eos106110.x

[75] Givan DA, Fitchie JG, Anderson L, Zardiackas LD. Tensile fatigue of 4-META cement bonding three base metal alloys to enamel and comparison to other resin cements. Journal of Prosthetic Dentistry 1995; 73: 377–385.

[76] Frankenberger R, Krämer N, Petschelt A. Fatigue behaviour of different dentin adhesives. Clinical Oral Investigation 1999; 3: 11–17. DOI: 10.1007/s007840050072

[77] Fissore B, Nicholls JI, Yuodelis RA. Load fatigue of teeth restored by a dentin bonding agent and a posterior composite resin. Journal of Prosthetic Dentistry 1991; 65: 80–85. DOI: 10.1016/0022-3913(91)90054-Z.

[78] De Munck J, Braem M, Wevers M, Yoshida Y, Inoue S, Suzuki K, Lambrechts P, Van Meerbeek B. Micro-rotary fatigue of tooth-biomaterial interfaces. Biomaterials. 2005; 26: 1145–1153. DOI: 10.1016/j.biomaterials.2004.04.013.

[79] Wiskott HW, Nicholls JI, Belser UC. Fatigue resistance of soldered joints: a methodological study. Dental Materials 1994; 10: 215–220. DOI: 10.1016/0109-5641(94)90035-3.

[80] Roydhouse R H. Punch-shear test for dental purposes. Journal of Dental Research 1970; 49: 131–136. DOI: 10.1177/00220345700490010501

[81] Frankenberger R, Sindel J, Krämer N, Petschelt A. Dentin bond strength and marginal adaptation: direct composite resins vs ceramic onlays. Operative Dentistry 1999; 24: 147–155.

[82] Goracci C, Tavares AU, Fabianelli A, Monticelli F, Raffaelli O, Cardoso PC, Tay F, Ferrari M. The adhesion between fiber posts and root canal walls: comparison between microtensile and push-out bond strength measurements. European Journal of Oral Science 2004; 112: 353–361. DOI: 10.1111/j.1600-0722.2004.00146.x.

[83] Castellan CS, Santos-Filho PC, Soares PV, Soares CJ, Cardoso PE. Measuring bond strength between fiber post and root dentin: a comparison of different tests. Journal of Adhesive Dentistry 2010; 12: 477–485. DOI: 10.3290/j.jad.a17856.

[84] Ebert J, Leyer A, Günther O, Lohbauer U, Petschelt A, Frankenberger R, Roggendorf MJ. Bond strength of adhesive cements to root canal dentin tested with a novel pull-out approach. Journal of Endodontics 2011; 37: 1558–1561. DOI: 10.1016/j.joen.2011.08.009.

[85] Daghyani HR, Ye L, Mai YW. Mode-l fracture behaviour of adhesive joints. Part I. Relationship between fracture energy and bond thickness. The Journal of Adhesion 1995; 53: 149–162. DOI:10.1080/00218469508009935.

[86] Yan C, Mai YW, Yuan Q, Ye L, Sun J. Effects of substrate materials on fracture toughness measurement in adhesive joints. International Journal of Mechanical Sciences 2001; 43: 2091–2102. DOI: 10.1016/S0020-7403(01)00030-3.

[87] Tam LE, Pilliar RM. Fracture toughness of dentin/resin-composite adhesive interfaces. Journal of Dental Research 1993; 72: 953–959. DOI: 10.1177/00220345930720051801.

[88] Lin CP, Douglas WH. Failure mechanisms at the human dentin-resin interface: a fracture mechanics approach. Journal of Biomechanics 1994; 27: 1037–1047. DOI: 10.1016/0021-9290(94)90220-8.

[89] Tantbirojn D, Cheng YS, Versluis A, Hodges JS, Douglas WH. Nominal shear or fracture mechanics in the assessment of composite-dentin adhesion? Journal of Dental Research 2000; 79: 41–48. DOI: 10.1177/00220345000790010601.

[90] Ruse ND, Troczynski T, MacEntee MI, Feduik D. Novel fracture toughness test using a notchless triangular prism (NTP) specimen. Journal of Biomedical Materials Research 1996; 31: 457–463. DOI: 10.1002/(SICI)1097-4636(199608)31:4<457::AID-JBM4>3.0.CO;2-K.

[91] Far C, Ruse ND. Effect of bleaching on fracture toughness of composite-dentin bonds. Journal of Adhesive Dentistry 2003; 5: 175–182. DOI: 10.1177/00220345000790010601.

[92] Scherrer SS, Cesar PF, Swain MV. Direct comparison of the bond strength results of the different test methods: a critical literature review. Dental Materials 2010; 26: e78–e93. DOI: 10.1016/j.dental.2009.12.002.

[93] Van Noort R, Cardew GE, Howard IC, Noroozi S. The effect of local interfacial geometry on the measurement of the tensile bond strength to dentin. Journal of Dental Research 1991; 70: 889–893. DOI: 10.1177/00220345910700050501.

[94] Armstrong SR, Keller JC, Boyer DB. Mode of failure in the dentin-adhesive resin-resin composite bonded joint as determined by strength-based (muTBS) and fracture-based (CNSB) mechanical testing. Dental Materials 2001; 17: 201–210. DOI: 10.1016/S0109-5641(00)00070-1.

[95] Tam LE, Pilliar RM. Fracture surface characterization of dentin-bonded interfacial fracture toughness specimens. Journal of Dental Research 1994; 73: 607–619. DOI: 10.1177/00220345940730030601.

[96] Ausiello P, Apicella A, Davidson CL. Effect of adhesive layer properties on stress distribution in composite restorations – a 3D finite element analysis. Dental Materials 2002; 18: 295–303. DOI: http://dx.doi.org/10.1016/S0109-5641(01)00042-2.

[97] Magne P, Versluis A, Douglas WH. Rationalization of incisor shape: experimental-numerical analysis. Journal of Prosthetic Dentistry 1999; 81: 345–355. DOI: 10.1016/S0022-3913(99)70279-9.

[98] Maurer P, Holweg S, Knoll WD, Schubert J. Study by finite element method of the mechanical stress of selected biodegradable osteosynthesis screws in sagittal ramus osteotomy. British Journal of Oral and Maxillofacial Surgery 2002; 40: 76–83. DOI: 10.1054/bjom.2001.0752.

[99] Trivedi S. Finite element analysis: a boon to dentistry. Journal of Oral Biology and Craniofacial Research 2014; 4: 200–203. DOI: 10.1016/j.jobcr.2014.11.008.

[100] Doblaré M, García JM, Gómez MJ. Modelling bone tissue fracture and healing: a review. Engineering Fracture Mechanics 2004; 71: 1809–1840. DOI: 10.1016/j.engfracmech.2003.08.003.

[101] Liu S, Liu Y, Xu J, Rong Q, Pan S. Influence of occlusal contact and cusp inclination on the biomechanical character of a maxillary premolar: a finite element analysis. Journal of Prosthetic Dentistry 2014; 112: 1238–1245. DOI: 10.1016/j.prosdent.2014.04.011.

[102] O'Mahony AM, Williams JL, Spencer P. Anisotropic elasticity of cortical and cancellous bone in the posterior mandible increases peri-implant stress and strain under oblique loading. Clinical Oral Implants Research 2001; 12: 648–657. DOI: 10.1034/j.1600-0501.2001.120614.x.

[103] Pegoretti A, Fambri L, Zappini G, Bianchetti M. Finite element analyses of a glass fibre reinforced composite endodontic post. Biomaterials 2002; 23: 2067–2082. DOI: 10.1016/S0142-9612(01)00407-0.

3

Toughness Assessment and Fracture Mechanism of Brittle Thin Films Under Nano-Indentation

Kunkun Fu, Youhong Tang and Li Chang

Abstract

The nano-indentation technique is an effective tool for assessing the fracture toughness of brittle thin film. In this chapter, a comprehensive and systematic review of toughness measurement methods and fracture mechanisms in brittle film by indentation are presented. The classic method, the energy method, and the stress-based method are three major approaches for determining fracture toughness using the nano-indentation technique. The limits and application ranges of these methods are discussed in detail. In particular, the stress-based method depends highly on the fracture mechanism of cracking. This chapter also reviews different types of crack patterns induced by nano-indentation, such as radial cracks, ring cracks, picture-frame cracks, spiral cracks, and spalling. The possible mechanisms of the crack patterns are investigated considering the substrate effect, the indenter shape effect, and the load level effect. Understanding the fracture mechanism provides guidance in developing a more accurate stress-based model.

Keywords: fracture toughness, fracture mechanism, brittle thin film, nano-indentation

1. Introduction

Over decades, hard thin film has been extensively utilized as protective layers to minimize detrimental influences of the environment. However, hard film usually has a brittle character, and hence fracture failure often occurs in hard thin film when it is subjected to high stress. In designing a high-performance film, it is of great importance to learn the fracture properties of the coating.

Because of the size limitation, the fracture behavior of brittle thin films cannot be easily determined by the standard linear elastic fracture mechanics tension test or the three-point bending test. In this case, nano-indentation may be the only effective technique to quantitatively characterize the fracture toughness of the film. Anstis et al. [1] were the first to propose an indentation method to evaluate the fracture toughness of brittle materials by measuring the length of radial cracks. In their approach, the assumption was that radial cracks can develop well without any confinement during loading. This classical method has been successfully applied to some "thick" film/substrate systems where radial cracks are well propagated [2–5]. However, it is invalid for brittle thin films/substrate systems because the substrate effect is not negligible and has an effect on crack propagation. For brittle thin films, the energy method proposed by Li et al. [6] has been widely used in recent years for the measurement of fracture toughness although it suffers from some deficiencies that require further research. Subsequently, a number of studies [7–10] have proposed improvements to this energy method. An alternative approach for assessing the fracture toughness of brittle thin films is to use a stress-based model. The most important prerequisite for a stress-based model is understanding the stress distribution in brittle films. Finite element (FE) analysis [11–13] and simplified analytical solutions [14, 15] are two efficient ways to predict stress distribution under indentation. It is evident that stress distribution depends strongly on the indenter shape and the substrate. Different indenter shapes and substrates can result in various crack patterns. A number of indentation-induced crack patterns have been reported, such as radial cracks [16–18], ring cracks [7, 19, 20], picture-frame cracks [21, 22], spiral cracks [23, 24], and spalling [25, 26]. Understanding the mechanism of the above crack patterns by indentation would greatly support the development of a stress-based model.

This chapter proposes a comprehensive and systematic view of fracture toughness assessment methods and fracture mechanisms for brittle thin films under nano-indentation. First, we present a review of the current indentation methods for characterizing the fracture toughness of brittle films, namely the classical method, energy method, numerical method, and stress-based method. The limits and application range of each method are discussed. Next, we review the observation of crack patterns in brittle thin films/substrates under nano-indentation. The various effects of indenter shape, substrate effects, and load levels on crack formation are discussed to gain an understanding of the fracture mechanism of cracking. The chapter concludes with a summary and a roadmap for future trends.

2. Determination of fracture toughness of brittle thin films using nano-indentation

2.1. Classical method

It is agreed that a Vickers' pyramid indenter may produce a median/radial crack pattern in brittle materials. **Figure 1(a)** shows a schematic representation of the median/radial crack pattern induced by a Vickers indenter. The average length of the radial cracks is c_m, and a is the impression length. Lawn et al. [27] reported that the elastic/plastic field in material under indentation can be considered as a residual field in an unloaded solid and an ideal elastic field.

Then the fracture toughness of materials can be obtained by measuring the length of the radial crack and the critical indentation load, *P*, as follows:

$$K_c = \chi \frac{P}{c_m^{3/2}} \quad (1)$$

where χ is a factor relating to the indenter geometry and material properties of the tested specimen and is expressed in the form of

$$\chi = \xi \left(\frac{E}{H} \right)^{2/5} \quad (2)$$

where *E* and *H* are the elastic moduli and hardness of the material, respectively. ξ is a constant depending on the indenter geometry and can be determined by linear fitting of the relation between *P* and $c_m^{3/2}$. A value of 0.016 ± 0.004 for ξ was reported in [1] by experimentally fitting *P* and $c_m^{3/2}$ for a great variety of brittle materials.

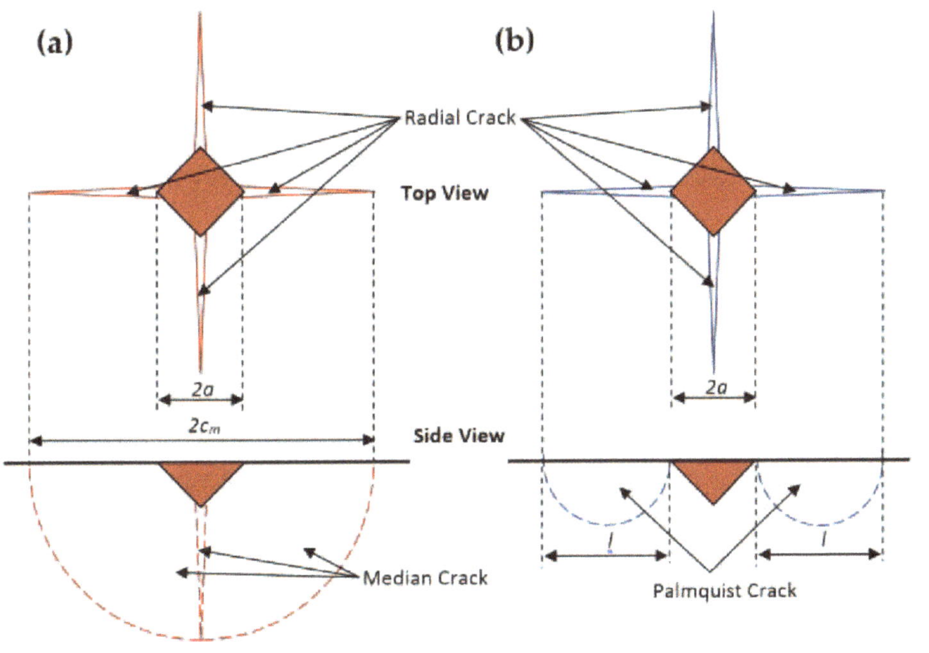

Figure 1. Schematic representation of the crack pattern induced by Vickers indenter: (a) the radial/median crack and (b) the Palmquist crack, based on [28].

The Palmquist crack is another commonly observed crack pattern in brittle materials caused by a Vickers indenter, as shown in **Figure 1(b)**. To assess the fracture toughness of materials in which the Palmquist crack pattern is detected, Eq. (1) needs to be modified as

$$K_c = \xi_v \left(\frac{a}{l}\right)^{1/2} \left(\frac{E}{H}\right)^{2/3} \frac{P}{c_m^{3/2}} \tag{3}$$

where ξ_v is a material constant and has the value of 0.015 reported in [29]. l is the average length of a Palmquist crack.

Although Eqs. (1) and (3) were initially derived from the stress distribution in material under a Vickers indenter, they can be extended to other types of pyramidal indenters, such as a Berkovich indenter and a cube-corner indenter. Comparison of the toughness results of various brittle materials obtained by a Berkovich indenter and a Vickers indenter in [29] showed that the Berkovich indenter could provide more accurate results at a low load range than those of the Vickers indenter, as the shape of the Berkovich indenter is sharper than that of the Vickers indenter. Furthermore, a cube-corner indenter is generally considered the sharpest indenter. It was reported by Pharr [30], therefore, that the cube-corner indenter can significantly reduce the cracking threshold, and is most suitable for toughness measurement of brittle materials at small load compared to other types of pyramidal indenter.

It should be stated that the residual stress in materials has an effect on the elastic/plastic field, and thus on the results of the predicted toughness. If there is a considerable amount of residual stress in brittle materials, Eq. (1) should be changed as follows [31]:

$$K_c = \chi \frac{P}{c_m^{3/2}} + 2m\sigma_R \sqrt{\frac{c_m}{\pi}} \tag{4}$$

where m denotes a dimensionless factor and σ_R denotes the residual stress in the materials.

Eqs. (1)–(4) have been successfully utilized with "thick" film/substrate systems [2–5] where the radial cracks or Palmquist cracks could propagate well. However, the toughness measurement of a "thin" film can be affected by the substrate effect. As a rule of thumb, the substrate effect is negligible when the indentation depth is less than 10% of the film thickness for a soft film/hard substrate. If the film thickness is submicro, the maximum indentation depth should be lower than tens of nanometers in order to minimize the influence of the substrate. However, for most nano-indenters, it is nearly impossible to keep an ideal pyramidal shape on such a small scale. Furthermore, the "1/10 principle" is not adequate for a hard film/soft substrate, which results in an even lower peak depth in order to reduce the influence of the substrate. Therefore, other methods should be employed to evaluate the fracture toughness of brittle thin films.

2.2. Energy method

Theoretically, using the energy method to characterize fracture toughness could minimize the substrate effect. Therefore, the energy method may be the most efficient method for the toughness measurement of brittle thin films. Li et al. [6] were the first to propose an energy method for assessing the fracture toughness of brittle thin films, and fracture toughness is given by

$$K_c = \left[\left(\frac{E_f}{(1-v^2) l_m} \right) \left(\frac{U_{fr}}{t'} \right) \right]^{1/2} \tag{5}$$

where E_f is the film modulus, v is the Poisson's ratio of the coating, l_m is the crack length in the film plane, U_{fr} is the fracture energy, t' is the effective film thickness, and has the form of

$$t' = t / \sin(\theta) \tag{6}$$

where t is film thickness and θ is the angle of the crack edge.

The key point of the energy method is to identify the dissipated energy according to the indentation load-depth curve. First, the correspondence between the load-depth curve and crack formation needs to be determined. For a perfect "pop-in" as illustrated in **Figure 2(a)**, there is a definite step in the loading curve suggesting where the crack initiates and ends. During the crack formation, a significant amount of energy is dissipated, which causes the discontinuity in the loading curve.

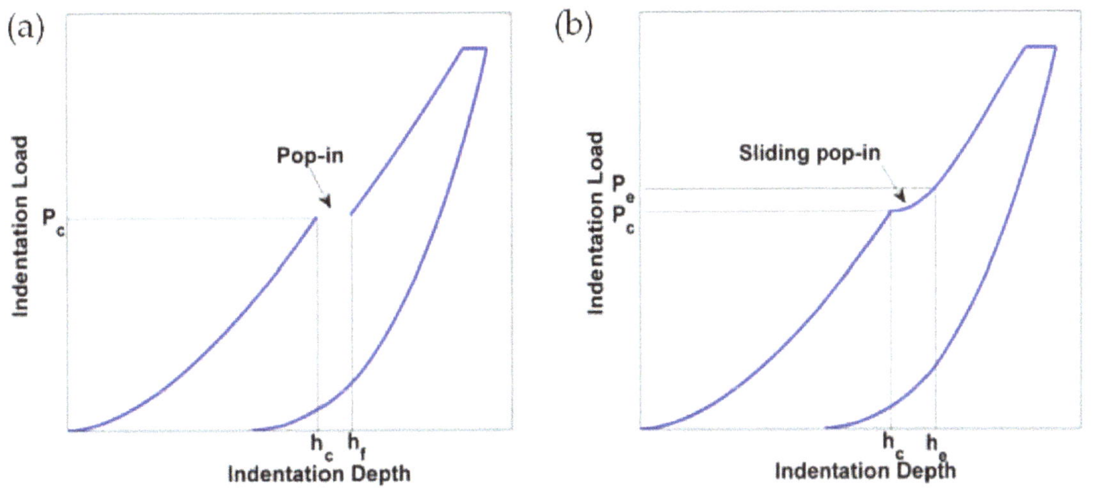

Figure 2. Schematic representation of (a) a perfect "pop-in" and (b) a "sliding pop-in" in load-depth curves [7].

A more general case is a "sliding pop-in" in the load-depth curve when a high-resolution transducer is used in an experiment. In **Figure 2(b)**, there is a clear starting point indicating the crack initiation. However, there is no end point in the load curve. Fu et al. [7] developed a method to define the end point of the "sliding pop-in" in a load-depth curve. They assumed that $\partial P/\partial h^2$ remains constant before and after a circumferential crack formation. **Figure 3** shows $\partial P/\partial h^2$ as a function of h^2. A quick drop of $\partial P/\partial h^2$ is observed, indicating the onset point of the "sliding pop-in" in the load-depth curve and also the initiation of the circumferential crack.

After the crack is complete, the derivative $\partial P/\partial h^2$ becomes stable again and the value is similar to that before the crack initiation. Accordingly, the end point of the "sliding pop-in" is identified.

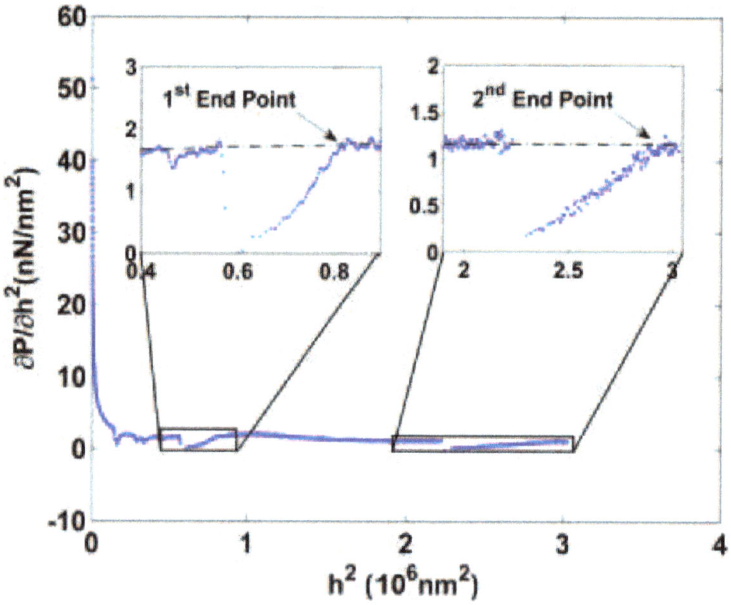

Figure 3. A method to define the end point of a "sliding pop-in" [7].

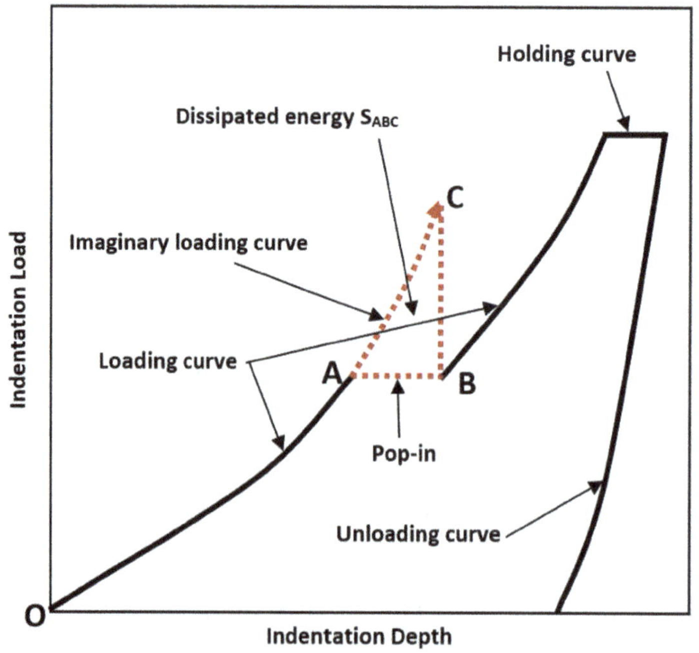

Figure 4. Schematic representation of a method to obtain the dissipated energy.

After the start point and end point of the crack in a loading curve have been obtained, the dissipated energy during crack formation may be calculated by different methods. For instance, a schematic of the load-depth curve in [6] is illustrated in **Figure 4**. The crack initiates at point A and ends at point B. If the crack does not occur, the loading curve follows the trend of the OA curve and reaches the point C. Hence, the enclosed area S_{ABC} is assumed to represent the dissipated energy during cracking [6]. This method is widely used to evaluate the fracture toughness of thin films, even though it ignores the change in elastic-plastic stress distribution of the film before and after crack propagation.

Given the shortcomings of the method in [6], Chen et al. [8] developed another method to define the dissipated energy by using the curve of total work as a function of displacement. **Figure 5** shows that a crack initiates at A and ends at D. The first work-displacement curve is extrapolated from A to C, and the second work-displacement curve is extrapolated from D to B. The work difference between CD and AB is considered to represent the dissipated energy during the crack formation. The negative or positive of AB relies on the film/substrate system. This method examines the work difference before and after cracking, and thus the energy caused by the change of elastic-plastic behavior is excluded. However, the accuracy of this method is still dependent on extrapolation of the imaginary work-displacement curves.

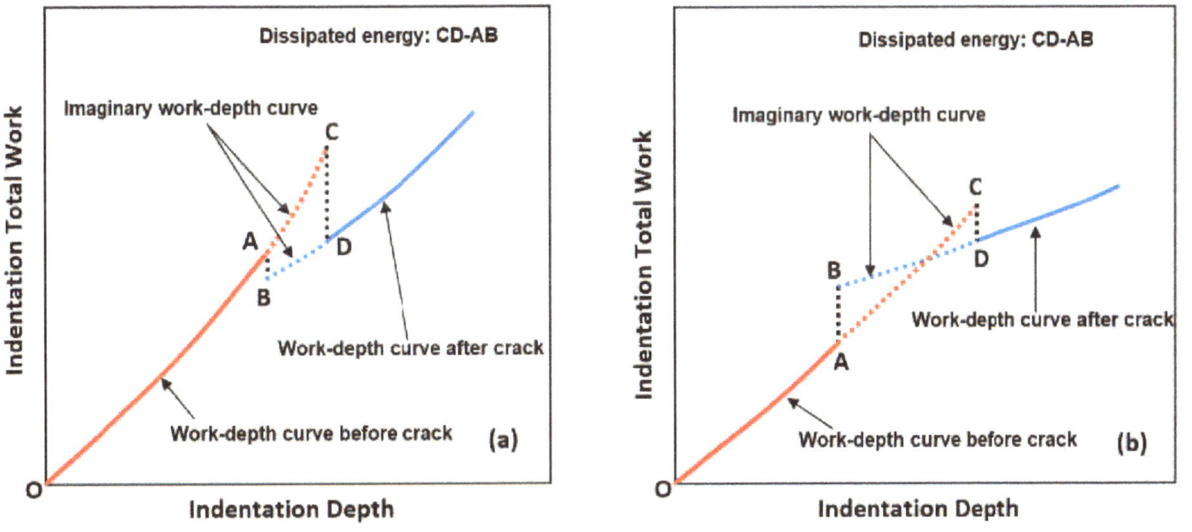

Figure 5. Schematic representation of a method to determine the dissipated energy CD-AB. Compared to CD, AB is positive in (a) or negative in (b), depending on the actual coated systems.

There are also a few studies focusing on estimating the bounds of the fracture energy. For instance, Toonder et al. [9] proposed an approach to evaluate the upper and lower limits of fracture energy by considering a material as pure elastic or perfect plastic. In **Figure 6(a)**, if the film/substrate system behaves as an elastic material, all the deformation will recover after unloading as the dashed curve BO. If the film/substrate system displays perfect plastic behavior, there is no elastic recovery. The unloading curves before and after the crack will be AC and BD, respectively. In practice, the film/substrate system is an elastic-perfect plastic material, and therefore, the dissipated energy should be between the enclosed area of ABO and ABDC. Hence, the upper and lower limits of the dissipated energy during cracking are given by

$$\frac{2}{3}P_{cr}(\delta_2-\delta_1) \le U_{fr} < P_{cr}(\delta_2-\delta_1) \tag{7}$$

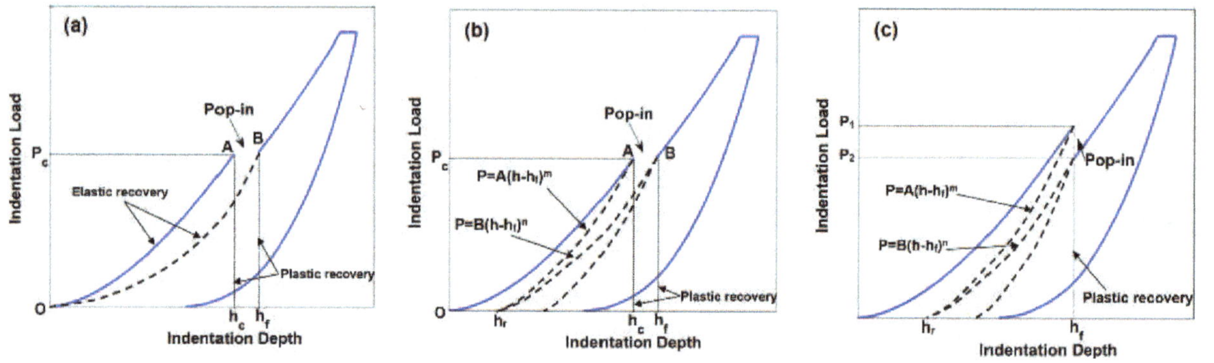

Figure 6. (a) A method and (b) an improved method to define dissipated energy using load control and (c) displacement control.

Further, Chen et al. [10] enhanced the method in [9] by performing a more reliable analysis of the unloading curves, as shown in **Figure 6(b)**. The bounds of the fracture toughness are expressed as

$$\int_{\delta_f}^{\delta_1} A(x-\delta_f)^m \mathrm{d}x + P_{cr}(\delta_2-\delta_1) \ge U_{fr} > \int_0^{P_{cr}} \left(\left(\frac{y}{B}\right)^{1/n} - \left(\frac{y}{A}\right)^{1/m} \right) \mathrm{d}y \tag{8}$$

Similarly, an approach to calculate the bounds of fracture energy of the film system under a displacement control is shown in **Figure 6(c)**. The dissipated energy is given by [10]

$$\int_{\delta_f}^{\delta_{cr}} A(x-\delta_f)^m \mathrm{d}x \ge U_{fr} > \int_{\delta_f}^{\delta_{cr}} \left(A(x-\delta_f)^m - B(x-\delta_f)^n \right) \mathrm{d}x \tag{9}$$

The energy method is efficient for obtaining the fracture toughness of brittle thin films because there is no need for information about crack propagation, such as crack size. However, the accuracy of the energy method depends highly on either the extrapolated imaginary load-depth curve or the imaginary total work-displacement curve. In experiments, it is impossible to know the accuracy of the imaginary extrapolated curve as the trend of the loading curve can be affected by many factors, such as roughness of the surface, defects in the films and the ratio of indentation depth to film thickness.

2.3. Stress-based method

The stress-based method is a straightforward approach for assessing the fracture toughness of brittle thin films. It is based on stress distribution in the brittle thin film under an indenter. To date, some analytical solutions have been derived for stress distribution in an incompressible elastic coating [32], compressible elastic coating/rigid substrate [33], and elastic film/elastic substrate [34, 35]. However, due to the complex elastic-plastic properties of film and substrate and the boundary condition of the indenter and film, it is not feasible to derive an analytical solution for evaluating the stress distribution in the brittle film/elastic-plastic substrate system. FE analysis is a useful tool for obtaining the fracture stress in film under indentation. Hence, stress distribution in a film is usually determined using FE analysis. For example, for a brittle film on a ductile substrate, the stress distribution under a conical indenter is obtained by FE analysis and is shown in **Figure 7**. A high tensile radial stress occurs at the surface of a film, which can open a crack.

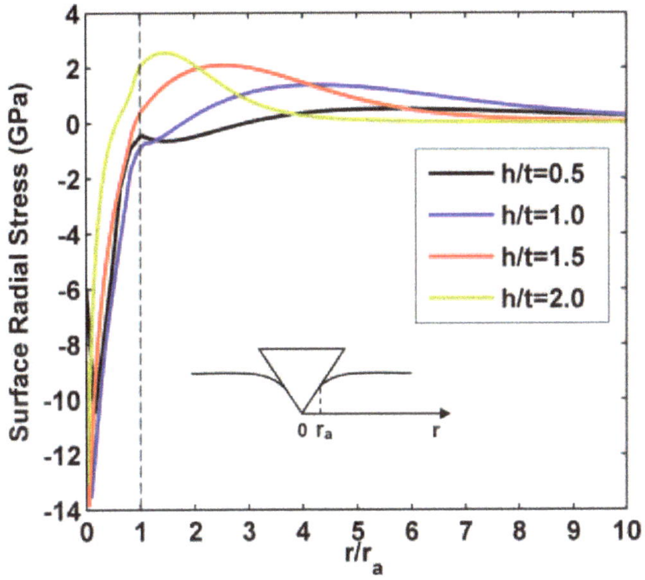

Figure 7. Radial stress in film at the surface for various ratios of indentation depth *h* and film thickness *t* [7].

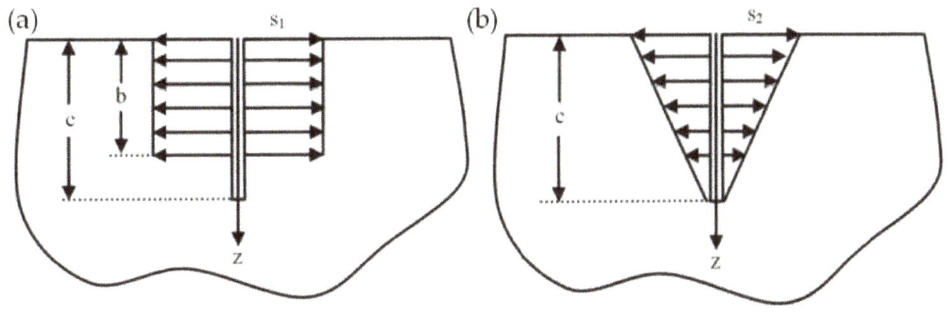

Figure 8. Stress applied to a crack during indentation: (a) constant stress and (b) linear stress [15].

When the stress distribution in a film has been obtained, the fracture toughness can be obtained by assuming the stress on pre-existing cracks as a uniformly distributed crack and a linearly distributed crack as shown in **Figure 8(a)** and **(b)**. Based on the linear fracture mechanics, fracture toughness is expressed as

$$K_{\mathrm{I1}} = 1.12 s_1 \sqrt{\pi c} + 0.439 s_2 \sqrt{\pi c} \tag{10}$$

where s_1 is a constant stress, s_2 is the highest stress in a linear stress distribution, b is the crack length, and 1.12 and 0.439 are the geometric factors in **Figure 8(a)** and **(b)**, respectively.

An initial short crack may grow to a circumferential channel crack by self-adjusting to a curved shape when the energy release rate G_{ps} by linear fracture mechanics is the same at each point, as reported by Steffensen et al. [11]. The energy release rate for the circumferential channel crack, G_{ss}, is given by

$$G_{\mathrm{ss}}(c) = \frac{1}{c}\int_0^c G_{\mathrm{ps}}(\tilde{c})\mathrm{d}\tilde{c} \tag{11}$$

G_{ps} is expressed as

$$G_{\mathrm{ps}} = K_{\mathrm{I}}^2 \left(\frac{1-v^2}{E_{\mathrm{f}}} \right) \tag{12}$$

Then the fracture toughness can be obtained by calculating the maximum of G_{ss} as proposed by Madsen et al. [12],

$$G_{\mathrm{c}} = \max(G_{\mathrm{ss}}(c)) \tag{13}$$

An alternative method for determining the fracture stress and thus the fracture toughness is to use a simplified stress-based model. The simplified model can give an explicit expression of fracture stress for a particular type of coated system. For example, Morasch and Bahr [14] developed a method to identify stress distribution of a brittle film/ductile substrate under an axisymmetric indenter. In their model, they believed that the ring crack was mainly induced by the film bending. Thus, the indentation load is assumed as a pressure with a radius of a_{c}, and the plastic zone is considered as a uniform distributed pressure. The bending moment by these two types of pressure is given by

$$M_\theta = \frac{P}{4\pi}(1+v)\left(\frac{a_{\mathrm{c}}^2}{4c^2} - \ln\frac{a_{\mathrm{c}}}{c} \right) - \frac{p_0 c^2}{8} \tag{14}$$

Then the bending stress can be obtained as

$$\sigma_{\mathrm{rr}} = \frac{12 M_\theta}{t^3} z \tag{15}$$

where z is the distance from the middle surface of the circular plate. It is clear that the maximum radial stresses occur on the surfaces $z = \pm t/2$ of the plate. This coincides with observations from the literature [7] that circumferential cracks initiate from the top surface of the brittle film. When the radial stress on top of the film surface reaches the strength, circumferential cracks initiate and begin to propagate along the interface between the film and the substrate. **Figure 9** shows the pressure distribution in the coated system with a modulus ratio of 14.0–69.8 under a spherical indenter and a conical indenter by FE analysis; however, it has been found that the pressure caused by the plastic zone is clearly not uniform.

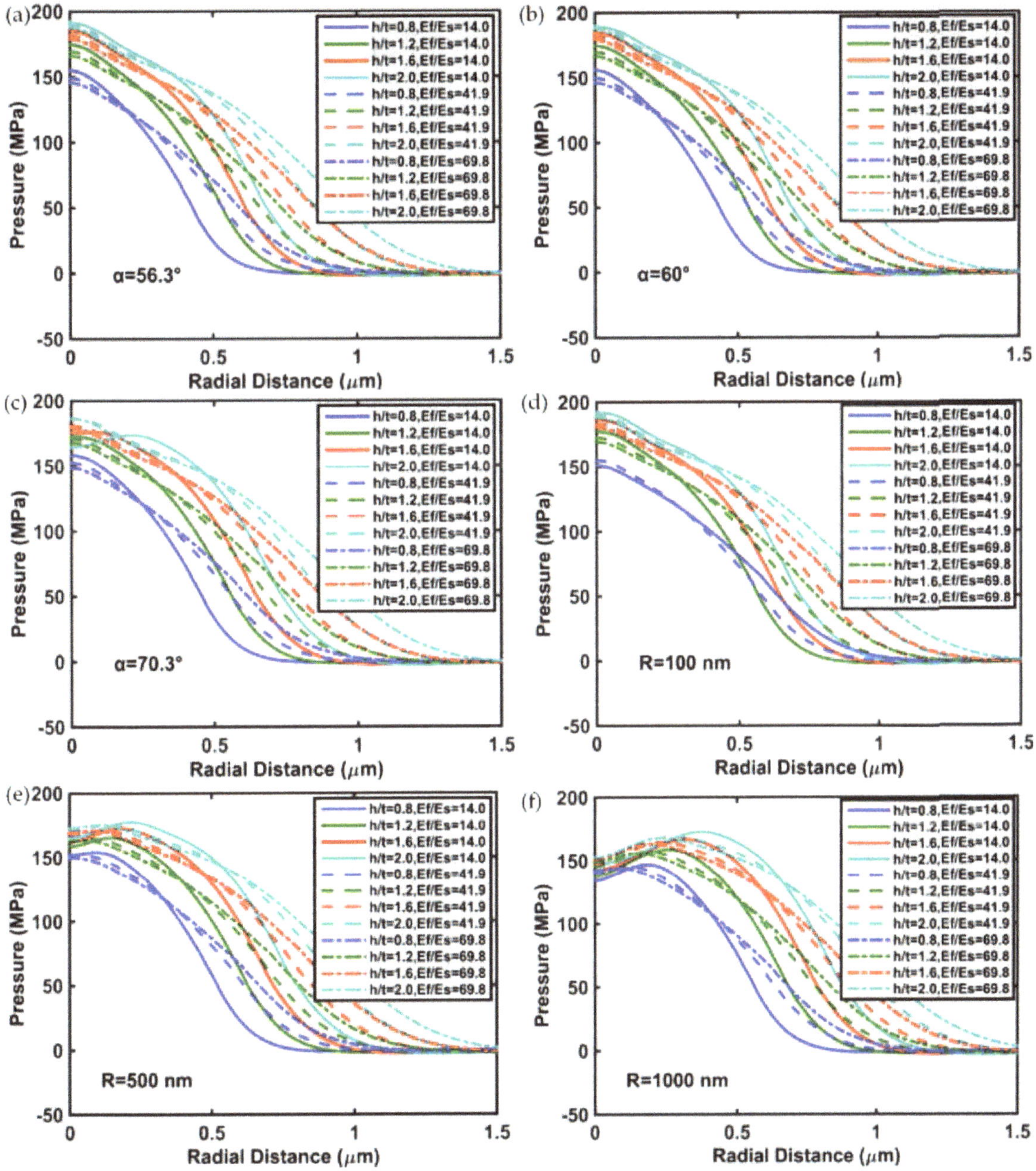

Figure 9. Pressure distribution in the interface under a conical indenter with a half-included angle α of (a) 56.3°, (b) 60.0°, and (c) 70.3°, and a spherical indenter with a radius R of (d) 100 nm, (e) 500 nm, and (f) 1000 nm [15].

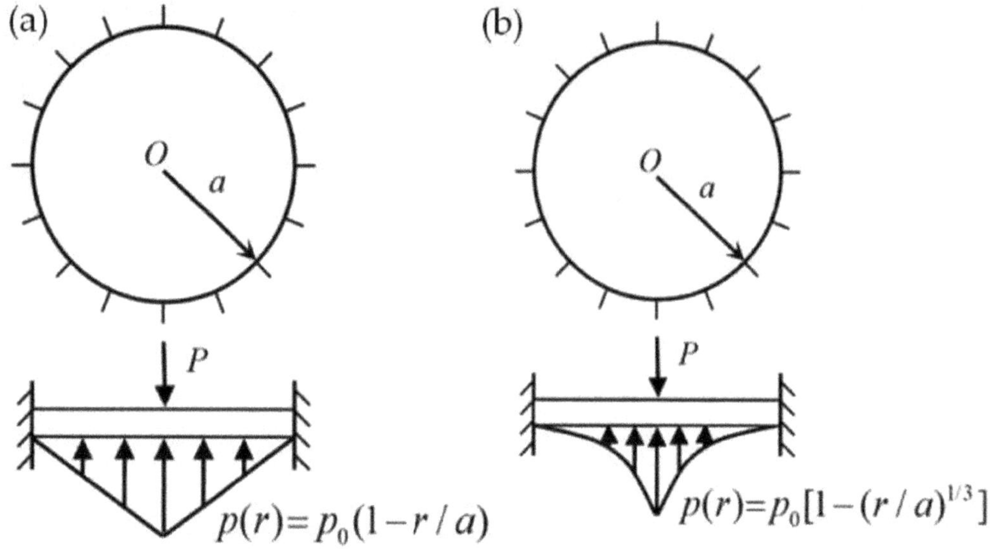

Figure 10. Schematic representation of a stress-based (a) model I and (b) model II.

In view of that, Fu et al. [15] proposed two stress-based models, as shown in **Figure 10**. The film is modeled as a circular elastic plate clamped at its edge. The radius of the circular plate, a, equals that of the plastic zone of the substrate beneath an axisymmetric indenter. A plastic zone of the ductile substrate beneath the film is considered as distributed pressure. Then the bending moments in films for model I and model II are given by

$$M_{\theta 1}=\frac{P}{240\pi a^3}\left(29a^3+29a^3v-135ar^2-45ar^2v+16r^3v+64r^3\right)-\frac{P}{4\pi}\left((1-v)\ln\frac{a}{r}-1\right) \tag{16}$$

$$\begin{aligned}M_{\theta 2}=&\frac{P}{1456\pi a^{7/2}}\left(205a^{7/2}+205a^{7/2}v-1911a^{1/3}r^2-637a^{1/3}r^2v+432r^{7/2}v+1440r^{7/2}\right)\\&-\frac{P}{4\pi}\left((1-v)\ln\frac{a}{r}-1\right)\end{aligned} \tag{17}$$

where P is the indentation load and r is the radial distance from the center. Subscripts 1 and 2 indicate models I and II, respectively.

The radial stress can be predicted by substituting the bending moment into Eq. (15). The normalized stresses (predicted radial stress divided by the maximum radial stress) predicted by models I and II are shown in **Figure 11**. It is clear that there is a high tensile stress outside the contact radius. Moreover, the highest compressive stress occurs at the center, which causes the radial crack. The stress obtained by the proposed models is comparable to the FE results and experimental observations [15].

After the stress distribution has been obtained, fracture toughness is obtained using linear fracture mechanics and a crack channeling criterion according to Eqs. (10)–(13) and Eq. (15).

It should be worth noting that the stress-based model is valid for a particular coated system only, that is, the brittle film/ductile substrate. In the next section, we introduce the mechanism of cracking in a brittle film on a ductile substrate to support the stress-based model.

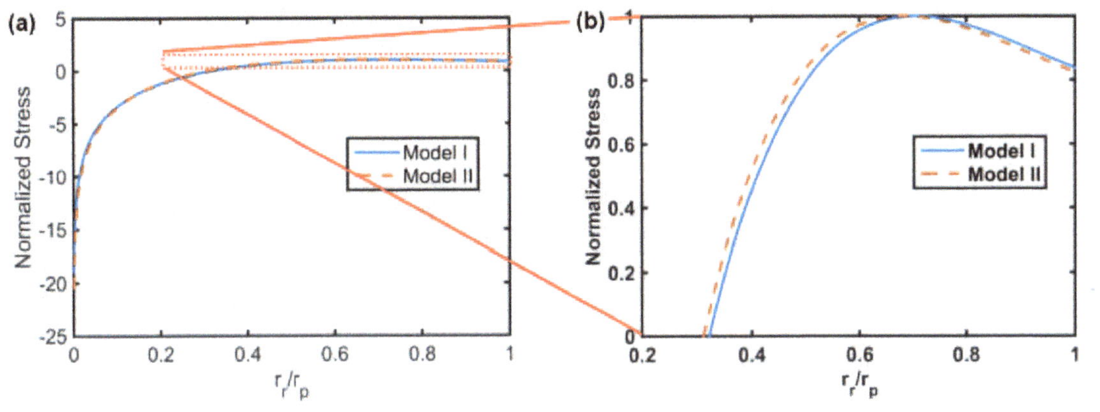

Figure 11. Normalized stress as a function of r_r/r_p (ring crack radius/plastic zone radius) predicted by the model I and model II (Poisson's ratio is assumed to be 0.21).

3. Fracture mechanism of brittle thin films under nano-indentation

Toughness measurement using a stress-based method depends strongly on the fracture mechanism of cracking. Understanding the mechanism of the crack pattern in brittle thin films could provide guidance in developing a more accurate stress-based model for toughness assessment of the film. In this section, the fracture mechanisms of various crack patterns under indentation are discussed in detail.

A number of types of crack pattern induced by an indentation in brittle thin film have been reported, such as radial cracks [16–18], ring cracks [7, 19, 20], picture-frame cracks [21, 22], spiral cracks [23, 24], and spalling [25, 26]. The crack patterns are dependent on the substrate effect, indenter shape, and the indentation load level. Regarding the substrate effect, the coated system can be divided into two types, namely the brittle film/hard substrate system and the brittle film/ductile substrate system, depending on the effects of the substrate on cracking. Accordingly, we discuss the fracture mechanisms in these two types of coated system under different types of indenter.

3.1. Crack patterns in brittle film/hard substrate

In a brittle film on a hard substrate, it has been found that cracks often occur at the contact edge of the indenter due to the stress concentration. For example, three radial cracks were detected in a diamond-like carbon (DLC) film/silicon substrate with a film thickness of 115 nm under a Berkovich indenter, as shown in **Figure 12(a)**. The radial cracks were mainly caused by the stress concentration at the contact edge. Also, a significant pileup was found around the impression. Due to the confinement of the substrate and the pileup, the radial cracks did not propagate along the edge. Therefore, it would be invalid to use a classical method to characterize the fracture toughness of the film. For a ZnO film on an architectural glass substrate, picture-frame cracks were detected inside the impression, as shown in **Figure 12(b)**. Meanwhile, radial cracks along the indenter edge were also detected. Bull [36] argued that the picture-frame cracks were caused by the high contact stress. In addition, the FE result in **Figure 18(a)** shows that a high stress occurs around the edge of the indenter, which supports the argument of Bull [36].

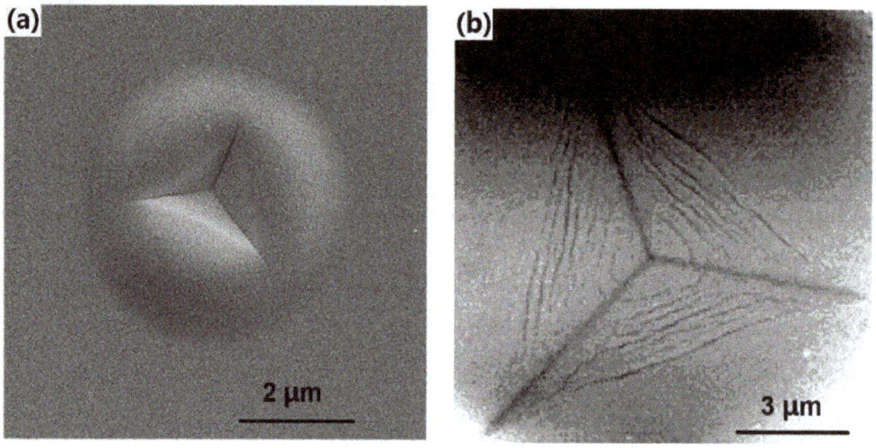

Figure 12. (a) Radial cracks and (b) picture-frame cracks [21] in a brittle film/hard substrate.

With an increase in the peak indentation load, the film on a hard substrate usually ends up with a failure of spalling, as reported in [25]. In their work, ring-like cracks together with spalling failure were observed in a DLC film caused by a conical indenter. The film outside the ring-like cracks was detached from the silicon substrate. Also, the crack pattern in a DLC film/silicon substrate caused by a cube-corner indenter was also shown in [25]. It is evident that radial cracks, ring-like cracks, and spalling occur in the films. Around the impression, the film is not detached from the substrate because of the high compressive contact stress. The spalling occurs outside the impression. It is shown that spalling occurs with a relatively higher load, regardless of the type of indenter.

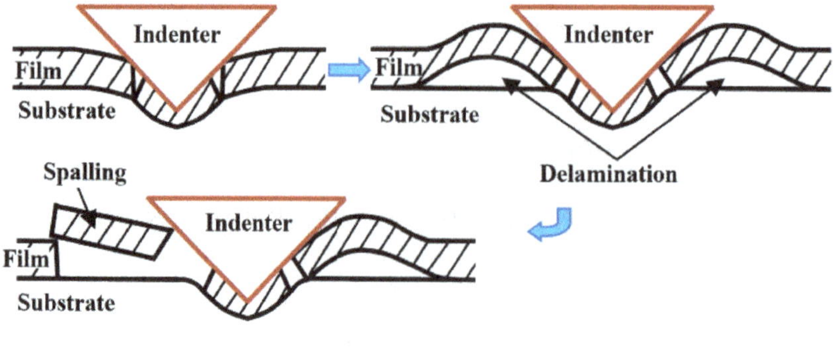

Figure 13. Schematic representation of various stages of indentation-induced cracking in a brittle film/hard substrate system.

The spalling that usually occurs in a brittle film/hard substrate follows three stages, as illustrated in **Figure 13**. First, a ring-like crack initiates and propagates toward the interface due to the high contact stress. In the second stage, delamination occurs owing to the weak bonding between film and substrate. Then the film begins to buckle because of the delamination caused by the extrusion of the indenter. In the last stage, the buckled film results in a high tensile stress. When the tensile stress reaches the film strength, the film begins to break, and spalling forms. During the formation of spalling, a significant energy release occurs, which leads to a perfect "pop-in" in the load-depth curve. Then we could measure the released energy according to the "pop-in" in the load-depth curve as described in Section 2.2.

3.2. Crack patterns in brittle film/ductile substrate

In a brittle film on a ductile substrate, the crack pattern and fracture mechanism are quite different from those in a brittle film/hard substrate system. For example, **Figure 14(a)** and **(b)** shows the crack patterns in a DLC film/polyether ether ketone (PEEK) substrate under a conical indenter and a Berkovich indenter, respectively. In **Figure 14(a)**, a few radial cracks and a ring cracks are observed. The radial cracks do not reach the ring crack, which indicates that the two types of crack are independent of each other. Examination of the stress distribution in the films in **Figure 15** shows that there is a high tensile stress on the surface of the film, and a plastic zone occurs in the substrate underneath the indenter. In the FE analysis, we found that the position of the ring crack was between the contact radius and the plastic zone, as shown in **Figure 16**. The stiffness difference between plastic zones caused the film to bend. As a result, a high tensile radial stress, which caused the ring crack formation, occurred on the surface of the indenter. It should be noted that the radial tensile stress consisted mainly of a bending stress, as well as a stretching stress in the thin film due to the contact load. If the modulus ratio of film and substrate, E_f/E_s, was significantly high, the stretching stress was negligible. In other words, the ring crack was caused mainly by film bending, which made the position of the ring crack move toward the plastic zone side. On the other hand, if E_f/E_s was low, the ring crack ran to the indenter edge. These conclusions confirmed the feasibility of the present stress-based model in Section 2.

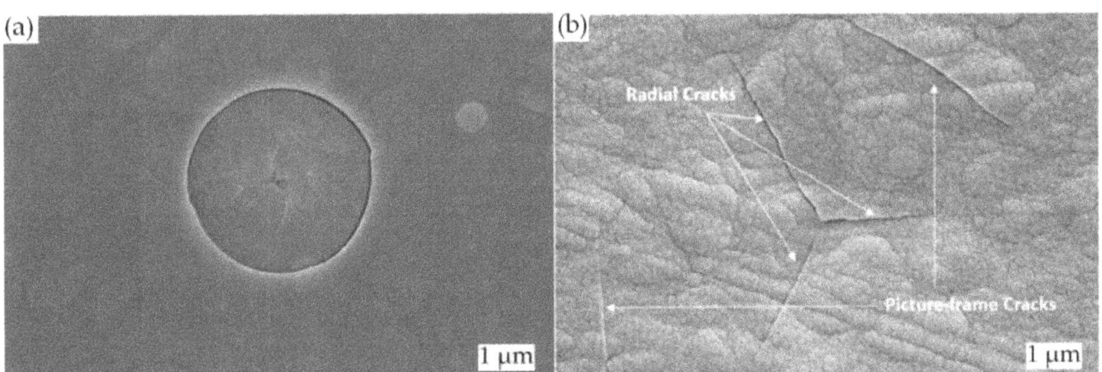

Figure 14. (a) Ring crack and (b) picture-frame crack [22] in a DLC film/PEEK substrate.

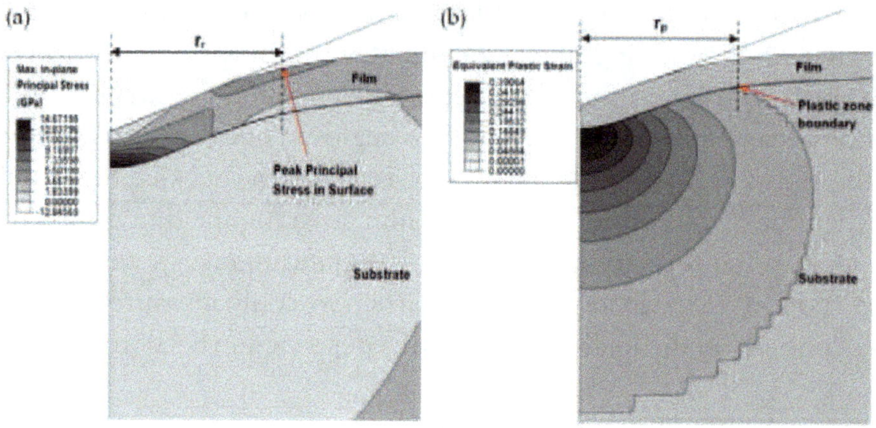

Figure 15. FE results of (a) maximum in-plane principal stress and (b) equivalent plastic strain distribution [7].

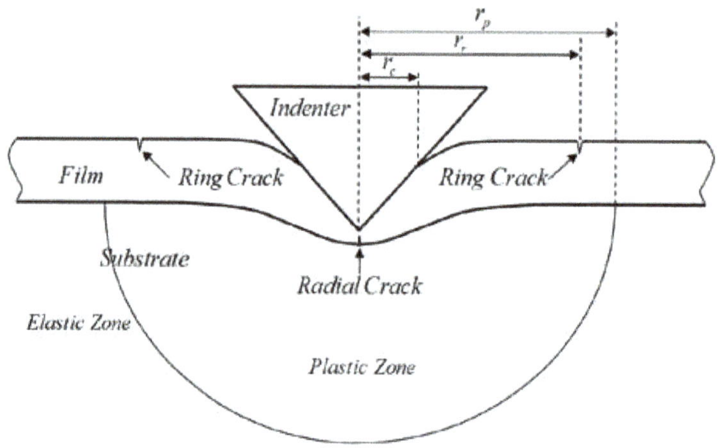

Figure 16. Fracture mechanism of a brittle film on a ductile substrate by indentation [15].

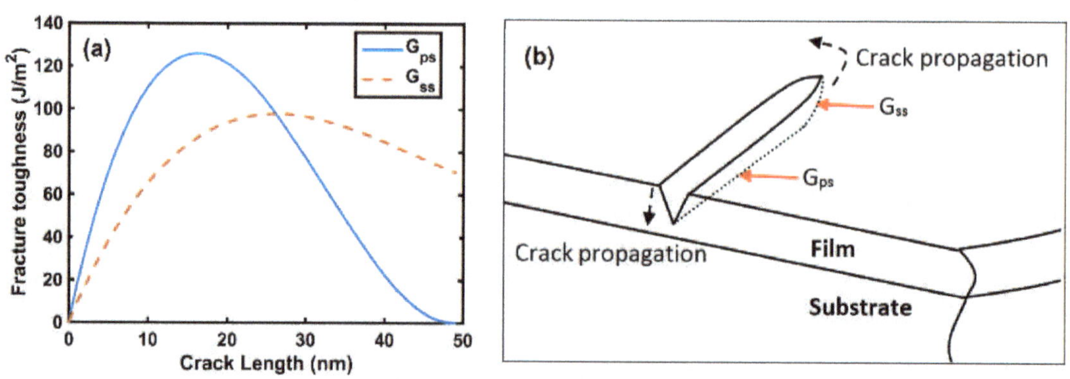

Figure 17. (a) Energy release rate of the plane strain crack G_{ps} propagating through the film and energy release rate of the channel front G_{ss} for indentation depth, and (b) illustration of a channel crack propagating through the film.

The FE analysis gives the stress distribution of film under an axisymmetric indenter, which can provide information about crack initiation. To fully understand the formation process of a ring crack, more detail is necessary. Prior to the indentation, it is highly improbable that a circular defect exists in the film. The crack channeling criterion may be the best explanation for the formation of a ring crack. **Figure 17(a)** shows the energy release rate by a linear fracture mechanics and a crack channeling energy release rate. In the first stage, a pre-existing short crack exists as shown in **Figure 17(b)**. As the stress in the film increases, the crack begins to propagate toward the interface when the energy release rate by linear fracture mechanics reaches the fracture toughness. With the increase in crack length, the energy release rate by the linear fracture mechanics decreases. When it reaches the critical energy release rate for a channeling crack, the channeling ring crack initiates. The short crack adjusts its own curvature to form a ring crack because the G_{ps} at each point around the circle is the same. After that, the ring crack propagates and stops at the interface. Hence, we could obtain the fracture toughness of the brittle film on a ductile substrate if a ring crack is observed using Eq. (13).

In **Figure 14(b)**, three radial cracks were detected and two picture-frame cracks were observed outside the contact region. Also, no delamination occurred. The FE results in **Figure 18** show that the radial crack is induced by the high compressive stress due to the contact stress, which is the same as the observation in a brittle film/hard substrate. However, a high tensile stress occurs outside the contact region. With an increase in E_f/E_s, the high stress in a film surface moves outward to the contact area. The film bending by a Berkovich indenter is also controlled by the plastic zone below. After the critical stress is reached, the crack may propagate to the interface or parallel to the edge of the indenter, depending on the value of the energy release rate using the channeling crack criterion or linear fracture mechanics.

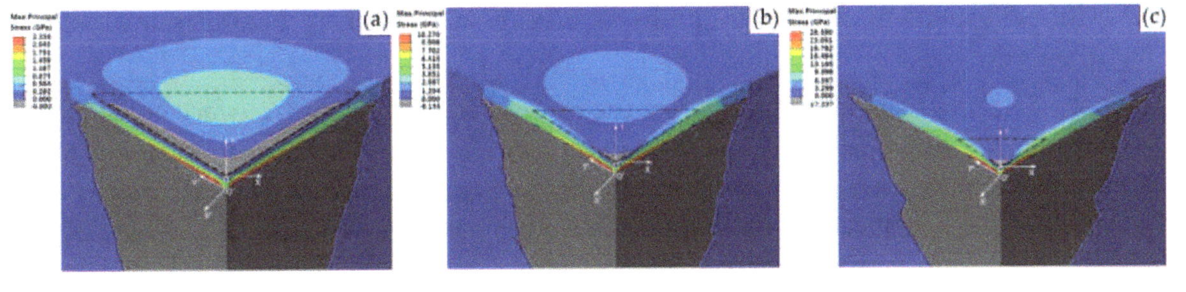

Figure 18. Maximum principal stress distribution in a brittle film on a ductile substrate (dashed lines indicate projected contact edges) with a modulus ratio of (a) $E_f/E_s = 2.3$, (b) $E_f/E_s = 14.0$, and (c) $E_f/E_s = 41.9$ [22].

If we keep increasing the indentation load in a brittle film/ductile substrate system, multiple ring cracks sometimes appear on the surface. **Figure 19** shows multiple ring cracks in a DLC film/PEEK substrate. The film thickness is 140, 400, and 1300 nm, respectively in **Figure 19(a–c)**. The peak indentation load is 10 mN. It is evident that, in all the films, there are a few radial cracks in the middle and a few ring cracks outside the radial crack. When the thickness increases to 400 nm, we can also observe a similar radial crack and ring crack pattern. In a brittle film with a high thickness, however, the crack pattern changes to a combination of both ring crack and spalling in **Figure 19(c)**. In other words, ratio of indentation depth and film thickness also affects the crack patterns. In a thin film, the ring crack is controlled by the plastic zone in the substrate under the indenter. The contribution of the plastic zone is less when a thick film is tested because the critical depth is much smaller than the film thickness. The substrate effect might not be dominant in crack formation, and the fracture mechanism is similar to that in the bulk brittle material.

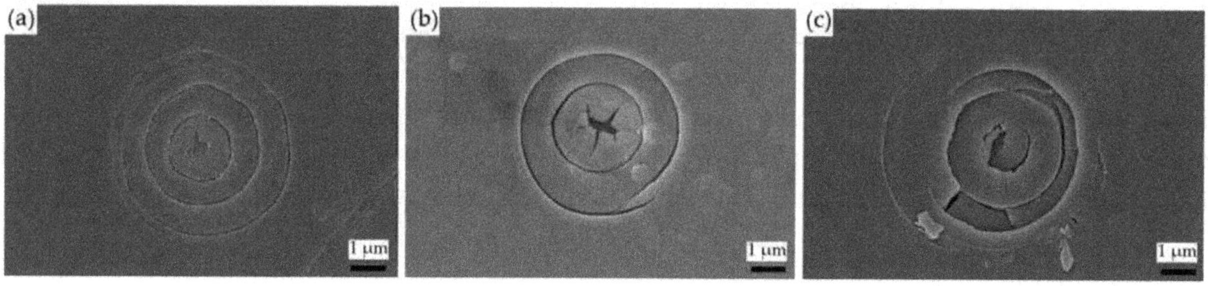

Figure 19. Multiple ring cracks in a DLC film/PEEK substrate with film thickness of (a) 140 nm, (b) 400 nm, and (c) 1300 nm.

Given the mechanism of ring crack formation, FE analysis integrated with cohesive elements can be used to simulate the formation of a ring crack. We ignore the channel crack formation here. The simulation steps are as follows:

Step 1. An FE analysis is performed without using cohesive elements. When the tensile stress in a surface reaches the tensile strength, the analysis stops, and the position of the peak tensile stress is recorded.

Step 2. Cohesive elements are inserted into the FE model at the recorded position along the thickness.

Step 3. The FE analysis is rerun, and when the tensile stress in the film surface equals the strength, a ring crack (represented by the cohesive elements) begins to initiate and propagates toward the interface. When the stress outside the first ring crack reaches the tensile strength, the position of the peak tensile stress is recorded again.

Step 4. Repeat Step 2 to insert another group of cohesive elements for the second ring crack formation.

It is seen that the FE results agree well with the experimental observations in **Figure 20**. However, it is noted that the drawback of this FE model is that it cannot simulate the propagation of a channeling crack.

Another specimen is a 100 nm Al_2O_3 film on a PEEK substrate. After indentation, there was still a multiple ring crack pattern under a conical indenter as shown in **Figure 21(a)**. We recorded the critical load and the ring crack diameter. We found the linear relation between the critical load and the diameter of the ring crack shown in **Figure 22**. In a bi-layer film, a much more complex crack pattern, a flower-shaped crack, was induced. For instance, we found that the crack pattern became a flower-shaped crack in the surface of a TiO_2/Al_2O_3 film/PEEK substrate system, as illustrated in **Figure 21(b)**. That crack is induced by the combination of substrate effect and the interaction of the two brittle films. For instance, the confinement of TiO_2 film causes uncertainty of channeling crack formation.

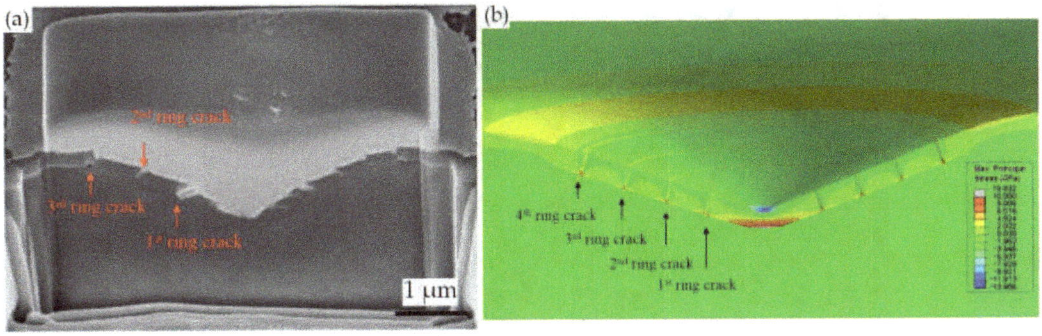

Figure 20. Focused ion beam observation of multiple ring crack and (b) simulation results of maximum principal stress, modified from [37].

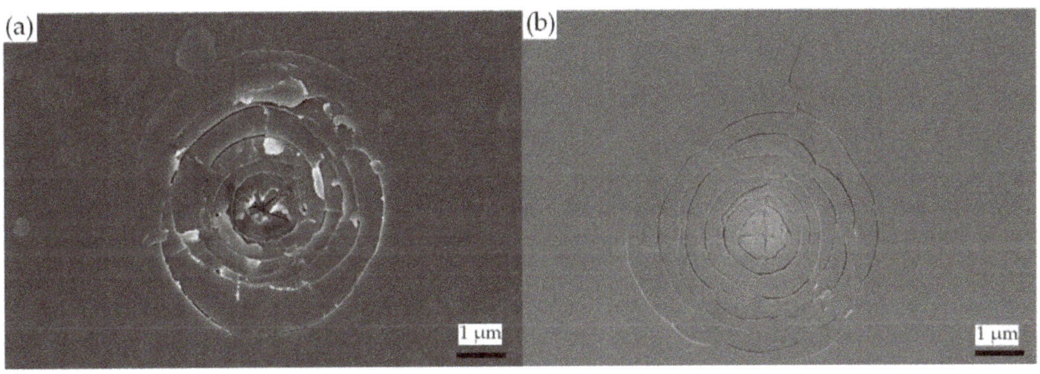

Figure 21. Crack pattern in (a) an Al_2O_3 film/PEEK substrate, and (b) a TiO_2/Al_2O_3/PEEK substrate.

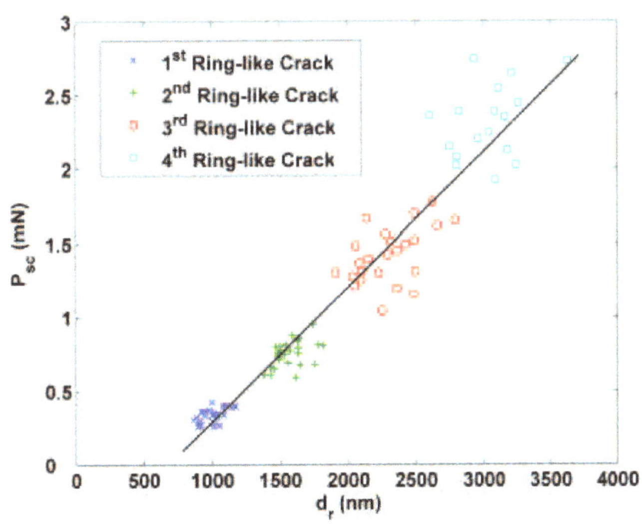

Figure 22. Critical load P_{sc} versus the diameter of ring-like crack d_r (solid lines indicate the trend) [38].

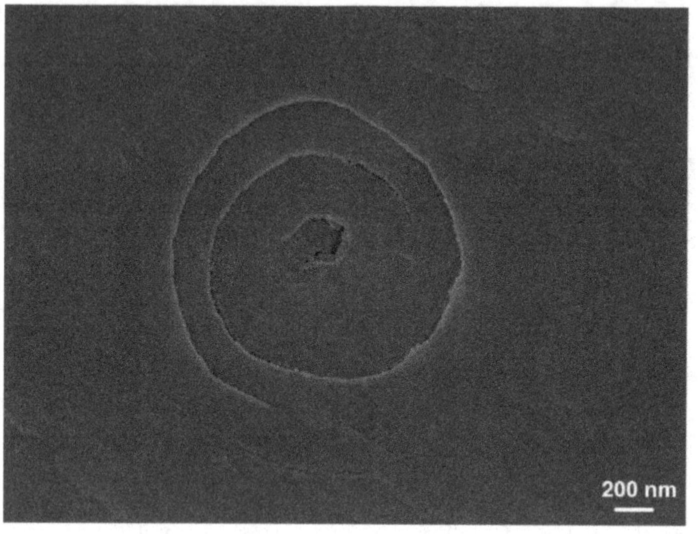

Figure 23. A spiral crack after unloading.

The spiral crack is another crack pattern induced by indentation. In **Figure 23**, a spiral crack is shown in a DLC film on a PEEK substrate. Under a conical indenter, a high equivalent stress arises near the interface owing to the film bending. Then a small defect grows to a spiral crack in the film. Once the indenter begins to withdraw, the increasing equi-biaxial stress field provides the driving force for a spiral crack extension due to the film bending curvature effect as reported in [23].

4. Summary

In this chapter, several methods for the measurement of fracture toughness of brittle film/ substrate systems were presented and detailed. Fracture mechanisms for some particular crack patterns were introduced to give a better understanding of the stress-based model. The conclusions are as follows:

The classic method was a widely used method for toughness measurement in "thick" films where a radial crack was well developed. The energy method was an efficient method for determining the toughness of thin films and depends on obtaining the dissipated energy during cracking. The stress method was a straightforward method based on stress distribution and the fracture mechanism in brittle films under indentation. In particular, the fracture mechanism in brittle films depends on the substrate effect, indenter shape, and load level, which can result in different crack patterns. The FE method may be the most efficient tool to characterize the evolution of stress under indentation.

To date, there is still no generally accepted method for the measurement of brittle thin films using nano-indentation. Each method mentioned above has its limitations. In future, with a better understanding of each crack pattern, it is expected that a general stress-based solution for the toughness measurement of thin film will be developed.

In the meantime, a more detailed numerical model is needed to investigate the complex experimental observation. An example might be an FE model that can simulate the indentation process considering the combination of a channeling crack and delamination.

Acknowledgements

Youhong Tang is grateful for the research support of the Premier's Research and Industry Fund (PRIF) with a Catalyst Research Grant (Grant No.: CRG 65) for the research work. Li Chang is grateful for the funding from the Faculty of Engineering and Information Technologies, the University of Sydney, under the Faculty Research Cluster Program. The authors greatly acknowledge Ahmadzada for the samples of Al_2O_3 film/PEEK substrate and TiO_2/Al_2O_3/PEEK substrate.

Author details

Kunkun Fu[1], Youhong Tang[2*] and Li Chang[1]

*Address all correspondence to: youhong.tang@flinders.edu.au

1 School of Aerospace, Mechanical and Mechatronic Engineering, University of Sydney, NSW, Australia

2 School of Computer Science, Engineering and Mathematics, Flinders University, SA, Australia

References

[1] Anstis G, Chantikul P, Lawn BR, Marshall D. A critical evaluation of indentation techniques for measuring fracture toughness: I, direct crack measurements. Journal of the American Ceramic Society. 1981;64(9):533–8. DOI: 10.1111/j.1151-2916.1981.tb10321.x

[2] Ucisik AH, Bindal C. Fracture toughness of boride formed on low-alloy steels. Surface and Coatings Technology. 1997;94(5):561–5. DOI: 10.1016/s0257-8972(97)00466-0

[3] Feng W, Yan D, He J, Zhang G, Chen G, Gu W, Yang S. Microhardness and toughness of the TiN coating prepared by reactive plasma spraying. Applied Surface Science. 2005;243(1–4):204–13. DOI: 10.1016/j.apsusc.2004.09.064

[4] Cantera EL, Mellor B. Fracture toughness and crack morphologies in eroded WC–Co–Cr thermally sprayed coatings. Materials Letters. 1998;37(4–5):201–10. DOI: 10.1016/s0167-577x(98)00092-5

[5] Xie X, Guo H, Gong S, Xu H. Lanthanum–titanium–aluminum oxide: a novel thermal barrier coating material for applications at 1300 °C. Journal of the European Ceramic Society. 2011;31(9):1677–83. DOI: 10.1016/j.jeurceramsoc.2011.03.036

[6] Li XD, Diao DF, Bhushan B. Fracture mechanisms of thin amorphous carbon films in nanoindentation. Acta Materialia. 1997;45(11):4453–61. DOI: 10.1016/s1359-6454(97)00143-2

[7] Fu KK, Yin YB, Chang L, Shou DH, Zheng BL, Ye L. Analysis on multiple ring-like cracks in thin amorphous carbon film on soft substrate under nanoindentation. Journal of Physics D: Applied Physics. 2013;46: 505314. DOI: 10.1088/0022-3727/46/50/505314

[8] Chen J, Bull SJ. Assessment of the toughness of thin coatings using nanoindentation under displacement control. Thin Solid Films. 2006;494(1–2):1–7. DOI: 10.1016/j.tsf.2005.08.176

[9] den Toonder J, Malzbender J, de With G, Balkenende R. Fracture toughness and adhesion energy of sol-gel coatings on glass. Journal of Materials Research. 2002;17(1):224–33. DOI: 10.1557/jmr.2002.0032

[10] Chen J, Bull SJ. Modelling the limits of coating toughness in brittle coated systems. Thin Solid Films. 2009;517(9):2945–52. DOI: 10.1016/j.tsf.2008.12.054

[11] Steffensen S, Madsen ND, Jensen HM. Numerical estimation of fracture toughness from indentation-induced circumferential cracking in thin films on ductile substrates. International Journal of Solids and Structures. 2013;50(20–21):3406–17. DOI: 10.1016/j.ijsolstr.2013.06.009

[12] Madsen ND, Steffensen S, Jensen HM, Bøttiger J. Toughness measurement of thin films based on circumferential cracks induced at conical indentation. International Journal of Fracture. 2015;193(2):117–30. DOI: 10.1007/s10704-015-0022-5

[13] Steffensen S, Jensen HM. Energy release rate for circular crack due to indentation in a brittle film on a ductile substrate. European Journal of Mechanics-A/Solids. 2014;43:133–41. DOI: 10.1016/j.euromechsol.2013.09.010

[14] Morasch KR, Bahr DF. Nanomechanical testing for fracture of oxide films. Journal of Materials Research. 2005;20(06):1490–7. DOI: 10.1557/jmr.2005.0188

[15] Fu KK, Chang L, Ye L, Yin YB. Indentation stress-based models to predict fracture properties of brittle thin film on a ductile substrate. Surface and Coatings Technology. 2016;296:46-57. DOI: 10.1016/j.surfcoat.2016.03.067

[16] Volinsky AA, Vella JB, Gerberich WW. Fracture toughness, adhesion and mechanical properties of low-K dielectric thin films measured by nanoindentation. Thin Solid Films. 2003;429(1–2):201–10. DOI: 10.1016/s0040-6090(03)00406-1

[17] Karimi A, Wang Y, Cselle T, Morstein M. Fracture mechanisms in nanoscale layered hard thin films. Thin Solid Films. 2002;420:275–80. DOI: 10.1016/s0040-6090(02)00944-6

[18] Zhang S, Zhang X. Toughness evaluation of hard coatings and thin films. Thin Solid Films. 2012;520(7):2375–89. DOI: 10.1016/j.tsf.2011.09.036

[19] Hainsworth SV, McGurk MR, Page TF. The effect of coating cracking on the indentation response of thin hard-coated systems. Surface and Coatings Technology. 1998;102(1–2): 97–107. DOI: 10.1016/s0257-8972(97)00683-x

[20] Thomsen NB, Fischer-Cripps AC, Swain MV. Crack formation mechanisms during micro and macro indentation of diamond-like carbon coatings on elastic–plastic substrates. Thin Solid Films. 1998;332(1–2):180–4. DOI: 10.1016/s0040-6090(98)01101-8

[21] Chen J, Bull SJ. Indentation fracture and toughness assessment for thin optical coatings on glass. Journal of Physics D: Applied Physics. 2007;40(18):5401. DOI: 10.1088/0022-3727/40/18/s01

[22] Fu KK, Chang L, Zheng BL, Tang YH, Yin YB. Analysis on cracking in hard thin films on a soft substrate under Berkovich indentation. Vacuum. 2015;112:29–32. DOI: 10.1016/ j.vacuum.2014.11.013

[23] Yonezu A, Liu L, Chen X. Analysis on spiral crack in thick diamond-like carbon film subjected to spherical contact loading. Materials Science and Engineering: A. 2008;496(1):67–76. DOI: 10.1016/j.msea.2008.04.069

[24] Xie ZH, Munroe PR, McGrouther D, Singh RK, Hoffman M, Bendavid A, Martin PJ, Yew S. Three-dimensional study of indentation-induced cracks in an amorphous carbon coating on a steel substrate. Journal of Materials Research. 2006;21(10):2600–5. DOI: 10.1557/jmr.2006.0313

[25] Li X, Bhushan B. Measurement of fracture toughness of ultra-thin amorphous carbon films. Thin Solid Films. 1998;315(1–2):214–21. DOI: 10.1016/s0040-6090(97)00788-8

[26] Evans AG, Hutchinson JW. On the mechanics of delamination and spalling in compressed films. International Journal of Solids and Structures. 1984;20(5):455–66. DOI: 10.1016/0020-7683(84)90012-x

[27] Lawn BR, Evans A, Marshall D. Elastic/plastic indentation damage in ceramics: the median/radial crack system. Journal of the American Ceramic Society. 1980;63:574–81. DOI: 10.1111/j.1151-2916.1980.tb10768.x

[28] Strecker K, Ribeiro S, Hoffmann MJ. Fracture toughness measurements of LPS-SiC: a comparison of the indentation technique and the SEVNB method. Materials Research. 2005;8:121–4. DOI: 10.1590/s1516-14392005000200004

[29] Dukino RD, Swain MV. Comparative measurement of indentation fracture toughness with Berkovich and Vickers indenters. Journal of the American Ceramic Society. 1992;75:3299–304. DOI: 10.1111/j.1151-2916.1992.tb04425.x

[30] Pharr G. Measurement of mechanical properties by ultra-low load indentation. Materials Science and Engineering: A. 1998;253(1–2):151–9. DOI: 10.1016/s0921-5093(98)00724-2

[31] Marshall D, Lawn B. An indentation technique for measuring stresses in tempered glass surfaces. Journal of the American Ceramic Society. 1977;60:86–7. DOI: 10.1111/j.1151-2916.1977.tb16106.x

[32] Yang F. Axisymmetric indentation of an incompressible elastic thin film. Journal of Physics D: Applied Physics. 2003;36(1):50. DOI: 10.1088/0022-3727/36/1/307

[33] Yang F. Asymptotic solution to axisymmetric indentation of a compressible elastic thin film. Thin Solid Films. 2006;515(4):2274–83. DOI: 10.1016/j.tsf.2006.07.151

[34] Sburlati R. Elastic indentation problems in thin films on substrate systems. Journal of Mechanics of Materials and Structures. 2006;1(3):541–57. DOI: 10.2140/jomms.2006.1.541

[35] Sburlati R. Adhesive elastic contact between a symmetric indenter and an elastic film. International Journal of Solids and Structures. 2009;46(5):975–88. DOI: 10.1016/j.ijsolstr.2008.10.006

[36] Bull SJ. Analysis methods and size effects in the indentation fracture toughness assessment of very thin oxide coatings on glass. Comptes Rendus Mecanique. 2011;339(7–8):518–31. DOI: 10.1016/j.crme.2011.05.009

[37] Fu KK, Zheng BL, Yin YB, Hu TY, Ye L, Shou DH. Fracture analysis of diamond-like carbon films under conical nanoindentation. Journal of University of Science and Technology Beijing. 2014;36:70–6. DOI: 10.13374/j.issn1001-053x.2014.02.011

[38] Fu KK, Chang Y, Chang L, Zheng BL. An improved indentation method for estimating limits of fracture toughness in brittle films. Advanced Materials Research. 2015;1095:598–602. DOI: 10.4028/www.scientific.net/amr.1095.598

4

Fracture Toughness of Ferritic Steels in the Ductile-to-Brittle Transition Region

Carlos Berejnoi and Juan E. Perez Ipiña

Abstract

Ferritic steels, as other materials, have different failure modes depending on the temperature. At elevated temperatures, they behave as ductile materials, while at low temperatures they are brittle. There is an intermediate temperature region where these alloys have a failure mode resulting from the competition between cleavage and ductile mechanisms. This region is known as the ductile-to-brittle transition zone. The characterization of fracture resistance of ferritic steels in the ductile-to-brittle transition region is problematic due to scatter in results, as well as size and temperature dependences. American Society for Testing and Materials (ASTM) has standardized the determination of a temperature reference (T_0) for the fracture toughness characterization of ferritic steels in this region. This chapter presents the evolution of the statistical treatment of fracture toughness data until the present, including some comments on T_0 determination, and some aspects that require a deeper analysis.

Keywords: ductile-to-brittle, fracture toughness, ferritic steels, weakest link, master curve

1. Introduction

The so-called ferritic steels are, as defined in American Society for Testing and Materials (ASTM) E1921-2015a$^{\varepsilon 1}$ [1], "typically carbon, low-alloy, and higher alloy grades. Typical microstructures are bainite, tempered bainite, tempered martensite, and ferrite and pearlite. All ferritic steels have body centered cubic crystal structures that display ductile-to-cleavage transition temperature fracture toughness characteristics."

Ferritic steels, as other materials, have different failure modes depending on the temperature. At elevated temperatures, they behave as ductile materials, while at low temperatures they are brittle. There is an intermediate temperature region where these alloys have a failure mode resulting from the competition between cleavage and ductile mechanisms. This region is known as the ductile-to-brittle transition zone, where fracture toughness decreases with decreasing temperatures. **Figure 1** shows the fracture behavior of different materials with temperature, including some that don't have a transition zone.

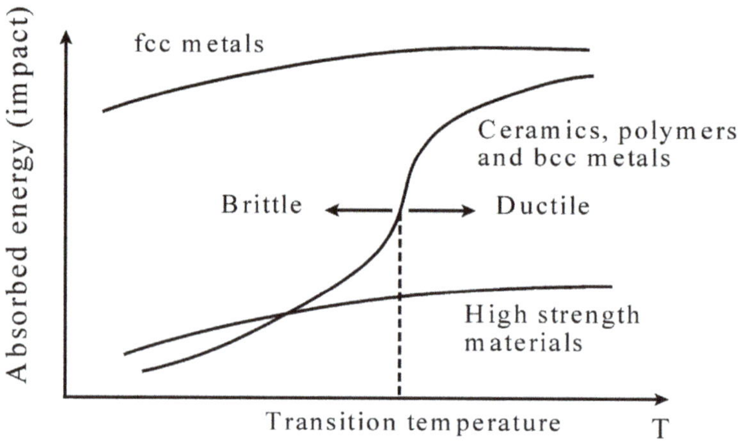

Figure 1. Fracture behavior of different materials as a function of temperature.

The characterization of fracture resistance of ferritic steels in this region is problematic due to scatter in results, as well as size and temperature dependences [2–9].

Size effects imply decreasing of the median value of fracture toughness and a larger scatter in small specimen than in larger ones. **Figure 2** describes schematically the failure probability density functions for a material in equal conditions but with two different sizes (thicknesses B_1 and B_N), using Jc as the fracture toughness parameter.

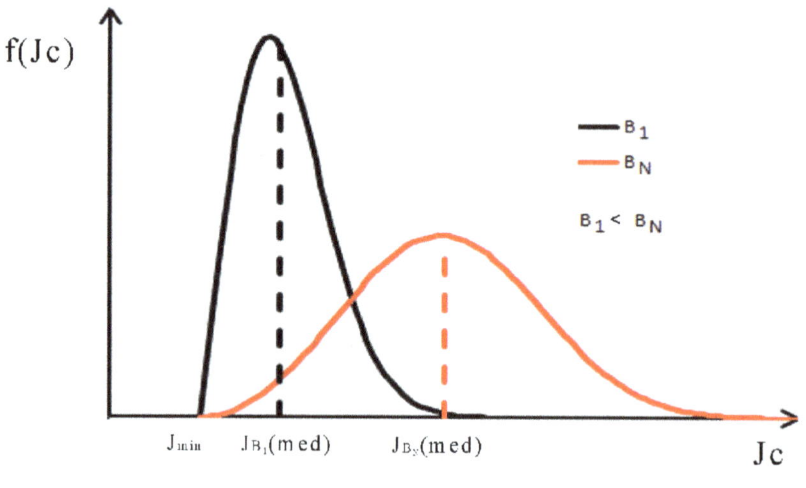

Figure 2. Weibull probability density function for two specimen sizes.

Originally, there were two explanations to the size effect. One was based on constraint effects, while the other made use of statistical weakest link concepts to explain the probability to find a cleavage initiator site at the crack front.

1.1. Constraint theory

This theory is based on the hypothesis that specimens with larger thickness present larger constraint than thinner specimens and that the average fracture toughness will be smaller for larger specimens than for smaller ones [10]. But this theory fails to explain the differences in scatter for different sizes.

1.2. Statistical theory

According to this theory, there are small areas of low toughness or weak links (possible initiators of cleavage) randomly distributed in the crack front, so that the brittle fracture would be a statistical event (Landes and Schaffer [11], Landes and McCabe [12]). The cleavage fracture is a local fracture process controlled by a critical stress, and it will occur when the critical stress is reached in one of these weak links. The load required to produce the fracture will depend upon the location of the weak link and its critical stress.

In addition to the scatter that occurs in the transition region, the weakest link model also explains the effect of specimen size, since an increase in the length of the crack front enlarges the highly stressed volume of material at the tip of the crack, also increasing the likelihood to find a weak link.

Based mainly on the works of Wallin [13–16], with the master curve (MC) methodology, the ASTM has standardized the determination of a temperature reference (T_0) for the fracture toughness characterization of ferritic steels in this region.

This chapter presents the evolution of the statistical treatment of fracture toughness data, including some comments on T_0 determination, and some aspects that require a deeper analysis.

2. Evolution of the statistical-based theory

2.1. Weakest link original proposal

Waloddi Weibull [17] proposed a probability distribution function (later called Weibull distribution). The probability of choosing an individual with a value X less than a given value x is given by Eq. (1):

$$P(X \leq x) = P(x) = 1 - \exp(-\varphi(x)) \tag{1}$$

The function $\varphi(x)$ can have any expression, although the simplest one is

$$\varphi(x)=\left(\frac{x-a}{b}\right)^{c} \tag{2}$$

where

a: threshold parameter

c: shape parameter or Weibull slope

b: scale parameter

Equation (3) results from replacing $\varphi(x)$ of Eq. (1) with Eq. (2):

$$P(x)=1-\exp\left(-\left(\frac{x-a}{b}\right)^{c}\right) \tag{3}$$

The probability density function is

$$f(x)=\frac{dP}{dx} \tag{4}$$

If we have a chain with **N** links, with each of them having a probability of failure P, the chain will fail when the weakest link fails, so the "non-failure" probability of the chain is

$$(1-P_N)=(1-P)^N \tag{5}$$

Then, the chain failure probability is

$$P_N(x)=1-\exp(-N\varphi(x)) \tag{6}$$

2.2. Landes proposal

In 1980, Landes and Shaffer [11] proposed that the cleavage fracture toughness of a metallic specimen is controlled by the point of minimum toughness (weakest link) at the crack front. According to these authors, it would be possible to predict the fracture toughness of large structures or specimens by testing small specimens. Using a two-parameter Weibull (2P-W) distribution, adapting the names of the variables and parameters, they proposed Eq. (7) as the failure probability for a specimen thickness B:

$$P = 1 - \exp\left(-\left(\frac{Jc}{J_0}\right)^b\right) \tag{7}$$

The non-failure probability is

$$1 - P = \exp\left(-\left(\frac{Jc}{J_0}\right)^b\right) \tag{8}$$

For a thickness B_N= N.B:

$$(1 - P_N) = (1 - P)^N = \exp\left(-N\left(\frac{Jc}{J_0}\right)^b\right) \tag{9}$$

The failure probability for a B_N thickness results

$$P_N = 1 - \exp\left(-N\left(\frac{Jc}{J_0}\right)^b\right) \tag{10}$$

The problem with this distribution is that the mean Jc value of the sample, Eq. (11), tends to zero when N tends to infinity (very large thicknesses), which is physically impossible since every material has a minimum value of toughness:

$$\overline{J_c} = \frac{J_0}{N^{\frac{1}{b}}} \tag{11}$$

In order to solve this problem, Landes and McCabe [12] proposed to use a three-parameter Weibull function (3P-W), where the third parameter J_{min} corresponds to a threshold parameter. **Figure 2** shows schematically the probability density functions derived from the probability distribution given by Eq. (12), where the same threshold for both distributions can be observed for different sizes. If the effect of size is incorporated, Eq. (13) would be applied but considering b and J_0 parameters from the distribution for B size. In this case, the probability density functions observed in **Figure 2** would be coincident.

For thickness B:

$$P = 1 - \exp\left(-\left(\frac{Jc - J_{\min}}{J_0 - J_{\min}} \right)^b \right) \tag{12}$$

For thickness B_N = N.B:

$$P = 1 - \exp\left(-N\left(\frac{Jc - J_{\min}}{J_0 - J_{\min}} \right)^b \right) \tag{13}$$

It should be highlighted that Landes and McCabe [12] found that the weakest link theory (expressed by a 3P-W function) described well the thickness effect in the fracture toughness observed in the ductile-to-brittle transition region. But this theory of weakest link does not have a theoretical basis to justify the relationship between the weakest link and the function of Weibull distribution, but it was stated that the latter is adjusted to the experimental data.

2.3. Kim Wallin proposal

Wallin [13–16] assumed that the crack-front material presents a random distribution of potential cleavage initiators. The cumulative probability distribution for a single critical site is a complex function that depends, among other things, on the size distribution of the initiators, stress, strain, temperature, loading rate, etc. It is considered that the shape and origin of the initiators distribution is not important for the case of sharp cracks, and no global interaction between initiators exists. It is considered that there is no interaction between initiators on a global scale. It may happen that there is a cluster of initiators to start the macroscopic fracture, and in this case the cluster is treated as a single site.

By means of theoretical assumptions of the probability distribution of volume elements near the crack tip and complex mathematical deductions, Wallin obtained the following expression for the probability of failure:

$$P = 1 - \exp\left[-\text{constant} \; . \; \text{B} \; . \; K_I^{\;4} \right] \tag{14}$$

Note that the shape parameter is fixed and equal to 4. The theoretical assumptions of Wallin, although different to those raised originally by Weibull, are also based on a weakest link failure mechanism.

Taking into account the existence of a minimum value of toughness, Eq. (14) is modified introducing a threshold parameter, and taking into account a conditional crack propagation criterion results in Eq. (15):

$$P = 1 - \exp\left[-\text{constant} \,.\, \text{B} \,.\, \left(K_I - K_{\min}\right)^4\right] \tag{15}$$

Equation (15) is expressed in the form of Eq. (16):

$$P = 1 - \exp\left(-\frac{B}{B_0}\left(\frac{K_{Jc} - K_{\min}}{K_0 - K_{\min}}\right)^4\right) \tag{16}$$

where K_0 and B_0 are normalization constants.

Wallin [18] concluded that, despite the fact that K_{min} depends on the temperature and material, the value that fits the data better using small sample sizes is 20 MPa.m$^{1/2}$.

2.4. Other proposals

According to Anderson et al. [19], the probability of failure based on the weakest link model corresponds to a 2P-W distribution of the type

$$P = 1 - \exp\left[-\frac{B}{B_0}\left(\frac{J_C}{J_0}\right)^2\right] \tag{17}$$

or, in terms of K:

$$P = 1 - \exp\left(-\frac{B}{B_0}\left(\frac{K_{Jc}}{K_0}\right)^4\right) \tag{18}$$

Regarding the type of distribution, 2P-W or 3P-W parameters, the use of both distributions is used in the classical literature [20–22]. The trend is to use a distribution with fixed parameters, which reduces the amount of specimens required to obtain the statistical distribution in a set of data, as would only have to determine a single parameter of the distribution (Eqs. (19) and (20)):

$$P = 1 - \exp\left[-\left(\frac{J_C}{J_0}\right)^2\right] \tag{19}$$

$$P = 1 - \exp\left(-\left(\frac{K_{Jc} - 20}{K_0 - 20}\right)^4\right) \tag{20}$$

Nowadays, 3P-W is well accepted in terms of K given by Eq. (20), due to the small sample size necessary for the estimation of K_0. According to McCabe [21], such a practice would only be suitable for establishing trends in mean toughness, however, because the tails of the fitted distribution curves would be quite unreliable and not usable to estimate lower-bound values.

3. Master curve

Kim Wallin [16] proposed that most ferritic steels tend to conform to one universal curve of median fracture toughness versus temperature for 1-inch thick specimens (Eq. (21), **Figure 3**). The temperature dependence of the 1 T-C(T) median fracture toughness is based on an empirical equation calibrated at the T_0 temperature that corresponds to a K_{Jc}(med) = 100 MPa.m$^{0.5}$ (Eq. (21)). T refers to the test temperature:

$$K_{Jc(med)} = 30 + 70\exp\left(0.019\left(T - T_0\right)\right) \tag{21}$$

This curve, named MC, was standardized in 1997 by ASTM, after several decades of scientific investigations, with the effort of researchers all over the world and the realization of some round-robin projects. The last version of the standard is ASTM E1921-15a$^{\varepsilon 1}$ [1].

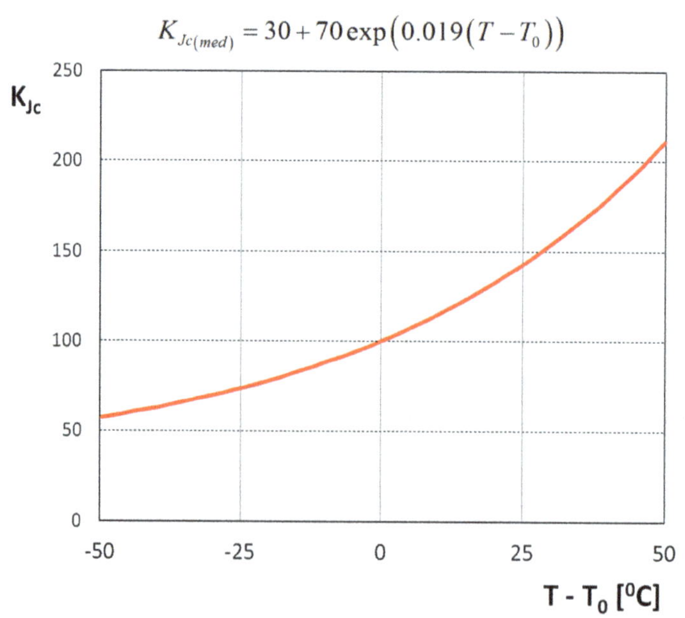

Figure 3. Master Curve.

The reference temperature (T_0) must be known to place the curve in the temperature axis and then have the fracture toughness characterization of ferritic steels in this region. The standardized procedure includes a size conversion equation for those situations where different specimen sizes are used, and some instructions for censoring data for excessive plasticity and ductile crack growth prior to fracture and for loss of constraint.

As the standard makes use of linear elastic fracture mechanics and the measurement of the fracture toughness is made by means of the elastic plastic *Jc* parameter, their values have to be converted to K_{Jc} equivalent values by means of Eq. (22), where E and ν are the Young and the Poisson modulus, respectively.

$$K_{J_C} = \sqrt{\frac{J_C.E}{1-\nu^2}} \tag{22}$$

The MC concept is based on a 3P-W distribution with shape parameter equal to 4 and threshold value equal to 20 MPa.m$^{0.5}$, for compact specimens of 1-inch size (Eq. (20)). In this way, only K_0 must be estimated.

ASTM E1921-15a$^{\varepsilon 1}$ [1] sets up a procedure for K_0 determination. It includes the conversion of the K_{Jc} values obtained for B thickness specimens to 1-inch size equivalent ($K_{Jc(1T)}$) by means of Eq. (23), as well as specifications for data censoring:

$$K_{Jc(1T)} = 20 + \left(K_{JC} - 20\right) B^{1/4} \tag{23}$$

This standard imposes two limits for K_{Jc} values: the first one is given by the condition of a high crack-front constraint at fracture (Eq. (24)):

$$K_{J\max} = \sqrt{\frac{Eb_0\sigma_{YS}}{30\left(1-\nu^2\right)}} \tag{24}$$

The second limit states that K_{Jc} values also shall be regarded as invalid for tests that terminate in cleavage after more than 0.05(W-a_o) or 1 mm (0.040 in.), whichever is smaller, of slow-stable crack growth.

The standardized procedure includes some instructions for censoring data for excessive plasticity and ductile crack growth prior to fracture and for loss of constraint.

K_0 is calculated by means of Eq. (25):

$$K_0 = \left[\sum_{i=1}^{N} \frac{\left(K_{Jc(i)} - 20\right)^4}{r} \right]^{1/4} + 20 \tag{25}$$

$K_{Jc(i)}$ corresponds to the individual K_{Jc} (originally 1 inch or converted to 1 inch equivalent), r is the quantity of non-censored tests, and N is the total number of tests.

The ASTM E1921-15a$^{\varepsilon 1}$ [1] standard also allows the T_0 determination for testing speeds other than static test and for different specimen configurations or geometries, besides the fact that single or multiple temperature tests are considered.

Prior to the MC, ASME Boiler and Pressure Vessel Code already established the lower-bound K_{IC} and K_{Ia} curves for the characterization of ferritic pressure vessel steels [23]. The reference temperature was RT_{NDT} instead of T_0 (**Figure 4**). The implementation of the MC has been a huge advance in the need to have adequate tools for treating the complexities related to temperature, size, and scatter in the ductile-to-brittle transition region for ferritic steels.

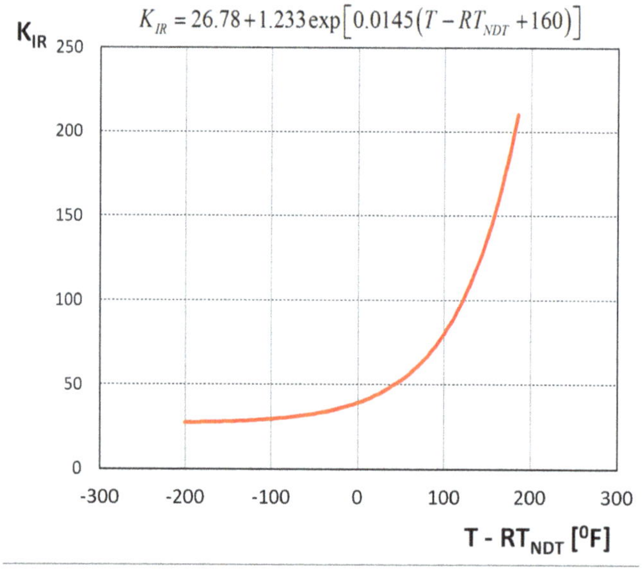

Figure 4. *K* variation with temperature (ASME code).

4. Some aspects on the statistical data fitting

The relationship between the parameters K and J given by Eq. (22) would lead one to believe that the Weibull slope in terms of K (b_K) is twice the slope (b_J) when J data are used.

It would appear correct to think that $b_K = 2b_J$, so if it is accepted that $b_K = 4$, the b_J value would be 2. But this is only valid when working with two-parameter distributions, without a threshold parameter.

An analysis about the relationship between Weibull distributions expressed in terms of J and K was presented in reference [7]. It was shown that if the J_C results follow a 3P-W, their equivalent K_{JC} values do not exactly fit a 3P-W function obtained by means of a simple transformation of the three parameters. Nevertheless, an approximated 3P-W function in K terms was proposed. It fits very well with the transformed values and their parameters are related to the ones expressed in J terms.

Equation (22) is applied to convert J_{min} and J_0 parameters, resulting in the new parameters of the K distribution from Eqs. (26) and (27):

$$K_{min} = K_{J_{min}} = \sqrt{\frac{E\,J_{min}}{(1-\nu^2)}} \tag{26}$$

$$K_0 = K_{J_0} = \sqrt{\frac{E\,J_0}{(1-\nu^2)}} \tag{27}$$

The shape parameter b_K must be calculated by means of Eq. (28):

$$b_K = 2\frac{K_0}{K_0 + K_{min}} b_J = \xi\ b_J \tag{28}$$

Figure 5 shows values of ξ as a function of different combinations of K_0 and K_{min}. As already expressed, for the particular situation of a 2P-W, there is an exact equivalence between the distributions in terms of J and K, being the Weibull slope in terms of K twice the slope in terms of J ($K_{min} = 0$).

Figure 6 shows an example where a dataset of K values, transformed from Jc, is fitted using Weibull-based statistical distributions. The differences among them are the way the three parameters were obtained and are as follows:

- parameters estimated from K_{Jc} values converted from Jc,
- parameters calculated using Eqs. (26)–(28),
- parameters K_{min} and K_0 obtained using Eqs. (26) and (27); $b_K = 2b_J$.

Clearly, the latter only coincides with the others in zone B, near a failure probability of $P = 0.63$. The first two distributions are quite similar for all the probability levels, including zone C (near the lower-bound zone) and for high probability levels (zone A).

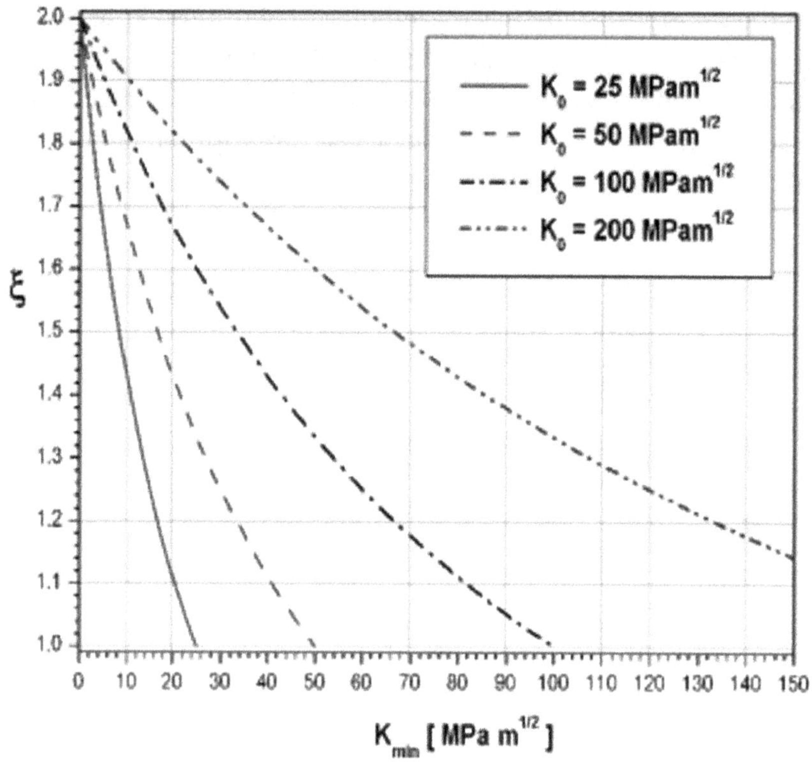

Figure 5. Dependence of ξ with K_0 and $K_{min.}$

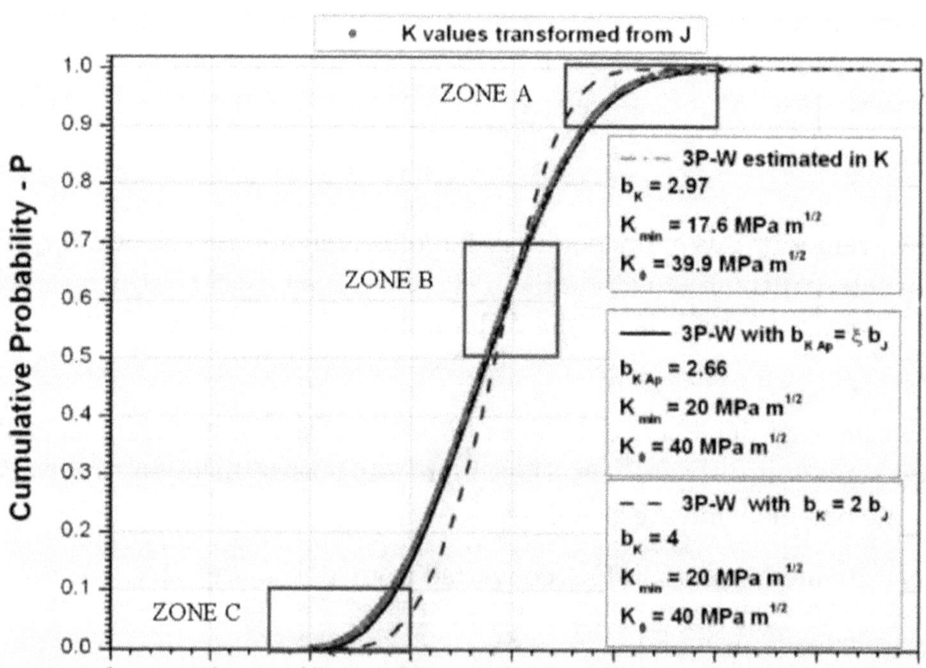

Figure 6. Comparison of cumulative probabilities obtained considering different options.

5. Unresolved aspects on the transition

The MC is a methodology to deal with the calculation of K_{Jmean} in the transition region using small datasets. Despite the fact that it is a good response to an engineering problem, there are some aspects of the transition region that must be investigated.

5.1. The real transition region

Perez Ipiña et al. [5] presented a ductile-to-brittle region reinterpretation based on experimental evidence (**Figure 7**). They proposed not a single curve but the area involving the scatter band. This area is limited by two curves, one corresponding to the toughness lower bound (thickness independent, i.e., material property) and the other fitting to the upper limit of the scatter band (thickness dependent). In this way, the area is larger for smaller thickness than for larger ones. Note that the scatter in ductile mechanisms is much lower than in cleavage.

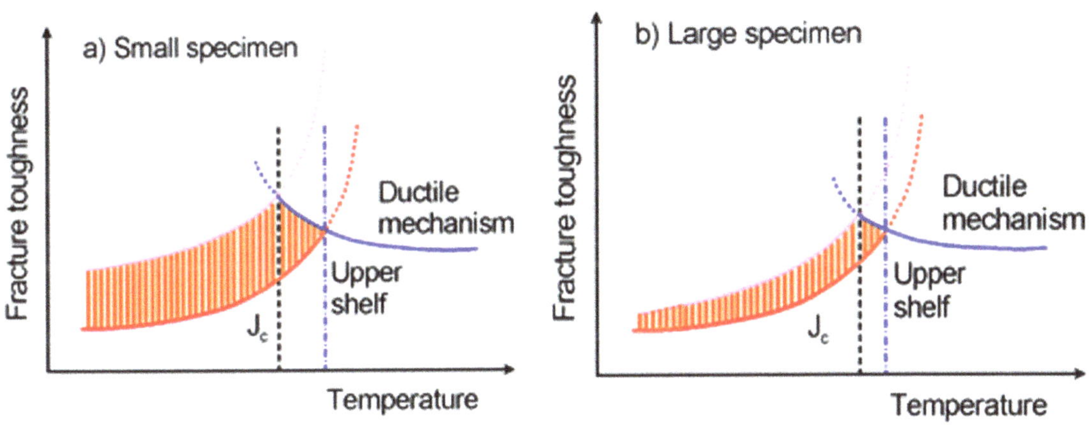

Figure 7. Ductile-to-brittle region scatter band for (a) small specimens and (b) large specimens.

Figure 8 shows that several subregions can be defined in the transition.

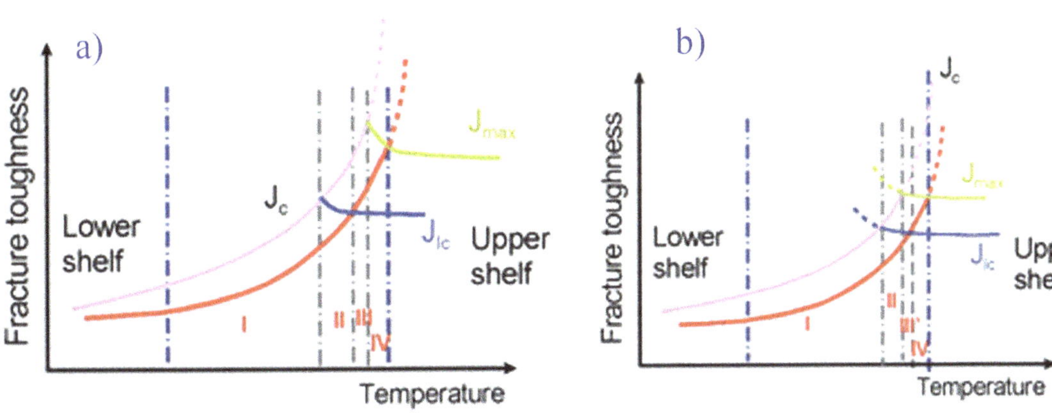

Figure 8. (a) Subregion III and (b) subregion III′ in the ductile-to-brittle transition curve.

I: All specimens fracture by cleavage without any stable crack growth.

II: Some specimens fracture by cleavage without any stable crack growth, while others fracture by cleavage after some amount of stable crack growth.

III: No cleavage without stable crack growth occurs. All the specimens fracture by cleavage after some amount of stable crack growth, or

III: some specimens fracture without stable crack growth, others with stable crack growth, and others reach the maximum load condition and do not present instability.

III or III′ will be present depending on the crossing of curves J_{IC} with lower-bound cleavage and the crossing of J_{max} with upper-bound cleavage curves. When the intersection of J_{IC} with the cleavage lower-bound curves and the intersection of J_{max} with the cleavage upper-bound curves occur at the same temperature, there will be no region III nor III′.

IV: Some specimens fracture after some amount of stable crack growth, while others reach the maximum load condition and do not present instability.

For higher temperatures, no cleavage occurs and this behavior corresponds to the upper shelf.

Maximum load toughness is size dependent: small specimens present the P_{max} plateau close past the stable crack growth initiation, while large specimens require more stable crack growth to reach this plateau, giving them larger J_{max} than small specimens. Maximum load curves intersect the cleavage curves—the upper cleavage curve is also size dependent—at different temperatures for different sizes, **Figure 8**. Region IV widens and displaces toward higher temperatures as size increases, making the beginning of the upper shelf also size dependent, as stated by Wallin [24].

5.2. Stable crack growth and loss of constraint limitations

In subregions where some or all specimens present ductile growth of cracks (DCG), the probability of failure of a set of data is affected by this stable growth. It may also happen that conditions of maximum J (J_{max}) are violated for small specimens and the fracture toughness of these samples will increase as a consequence of a loss of constraint. **Figure 9** [20] shows schematically these two effects in the cumulative failure probability.

Consequently, the weakest link model, and the toughness prediction for different thicknesses, seems not to work well in the superior third of the transition region, where ductile crack growth and/or loss of constraint are present. There are therefore conditions that must be met for its implementation. There are various proposals or explanations in the literature of the area of validity of the model weakest link. Among these is the existence of a single initiator of cleavage site, and limitations in the stable growth of cracks prior to cleavage (would affect the in-plane constraint) and in the thickness of the sample (would affect the out-of-plane constraint). The MC, as mentioned before, introduces censoring under conditions of stable crack growth or loss of constraint. This censoring scheme, when applied to 3P-W, gives artificially increased slope values.

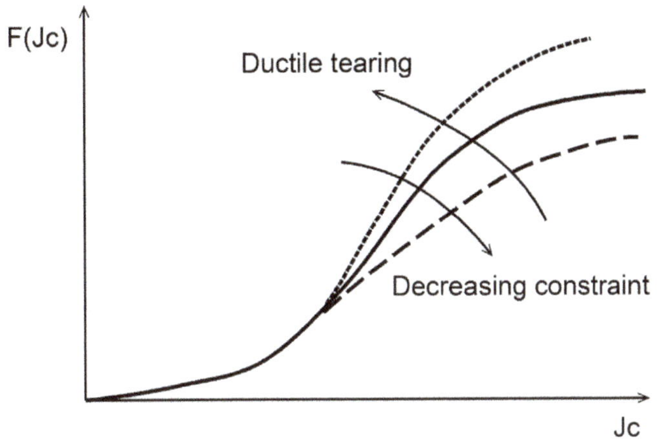

Figure 9. Effects of ductile tearing and decreasing constraint.

Some authors (Wallin [15] and McCabe et al. [22]) also stated that the model is not valid at temperatures corresponding to the lower shelf.

The use of small datasets does not allow a correct K_{min} and b_K estimation. Berejnoi and Perez Ipiña [4] have shown that the threshold parameter K_{min} and Weibull slope b_K are clearly dependent on temperature and different from the values of 20 MPa.m$^{0.5}$ and 4, which are considered in the MC. This was found even for 1-inch size sets with all valid data. **Figure 10** shows the MC from the material of the Euro round-robin [25] obtained with 1-inch specimens, corresponding to a $T_0 = -96$°C. The values of K_{Jmean} obtained using MC methodology at different temperatures and sizes are also shown, being clearly different from the MC.

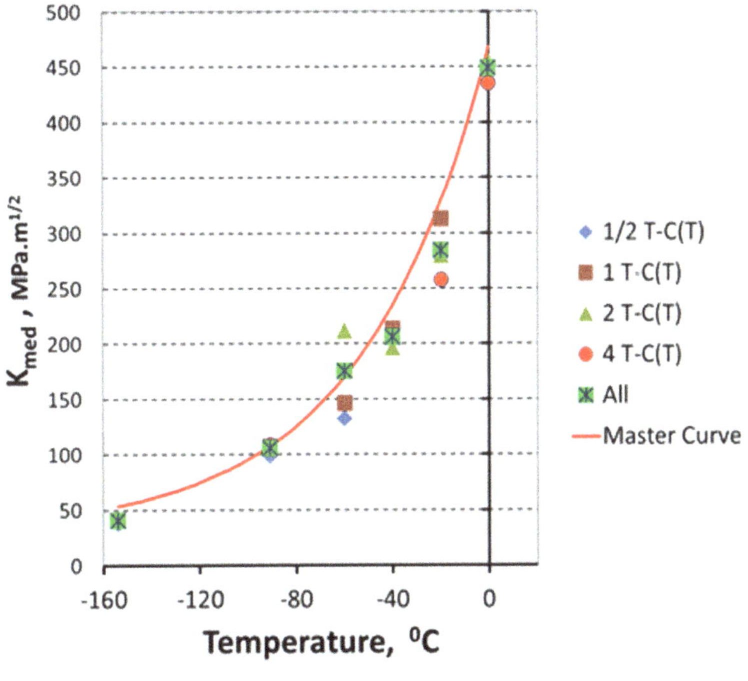

Figure 10. K_{med} versus T for different sizes and master curve according to ASTM E1921.

When only valid datasets of 1-inch size are considered [4], the values of K_0 and K_{med} obtained using a 3P-W distribution are in concordance with those obtained using ASTM E1921-15a$^{\varepsilon 1}$[1], although b_k and K_{min} were different. The fixed values stated in ASTM standard could not be appropriated when a size conversion criterion and/or some censoring procedure are included.

Figure 11 corresponds to datasets from the Euro round-robin, tested at $T = -60°C$ using 2-inch specimens. This temperature is within the range of $T_0 + 50°C$ of the material. The original data were converted to a 1-inch equivalent size, by means of Eq. (22). This procedure was also applied considering the 3P-W distribution with $b_K = 4$ and $K_{min} = 20$ (Eq. (20)).

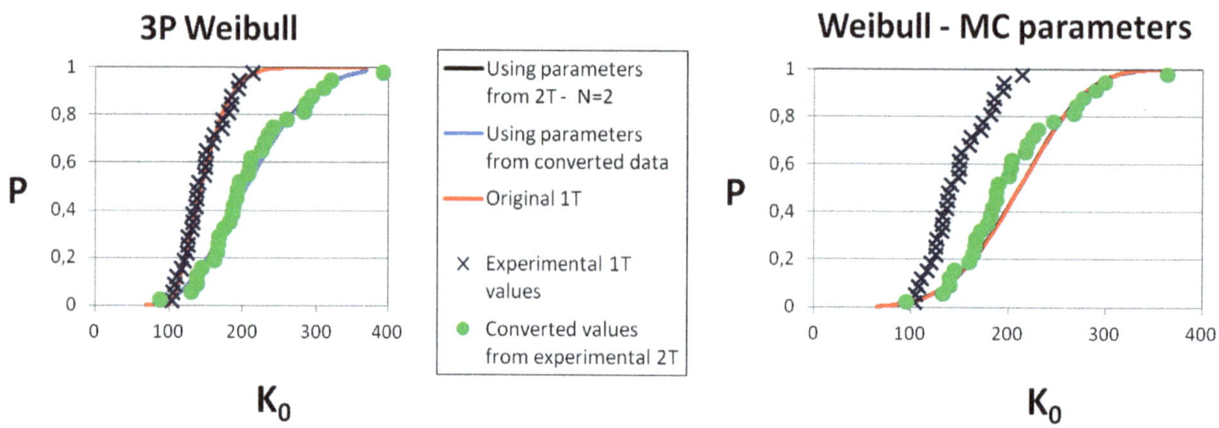

Figure 11. Probability distributions for $T = -60°C$ and $2T$ original dataset.

From **Figure 11**, it is seen that the conversion formula does not work properly, and it is not just as simple as using a factor N as the weakest link model states for converting toughness values, nor with 3P-W, nor with MC. The distributions (MC or 3P-W) do not fit the toughness values obtained experimentally using 1T size. Experimental converted-to-1T values (green circles in **Figure 11**) should be close to the original 1T values (black crosses).

6. Conclusions

The characterization of fracture resistance of ferritic steels in the ductile-to-brittle transition region is problematic due to scatter in results, as well as size and temperature dependences.

Originally, there were two explanations to the size effect. One based on constraint effects and another that made use of statistical weakest link concepts to explain the probability to find a cleavage initiator site at the crack front. The first fails to explain the observed scatter.

As the result of many years of investigations, which began in the 1980s decade, and based mainly on the works of Wallin, ASTM has standardized the determination of a temperature reference (T_0) for the fracture toughness characterization of ferritic steels in this region by means of a MC. This function gives the variation of K_{Jmean} with temperature, for 1T specimen size, and is based on a 3P-W probability distribution, with two of them fixed.

Despite the fact that this MC is a huge technological advantage, there are many aspects that need a deeper analysis, including

- the relationship between Weibull shape parameters when *J* or *K* results are used,
- size effect and validity of the specimen sizes conversion imposed in the ASTM standard,
- the validity of a model based only on statistical effect without taking into account the constraint.

Acknowledgements

The Consejo Nacional de Investigaciones Científicas y Tecnológicas de Argentina (CONICET) and Consejo de Investigación de la Universidad Nacional de Salta (CIUNSa) are gratefully acknowledged.

Author details

Carlos Berejnoi[1*] and Juan E. Perez Ipiña[2]

*Address all correspondence to: berejnoi@unsa.edu.ar

1 Faculty of Engineering, National University of Salta/CIUNSa, Avda. Bolivia, Salta, Argentina

2 Fracture Mechanics Group, National University of Comahue/CONICET, calle Buenos Aires, Neuquén, Argentina

References

[1] ASTM E1921-15a$^{\varepsilon 1}$. Standard test method for determination of reference temperature, T_0, for ferritic steels in the transition range. In: ASTM book of standards. 2015. DOI: 10.1520/E1921-15AE01.

[2] Berejnoi C, PerezIpiña JE. Evaluation of some methods for lower bound determination in the transition region of ferritic steels. Latin American Applied Research. 2004;34(4): 241–248.

[3] Berejnoi C, PerezIpiña JE. Ductile-to-brittle transition and upper shelf in steels: scatter analysis. In: 63° ABM Annual Congress, São Paulo, Brasil. 2008.

[4] Berejnoi C, PerezIpiña JE. Analysis of size and temperature effects in the ductile to brittle transition region of ferritic steels. Engineering Fracture Mechanics. 2015;148:180–191. DOI: 10.1016/j.engfracmech.2015.09.041

[5] PerezIpiña JE, Berejnoi C. Size effects in the transition region and the beginning of the upper shelf for ferritic steels. Fatigue & Fracture of Engineering Materials & Structures. 2010;33(3):195–202. DOI: 10.1111/j.1460-2695.2009.01432.x

[6] Perez Ipiña JE, Berejnoi C. Experimental validation of the relationship between parameters of 3P-weibull distributions based in JC or KJC.In: 13th International Conference on Fracture 2013 (ICF-13) Proceedings; 16–21 June 2013, International Congress on Fracture (ICF), Beijing, China. 2013.

[7] Larrainzar C, Berejnoi C, PerezIpiña JE. Comparison of 3P-Weibull parameters based on Jc and K_{Jc} values. Fatigue & Fracture of Engineering Materials & Structures. 2011;34(6):408–422. DOI: 10.1111/j.1460-2695.2010.01533.x

[8] Berejnoi C, Ruiz AI, PerezIpiña JE. Analysis of the influence of test temperature and specimen size in the determination of the master curve. In: 23rd ABCM International Congress of Mechanical Engineering; 6–11 December, Rio de Janeiro, Brazil. 2015.

[9] Perez Ipiña JE. Fracture Mechanics (in spanish). 1st ed. Buenos Aires: Ed. Alsina; 2004. ISBN: 950-5553-124-9

[10] Milne I, Chell GG. Effect of size on the J fracture criterion in elastic-plastic fracture. In: Landes JD, Begley JA, Clarke GA, editors. Elastic-plastic fracture symposium, ASTM Special Technical Publication 668; 16–18 Nov 1977, Atlanta, Baltimore. 1982. p. 358–377.

[11] Landes J, Shaffer DE. Statistical characterization of fracture in the transition region. In: Paris PC, editor. Fracture mechanics, 12th conference, ASTM Special Technical Publication 700; 21–23 May 1979, Washington University, Baltimore. 1980. p. 368–382.

[12] Landes JD, McCabe DE. Effect of section size on transition behavior of structural steels. Scientific paper 81-1D7-Metal-P2, Westinghouse R&D Center. Pittsburg, USA. 1982.

[13] Wallin K. The scatter in K_{IC} results. Engineering Fracture Mechanics. 1984;6:1085–1093.DOI: 10.1016/0013-7944(84)90153-X

[14] Wallin K. A simple theoretical charpy-V-K_{IC}correlation for irradiation embrittlement. In: Proc. ASME pressure vessels and piping conference, innovative approaches to irradiation damage and fracture analysis, PVP-Vol. 170, American Society of Mechanical Engineers, New York. 1989. p. 93–100.

[15] Wallin K. The effect of ductile tearing on cleavage fracture probability in fracture toughness testing. Engineering Fracture Mechanics. 1989;32(4):523–531. DOI: 10.1016/0013-7944(89)90186-0

[16] Wallin K. Recommendations for the application of fracture toughness data for structural integrity assessments. In: Proceedings of the joint IAEA/CSNI specialists meeting

on fracture mechanics verification by large-scale testing, NUREG/CP-0131 (ORNL/TM-12413); 26–29 Oct; Oak Ridge. 1992. p. 465–494.

[17] Weibull W. A statistical distribution function of wide applicability. Journal of Applied Mechanics, Trans. ASME. 1951; 18:293–297.

[18] Wallin K. Fracture toughness of engineering materials, estimation and application. United Kingdom: EMAS Publishing; 2011. 543 p. ISBN: 0-9552994-6-2

[19] Anderson TL, Stienstra D, Dodds RH. A theoretical framework for addressing fracture in the ductile-brittle transition region. In: Landes JD, McCabe DE, Boulet JAM, editors. Fracture mechanics: twenty-fourth volume, ASTM Special Technical Publication 1207; 30 June–2 July 1992; Gatlinburg, Tennessee. Ann Arbor, MI; 1994. p. 186–214.

[20] Heerens J, Zerbst U, Schwalbe KH. Strategy for characterizing fracture toughness in the ductile to brittle transition regime. Fatigue & Fracture of Engineering Materials & Structures. 1993;16(11):1213–1230. DOI: 10.1111/j.1460-2695.1993.tb00735.x

[21] McCabe DE. A comparison of Weibull and βIC analyses of transition range data. In: Chona R, editor. Fracture mechanics: twenty-third symposium, ASTM Special Technical Publication 1189; 18–20 June 1991; College Station, Texas. Baltimore; 1993. p. 80–94.

[22] McCabe DE, Zerbst U, Heerens J. Development of test practice requirements for a standard method on fracture toughness testing in the transition range. GKSS Publications, 93/E/81. GKSS Forschungzentrum, Geeshscht, Germany. 1993.

[23] Rules for construction of nuclear power plant components. ASME Boiler and Pressure Vessel Code Section III. 1983.

[24] Wallin K. Master curve analysis of the "Euro" fracture toughness dataset. Engineering Fracture Mechanics. 2002;69(4):451–481. DOI: 10.1016/S0013-7944(01)00071-6

[25] Heerens J, Hellmann D. Development of the fracture toughness dataset. Engineering Fracture Mechanics. 2002;69(4):421–449. DOI: 10.1016/S0013-7944(01)00067-4

5

The Strain Energy Release Rate for Stack of Coated Conductors with Interface Crack in Perpendicular Magnetic Field

Meng Zhao, Huadong Yong and Youhe Zhou

Abstract

Due to the ability of transporting huge current in the stack of high-temperature superconducting conductors, the electromagnetic body force generated by the interaction of magnetic field and current may affect the mechanical stability of structure. In this paper, the fracture behavior of the stack of coated conductors which contains an interface crack is studied for increasing field and decreasing field. The body forces are obtained with variational formulation for the Bean's critical state model. Based on the virtual crack closure technique (VCCT), the strain energy release rate of the stack of coated conductors with an interface crack is determined. The strain energy release rates are compared for different crack positions, crack lengths, magnetic fields and the thicknesses of substrate, respectively. These results may be useful for the practical application.

Keywords: strain energy release rate, stacked conductors, interface crack, magnetic field, superconductor

1. Introduction

Due to the ability of transporting high critical current density and trapping magnetic field, high-temperature superconductors have been expected to be used in the transmission cables, transformers, motors and maglev system [1–5]. A typical application is the stack of high-temperature superconductor coated tapes for power engineering [6]. However, the application of tape is limited for the reason that the superconductor is a brittle material, which cannot withstand large mechanical loadings [7, 8]. Meanwhile, the manufacturing process is complex for high-temperature superconductor. Thus, the cavity and microcracks are inevitable in superconductor [9, 10]. The electromagnetic body force induced by the interaction of electromagnetic field may result in the extension of crack in superconductor under high magnetic fields, which will finally give rise to the fracture of superconductor [11]. Cracking can reduce the critical current density of superconductor and mechanical stability [12]. Recently, significant efforts have been made to study the mechanical behavior of the superconductors with the development of critical current density.

Since the electromagnetic behavior in superconductor is complex, several different theoretical models were presented to describe the constitutive relation, such as Bean model, Kim model and E-J relation [13–15]. In recent years, the relationships between mechanical response and magnetic field have been concerned [16, 17]. The general procedure is that the current and magnetic field distributions are obtained based on the constitutive model. After deriving the electromagnetic body force, we will find the mechanical characteristics of superconductor. The superconductor will undergo mechanical deformation in magnetic field which is regarded as magnetostriction. The magnetostriction is dependent on the amplitude of magnetic field and critical state model [16, 17]. For the superconductors during the magnetization, the body force induced by the flux pinning evolves from initially compressive to tensile as the external field changes from increasing to decreasing. It was pointed out that the stress and strain induced by flux pinning show obvious irreversible behavior. In other words, the stresses are different for increasing field and decreasing field cases [18–20]. Moreover, two different cooling processes (zero field cooling and field cooling) are discussed. The quantitative analysis on the magneto-elastic problems in bulks and strips was made for different critical state models, such as the Bean model and Kim model [18–31]. For superconducting strip with transport current, the magnetostriction curve is not a closed loop [32]. The stress and magnetostriction in many complex structures also have attracted many attentions [33–37].

The crack problem is another challenge for the superconductors because the electromagnetic body force and thermal stress are important driving forces for crack growth. During the cycles of the applied magnetic field, the tensile and compressive stress will be generated alternately and cracking of superconductor usually occurs as the field is decreased to zero [38]. This is because the crack is prone to advance under the tensile stress. The linear elastic fracture theory was widely used to investigate the fractures behaviors of the thin film and bulk superconductors [39–47]. The collinear cracks, transverse crack and crack-inclusion problems in superconductors were also studied recently [48–50]. Superconductors are usually metal composites which consist of several components. Then, the crack problems in inhomogeneous superconductors were studied extensively [51–57]. Recently, the dynamic fracture behaviors in superconductors were analyzed with the finite element method [58].

Compared to the bulk superconductor, coated conductors have better mechanical strength [59]. As is known to us, the coated conductor structure is made up of several layers deposited onto the substrate. In the coated conductor structure, the interface crack may be generated by lattice mismatch, thermal expansion and chemical reactions during the deposition progress of the coated structure [60]. The delamination of superconducting layer is easy to take place at the interface between dissimilar materials [61]. Unlike the internal or edge crack problems, the compressive stress can also lead to the propagation of interface crack [62]. Thus, it is necessary to analyze the fracture behavior of interface crack to predict the mechanical instability in the coated conductor structure.

The fracture behavior of interface crack between superconducting film or tape and substrate was studied in magnetic field or with transport current [62–64]. However, there are few works on the crack problem in stack of conductors. In this paper, we consider the interface crack in the stack of high-temperature coated tapes. The effect of copper and substrate on the shielding current and magnetic field distributions in the superconducting layer is neglected. Based on the simplified geometrical and material assumptions, linear elastic equation is used. We carry out an in-depth investigation on the strain energy release rate for interface crack in different magnetic fields, crack lengths, substrate thicknesses with virtual crack closure technique (VCCT). The results may provide guidance for the design of stack of coated conductors.

This paper is organized as follows. In Section 2, we introduce VCCT to calculate the strain energy rate at the crack tip. Section 3 computes a bimaterial plate with a central interface crack to verify the accuracy of VCCT. In Section 4, we discuss the body force distribution in superconducting tape. In Section 5, the interface crack in stack of tapes is analyzed under perpendicular magnetic field. In Section 6, we draw a conclusion from the above work.

2. Strain energy release rate of interface crack using VCCT

In this section, we will first verify the accuracy of the numerical computation. Here, we only consider the linear elastic fracture behavior. For the interface crack problem, the stress intensity factor or strain energy release rate can be determined with the virtual crack closure technique (VCCT). VCCT is a well-established method for computing the crack problem [65], which was proposed by Rybicki and Kanninen for two-dimensional problems [66].

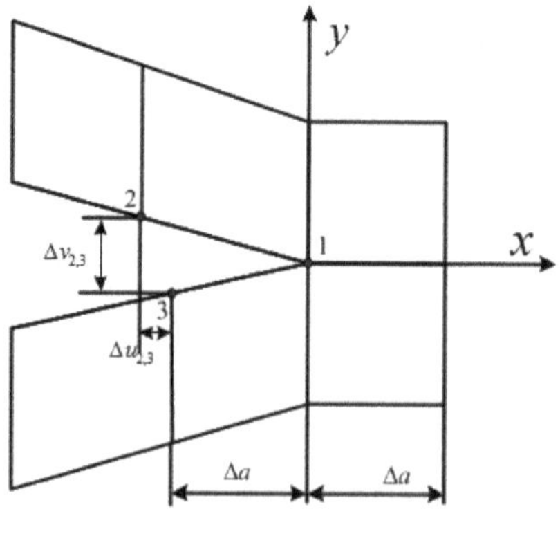

Figure 1. Calculating model of the strain energy release rate using VCCT.

Figure 1 shows the mesh model for computing the strain energy release rate with VCCT, where u and v stand for the displacement along x and y directions, respectively. Each mesh is a square and Δa is the side width. On the basis of assumption given in reference [66], we define the opening displacement behind the tip of virtual crack ($\Delta v_{2,3}$ and $\Delta u_{2,3}$) as the actual crack opening displacement for determining strain energy release rate with VCCT. The strain energy release rate represents the change of elastic strain energy per unit area of crack extension [67]. Then, the strain energy release rates for different fracture modes can be expressed as follows:

$$G_{\mathrm{I}} = \frac{F_y \Delta v_{2,3}}{2B\Delta a}, \ G_{\mathrm{II}} = \frac{F_x \Delta u_{2,3}}{2B\Delta a} \tag{1}$$

$$G = G_{\mathrm{I}} + G_{\mathrm{II}} \tag{2}$$

where F_x and F_y are the nodal forces and B is the thickness of the body with crack. VCCT is an effective method which does not need integration and it is insensitive to mesh.

3. Verification of numerical results

In order to verify the simulation results, we study the fracture behavior for bimaterial plate with an interface crack, as shown in **Figure 2**. It is well known that for the interface crack, there is the coupling between the tensile fracture mode and shear fracture mode even for mode I loadings. In addition, the oscillation of stress singularity will occur at the crack tip [65]. In this model, the interface crack is located in the center of the bimaterial plate and a uniform tension load σ_0 is applied on the top and bottom edges of the plate. The parameters used for numerical analysis are as follows [68]: $2w$ = 100 mm, $2l$ = 200 mm, a/w = 0.4, E_1 = 2.058 × 10^5 MPa, $\frac{E_1}{E_2} = 1 \sim 100$, $v_1 = v_2 = 0.3$, $\sigma_0 = 9.8$ MPa.

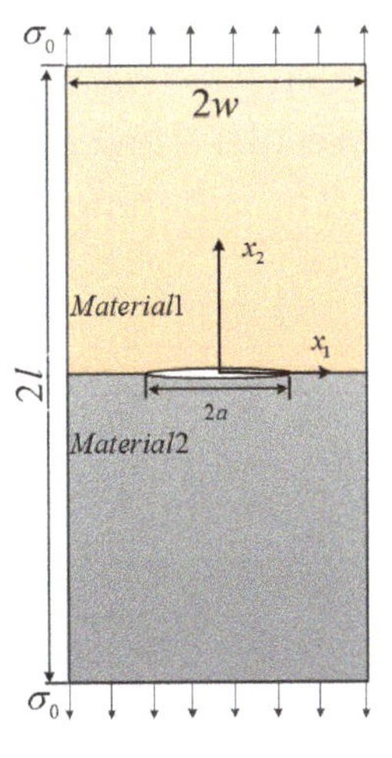

Figure 2. A finite bimaterial plate with interface crack under a uniform tension.

We use finite element software Abaqus to solve the interface crack problem. A 4-node bilinear plane stress quad-dominated structured element is used and mesh width is given as $\Delta a = 1\text{mm}$. It is convenient for us to use VCCT to compute the strain energy release rate. For the sake of comparing our results with those of Yu et al. [69], Nagashima et al. [70], Miyazaki et al. [71], we use the formulation proposed by Suo and Hutchinson [72]. For bimaterial interface crack problem, the relation between stress intensity factor and strain energy release rate is as follows [72]

$$G = \frac{c_1 + c_2}{16\cosh^2 \pi\varepsilon}|K|^2 \tag{3}$$

where subscripts 1 and 2 represent the material 1 and material 2, respectively, and

$$K^2 = K_1^2 + K_2^2 \tag{4}$$

$$c_1 = \frac{\kappa_1 + 1}{\mu_1}, \; c_2 = \frac{\kappa_2 + 1}{\mu_2} \tag{5}$$

$$\varepsilon = \frac{1}{2\pi}\ln\frac{1-\beta}{1+\beta}, \; \kappa = (3-\upsilon)/(1+\upsilon) \tag{6}$$

$$\beta = \frac{\mu_1(\kappa_2 - 1) - \mu_2(\kappa_1 - 1)}{\mu_1(\kappa_2 + 1) + \mu_2(\kappa_1 + 1)} \tag{7}$$

in which β is the parameter proposed by Dundurs to describe the mismatch of material 1 and material 2 [73], and μ is the shear modulus.

Table 1 shows the comparison of strain energy release rates for present work and the results given by others when $a/w = 0.4$. These results show that there is little difference between our work using VCCT and other methods. The difference is smaller than 1% and this verifies the accuracy of the numerical method.

E1/E2	Yu	Nagashima	Miyazaki	This work
1	0.0361	0.0361	0.0362	0.0358
2	0.0531	0.0529	0.0231	0.0526
3	0.0688	0.0685	0.0690	0.0683
4	0.0841	0.0835	0.0843	0.0834
10	0.1725	0.1699	0.1727	0.1711
100	1.4652	1.4264	1.4642	1.4529

Table 1. Comparison of the strain energy release rate G (N/mm).

4. Body force distribution in tape under magnetic field

Next, we turn our attention to the body force distributions in superconducting tape. **Figure 3** shows the configuration of the stack of superconducting conductors. The conductors consist of three tapes and these tapes are separated by the insulation layers (green). Generally, the superconducting layer (black) is attached to the substrate in each tape. In addition, copper layer is deposited on the superconducting layer which can also transport current. The thicknesses of the insulation layer, copper layer, superconducting layer and substrate are h_{Ins} = 25 μm, h_{Cop} = 40 μm, h_{SC} = 1 μm, h_{Sub} = 50 μm, respectively [74]. Then, the total thickness of stacked conductors is h = 373 μm and the width is 200 μm. The mechanical behavior is based on the electromagnetic body force and we will first calculate the field and current distributions. For simplicity, we use the Bean's critical state model, i.e., the critical current density is a constant which is independent of the magnetic field.

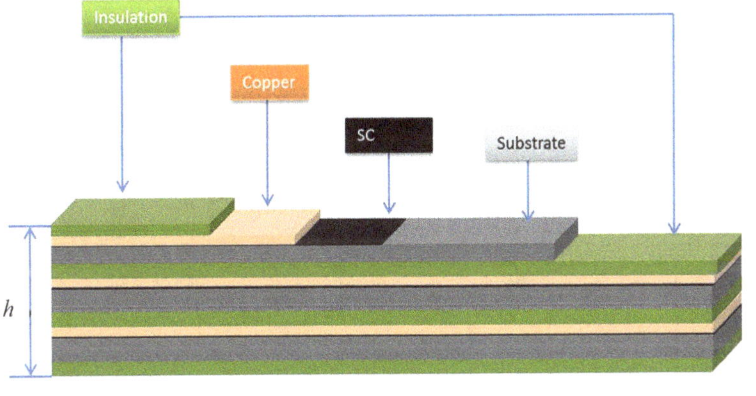

Figure 3. The configuration of the stack of superconducting conductors.

The distributions of shielding current are plotted in **Figure 4**. As the conductor is under magnetic field, the current will be induced to shield the external field. From **Figure 4 (a)**, we can find that the tape is divided into two parts and the current direction is opposite in two parts for increasing field case. As the external field is decreased, the shielding current will redistribute. The conductor can be divided into four parts for decreasing field case. However, it is interesting to see that the direction of shielding current is antisymmetric.

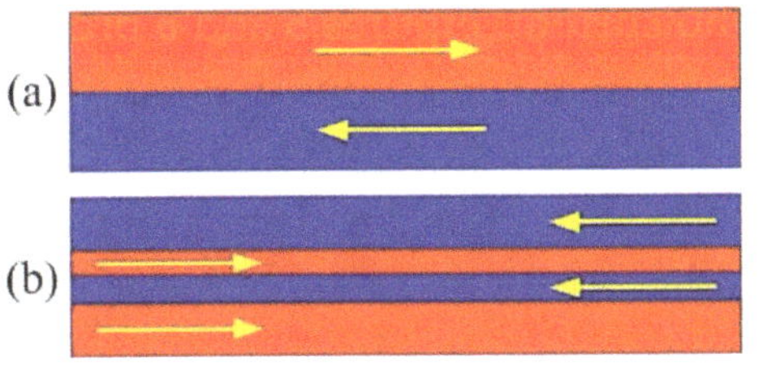

Figure 4. The current distribution in superconducting layer under magnetic field. (a) Increasing field case and (b) decreasing field case.

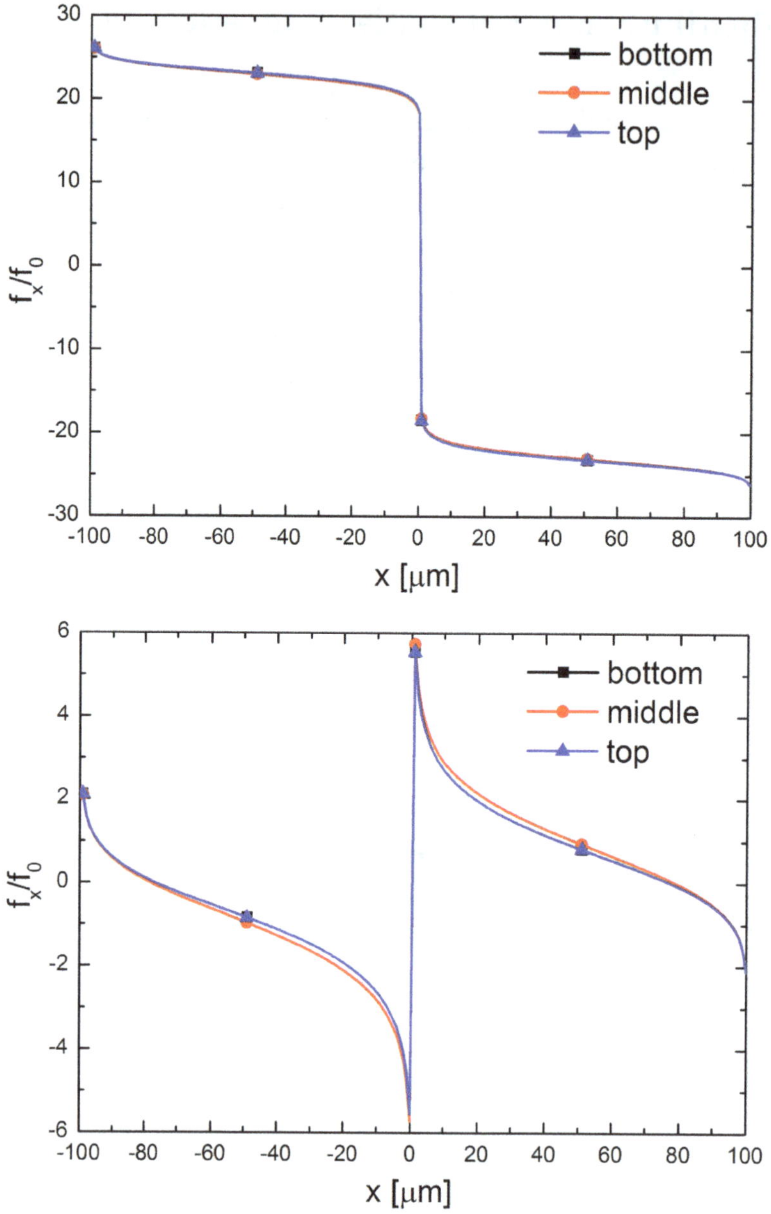

Figure 5. The distributions of body force along x direction. The magnetic field is increased from zero to B_e (upper). The field is decreased from B_e to zero (lower).

Using the variational formulation derived by Prigozhin [75], the electromagnetic responses of superconducting layer are obtained in the external field. The body force **f** = **j**×B can be determined with the flux density and current. **Figures 5** and **6** present the body forces f_x and f_y per unit volume for various stages of magnetization process. It is obvious that the body forces along x direction are very close for the bottom, middle and top layers. Moreover, the value of f_x for the increasing field is larger than that for decreasing field. The body force f_y is much smaller than f_x. This is due to the reason that the magnetic field is mainly along y direction. Thus, it can be expected that shear loading is important than the tensile loading. As external field B_e is larger than the full penetration field, the value of f_y for increasing and decreasing field is also very close.

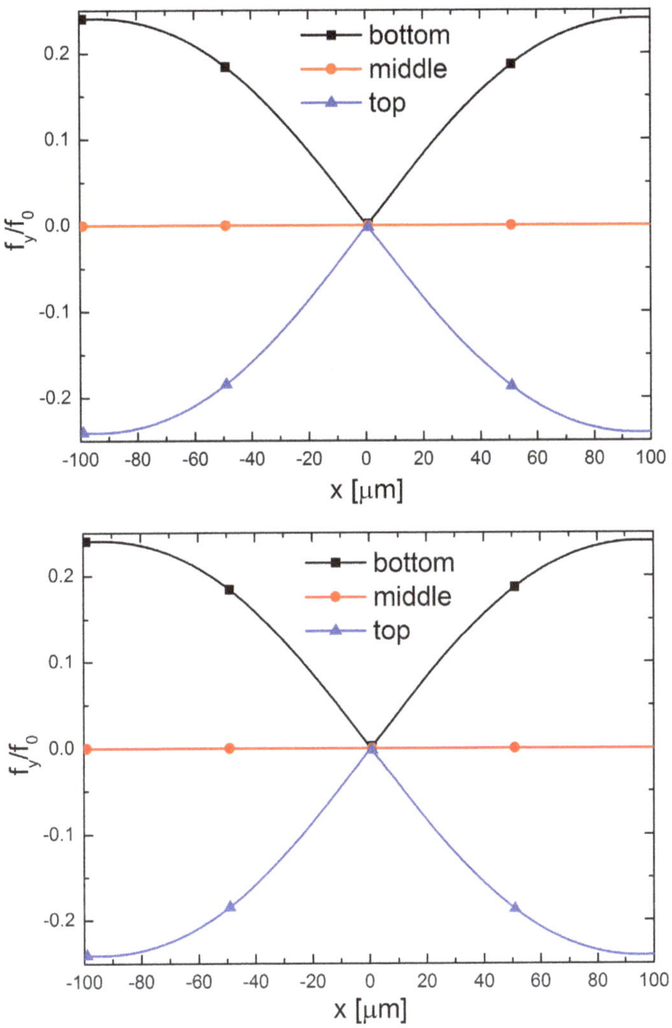

Figure 6. The distributions of body force along y direction. The magnetic field is increased from zero to B_e (upper). The field is decreased from B_e to zero (lower).

5. Interface crack problem for the stacked conductors structure

The structure of the stack of high-temperature superconducting conductors in perpendicular magnetic field is shown in **Figure 7**. The lower insulation layer which is located at the bottom is fixed. That is to say that the displacements of x and y directions at the bottom of conductors are zero. The stacked conductors consist of three superconducting tapes. The central interface crack is located between the superconducting layer and substrate, and is assumed to be through the length of conductor.

Although copper is an elasto-plastic material, we neglect the plastic deformation. Thus, the focus will be on the linear fracture mechanics problem. All constituent materials are isotropic, linearly elastic and homogeneous. The effects of copper and substrate on the distributions of magnetic field and current density in the superconductor layer can be ignored and the stack of superconducting conductors is placed in a magnetic field B parallel to the y direction. Since the interface crack is parallel to the direction of current flowing, the crack will not disturb the current density and field distributions. Moreover, the width of the stacked conductors is much smaller than the length of the tape along z direction. For the 2D dimensional analysis, a plane strain condition can be assumed.

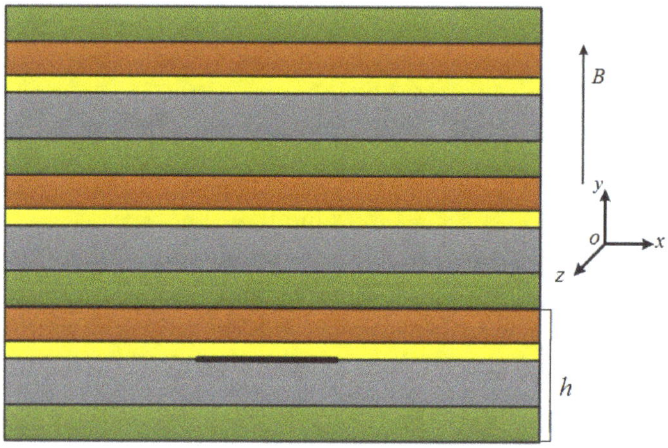

Figure 7. Schematic drawing of the stacked conductors with an interface crack in the perpendicular magnetic field.

During the following simulations, the used parameters are given as follows. We assume that the current density j is no larger than the critical current value $j_c = 3\times10^4$ A/mm^2. Besides, the applied maximum field is equal to $B_e = 8\mu_0 j_c t$, t is the thickness of the total superconducting layers and μ_0 is permeability in vacuum. The strain energy release is normalized by $G_0 = \frac{(1-\nu^2)\pi a}{E_{sc}}\left(\frac{\mu_0 j_c^2 tw}{\pi}\right)^2$.

	E(MPa)	Poisson's ratio
Copper	117×10^3	0.35
YBCO	178×10^3	0.3
Hastelloy	200×10^3	0.3
Insulation	3.5×10^3	0.3

Table 2. Parameters of the materials of the coated tapes [74].

We consider the magnetization process in which the field is increased from zero to maximum and subsequently decreased to zero. As discussed earlier, the body force direction will change for increasing field and decreasing field. Since the total strain energy release rate for increasing field is larger, we mainly discuss the increasing field case. We can assume $G_I = 0$ when the interface crack is closed [62]. For the sake of simplicity, the contact and friction effects between crack surfaces are not considered. The parameters used are given in **Table 2**.

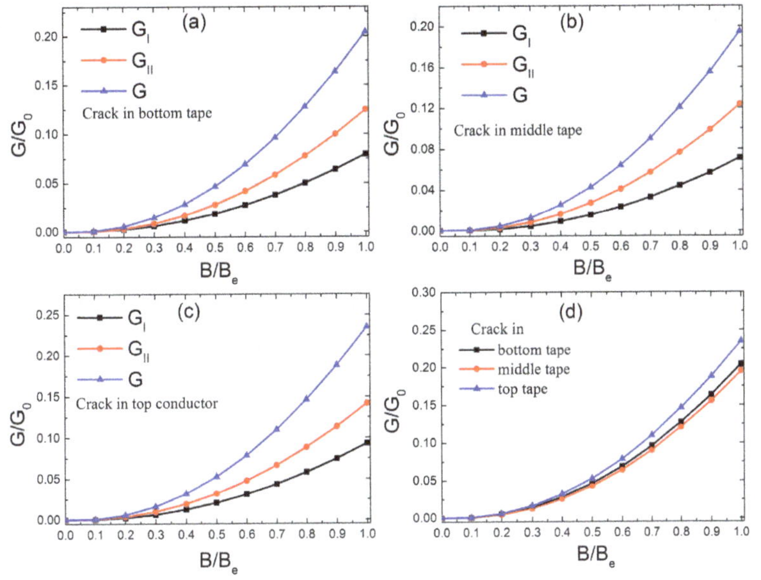

Figure 8. The variation of strain energy release rates with magnetic field as the interface crack is in (a) top tape, (b) middle tape and (c) top tape. The strain energy release rates are compared in (d).

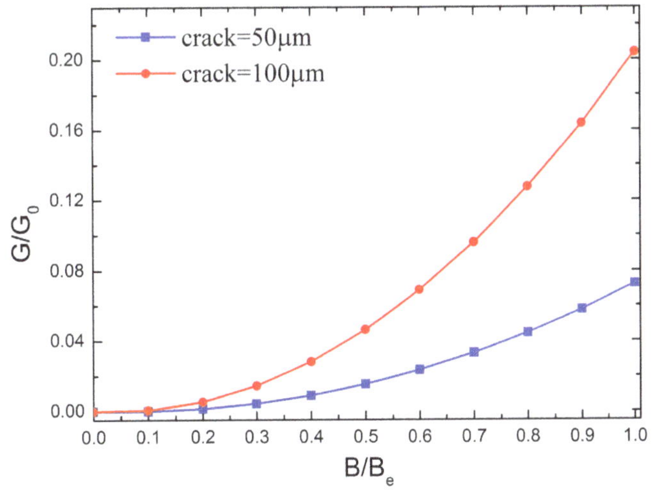

Figure 9. The strain energy release rates for the crack lengths of 50 μm and 100 μm during increasing magnetic field.

Next, we will calculate the strain energy release rates for the stacked conductors with the interface crack. **Figure 8(a)–(c)** shows the variation of strain energy release rate with the magnetic field, i.e., the field is increased from zero to maximum. The central interface crack is located at the bottom, middle or top tape. It can be found that all the strain energy release rates increase with the increasing field. Although the body force is compressive for increasing field case, G_I is smaller than G_{II} and this trend is similar to the results given in reference [62]. The strain energy release rates with interface crack in different tapes are compared in **Figure 8(d)**. It can be found that the strain energy release rate reaches the largest value as the interface crack is in the top tape during the increasing of field, while the value is the lowest as the crack is in the middle tape. We can draw a conclusion that the central interface crack in the top tape is most dangerous during an increasing magnetic field. This may be due to the reason that the penetration field is largest in the top tape. In **Figure 9**, it is to be noted that the strain energy release rate for the crack length of 100 μm is always larger than that of 50 μm when the magnetic field rises. However, since the shielding current reverses, the trend of strain energy release rate becomes complicated for decreasing field case, as shown in **Figure 10**. At the beginning of the decreasing field, both the strain energy release rates for the crack lengths of 100 μm and 50 μm reduce. It is obvious that the reduction for longer crack length is faster. With the decreasing external field, the strain energy release rate reaches the local minimum and begins to increase. The longer crack firstly reaches the local minimum at $B/B_e = 0.88$. Finally, the strain energy release rate will decrease again, and the value is very close to zero as external field vanishes. Comparing the values for increasing field and decreasing field (see **Figure 11**), we can find that the strain energy release rate is larger for increasing field case. Thus, the highest possibility of cracking may happen during the increasing field, which is opposite to the bulk superconductor.

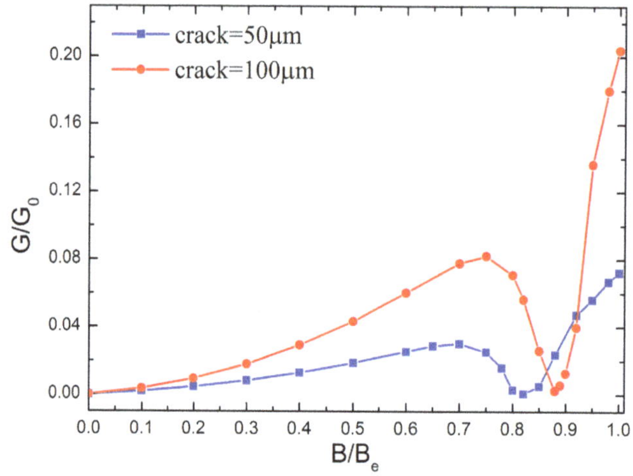

Figure 10. The strain energy release rates for the crack lengths of 50 μm and 100 μm during decreasing magnetic field.

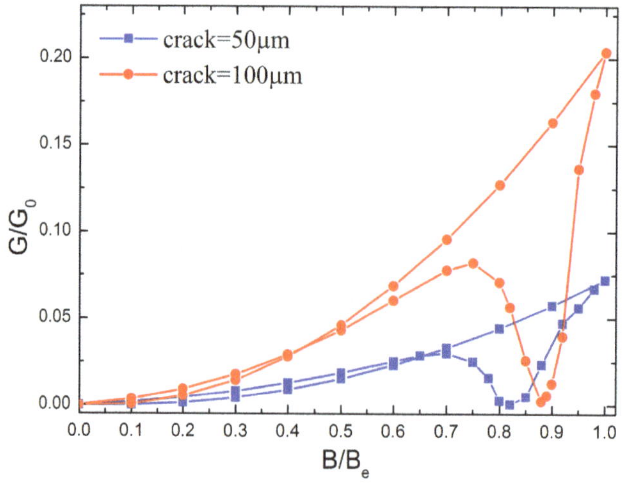

Figure 11. The comparison of strain energy release rates for increasing field and decreasing field.

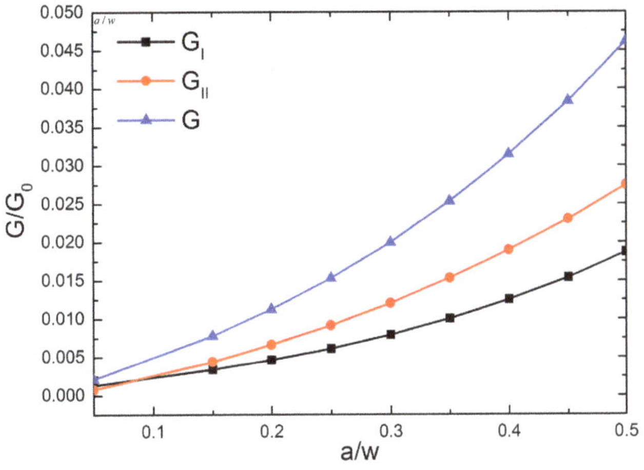

Figure 12. The change of strain energy release rate with the crack length and as the field is increased from zero to $0.5B_e$.

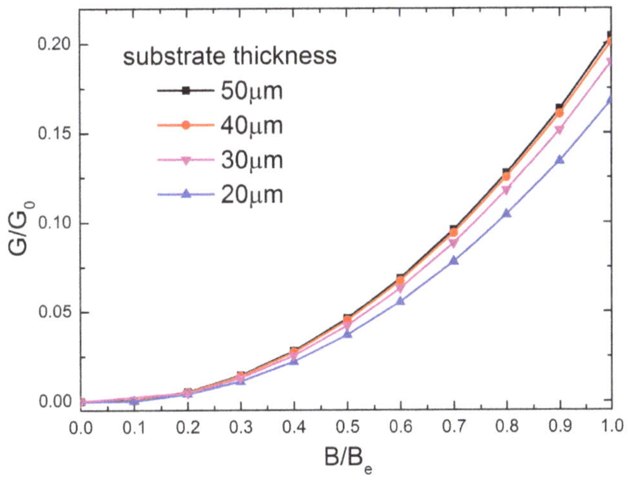

Figure 13. The strain energy release rates with different thicknesses of substrate.

For the central interface crack problem, it can be expected that the strain energy release rate increases with the length of the crack for a fixed magnetic field, as shown in **Figure 12**. The effect of substrate thickness on the strain energy release rate during increasing field is presented in **Figure 13**. One can find that with the increase of substrate thickness from 20 μm to 50 μm, the strain energy release rate also rises. In addition, the increasing strain energy release rate is more obvious in high field. However, the difference is very small as the thickness of substrate is larger than 40 μm. Thus, it is reasonable to reduce the thickness of substrate to improve the mechanical stability of stacked conductors structure.

6. Conclusions

In this paper, we considered the interface crack in the stack of high-temperature superconductor coated conductors with VCCT. A simple interfacial crack problem in bimaterial was calculated to verify the accuracy of numerical method. On the basis of the linear elastic fracture theory, the strain energy release rates for the stack of coated conductors with interface crack were presented for different magnetic fields, crack lengths, substrate thicknesses. The strain energy release rate increases monotonically with the increasing external field. However, the trend for decreasing field case is not monotonic. When the interface crack is located in the top tape, the strain energy release rate is the largest. In addition, the thickness of hastelloy substrate also has an effect on the strain energy release rate, and larger thickness leads to the higher energy release rate. When the thickness of the substrate is larger than 40μm, the effect of thickness becomes negligible. This study may provide a better understanding on the mechanical instability of the stack of coated conductor structure during the magnetization process.

Acknowledgements

We acknowledge the support from the Innovative Research Group of the National Natural Science Foundation of China (11421062), National Natural Science Foundation of China (Nos. 11202087 and 11472120), the National Key Project of Magneto-Constrained Fusion Energy Development Program (No. 2013GB110002), the National Key Project of Scientific Instrument and Equipment Development (No. 11327802) and New Century Excellent Talents in University of Ministry of Education of China (NCET-13-0266).

Author details

Meng Zhao, Huadong Yong* and Youhe Zhou

*Address all correspondence to: yonghd@lzu.edu.cn

Key Laboratory of Mechanics on Environment and Disaster in Western China, The Ministry of Education of China, Department of Mechanics and Engineering Sciences, College of Civil Engineering and Mechanics, Lanzhou University, Lanzhou, Gansu, PR China

References

[1] Bohno T, Tomioka A, Imaizumi M, et al. Development of 66 kV/6.9 kV 2 MVA prototype HTS power transformer. Physica C: Superconductivity and its Applications, 2005, 426–431: 1402–1407. DOI: 10.1016/j.physc.2005.03.080

[2] Igarashi M, Nakao H, Terai M, et al. Persistent current HTS magnet cooled by cryocooler (1)—project overview. IEEE Transactions on Applied Superconductivity, 2005, 15(2): 1469–1472. DOI: 10.1109/TASC.2005.849130

[3] Wiezoreck J, Schmidt F, Nick W, et al. Development of HTS power transmission cables. IEEE Transactions on Applied Superconductivity, 1999, 9(2): 406–411. DOI: 10.1109/77.783321

[4] Choi S, Kiyoshi T, Hahn S Y, et al. Stress analysis of a high temperature superconductor coil wound with Bi-2223/Ag tapes for high field HTS/LTS NMR magnet application. IEEE Transactions on Applied Superconductivity, 2009, 19(3): 2237–2240. DOI: 10.1109/TASC.2009.2018071

[5] Song H H, Brownsey P, Zhang Y F, et al. 2G HTS coil technology development at SuperPower. IEEE Transactions on Applied Superconductivity, 2013, 23(3): 4600806. DOI: 10.1109/TASC.2012.2233837

[6] Augieri A, De Marzi G, Celentano G, et al. Electrical characterization of ENEA high temperature superconducting cable. IEEE Transactions on Applied Superconductivity, 2015, 25(3): 1–4. DOI: 10.1109/TASC.2014.2364391

[7] Durrell J H, Dennis A R, Jaroszynski J, et al. A trapped field of 17.6 T in melt-processed, bulk Gd-Ba-Cu-O reinforced with shrink-fit steel. Superconductor Science and Technology, 2014, 27(8): 082001. DOI: 10.1088/0953-2048/27/8/082001

[8] Gruss S, Fuchs G, Krabbes G, et al. Superconducting bulk magnets: very high trapped fields and cracking. Applied Physics Letters, 2001, 79(19): 3131–3133. DOI: 10.1063/1.1413502

[9] Diko P, Krabbes G. Macro-cracking in melt-grown Y-Ba-Cu-O superconductor induced by surface oxygenation. Superconductor Science and Technology, 2002, 16(1): 90. DOI: 10.1088/0953-2048/16/1/316

[10] Diko P, Krabbes G. Formation of c-macrocracks during oxygenation of TSMG YBa2Cu3O7/Y2BaCuO5 single-grain superconductors. Physica C: Superconductivity, 2003, 399(3): 151–157. DOI: 10.1016/S0921-4534(03)01305-4

[11] Ren Y, Weinstein R, Liu J, et al. Damage caused by magnetic pressure at high trapped field in quasi-permanent magnets composed of melt-textured Y-Ba-Cu-O superconductor. Physica C: Superconductivity, 1995, 251(1): 15–26. DOI: 10.1016/0921-4534(95)00398-3

[12] Li Y X, Ta W R, Gao Y W, et al. Transport current distribution on Nb 3 Sn strand for TARSIS. Physica C: Superconductivity, 2013, 489: 25–31. DOI: 10.1016/j.physc.2013.03.052

[13] Bean C P. Magnetization of hard superconductors. Physical Review Letters, 1962, 8(6): 250. DOI: 10.1103/PhysRevLett.8.250

[14] Kim Y, Hempstead C, Strnad A. Magnetization and critical supercurrents. Physical Review, 1963, 129(2): 528. DOI: 10.1103/PhysRev.129.528

conductivity by finite element software. Superconductor Science and Technology, 2006, 19(12): 1246. DOI: 10.1088/0953-2048/19/12/004

[16] Ikuta H, Hirota N, Nakayama Y, et al. Giant magnetostriction in Bi2Sr2CaCu2O8 single crystal in the superconducting state and its mechanism. Physical Review Letters, 1993, 70(14): 2166. DOI: 10.1103/PhysRevLett.70.2166

[17] Ikuta H, Kishio K, Kitazawa K. Critical state models for flux-pinning-induced magnetostriction in type-II superconductors. Journal of Applied Physics, 1994, 76(8): 4776–4786. DOI: 10.1063/1.357249

[18] Johansen T H. Flux-pinning-induced stress and strain in superconductors: Case of a long circular cylinder. Physical Review B, 1999, 60(13): 9690. DOI: 10.1103/PhysRevB.60.9690

[19] Johansen T H. Flux-pinning-induced stress and strain in superconductors: Long rectangular slab. Physical Review B, 1999, 59(17): 11187. DOI: 10.1103/PhysRevB.59.11187

[20] Johansen T H, Shantsev D V. Magnetostrictive behaviour of thin superconducting disks. Superconductor Science and Technology, 2003, 16(9): 1109. DOI: 10.1088/0953-2048/16/9/324

[21] Eremenko V, Sirenko V, Szymczak H, et al. Magnetostriction of thin flat superconductor in a transverse magnetic field. Superlattices and Microstructures, 1998, 24(3): 221–226. DOI: 10.1006/spmi.1998.0583

[22] Nabiałek A, Szymczak H, Sirenko V, et al. Influence of the real shape of a sample on the pinning induced magnetostriction. Journal of Applied Physics, 1998, 84(7): 3770–3775. DOI: 10.1063/1.368555

[23] Çelebi S, Inanir F, LeBlanc M. Coexistence of critical and normal state magnetostrictions in type II superconductors: a model exploration. Journal of Applied Physics, 2007, 101(1): 3906. DOI: 10.1063/1.2407266

[24] Xue F, Yong H D, Zhou Y H. Effect of flux creep and viscous flux flow on flux-pinning-induced stress and magnetostriction in a long rectangular slab superconductor. Journal of Applied Physics, 2010, 108(10): 103910. DOI: 10.1063/1.3506704

[25] Huang C G, Yong H D, Zhou Y H. Magnetostrictive behaviors of type-II superconducting cylinders and rings with finite thickness. Superconductor Science and Technology, 2013, 26(10): 105007. DOI: 10.1088/0953-2048/26/10/105007

[26] Xue C, He A, Yong H, et al. Magneto-elastic behaviour of thin type-II superconducting strip with field-dependent critical current. Journal of Applied Physics, 2013, 113(2): 023901. DOI: 10.1063/1.4773483

[27] Yong H D, Zhou Y H. Flux pinning induced stress and magnetostriction in a long elliptic cylindrical superconductor. Journal of Applied Physics, 2013, 114(2): 023902. DOI: 10.1063/1.4811531

[28] Huang C G, Zhou Y H. Magnetic and magnetostrictive properties of finite superconducting cylinders containing a cavity. Journal of Applied Physics, 2014, 115(3): 033904. DOI: 10.1063/1.4862856

[29] Yang Y, Xiao L-Y, Li X-H. Impact of viscous flux flow on the stress in long rectangular slab superconductors. Journal of Applied Physics, 2010, 107(2): 3910. DOI: 10.1063/1.3284080

[30] Johansen T H. Flux-pinning-induced stress and magnetostriction in bulk superconductors. Superconductor Science and Technology, 2000, 13(10): R121. DOI: 10.1088/0953-2048/13/10/201

[31] Johansen T, Wang C, Chen Q, et al. Enhancement of tensile stress near a hole in superconducting trapped-field magnets. Journal of Applied Physics, 2000, 88(5): 2730–2733. DOI: 10.1063/1.1287123

[32] Yong H D, Zhou Y H. Stress distribution in a flat superconducting strip with transport current. Journal of Applied Physics, 2011, 109(7): 073902. DOI: 10.1063/1.3561366

[33] Yang Y, Wang X. Magnetization and magnetoelastic behavior of a functionally graded rectangular superconductor slab. Journal of Applied Physics, 2014, 116(2): 023901. DOI: 10.1063/1.4887138

[34] Yang Y, Wang X. Stress and magnetostriction in an infinite hollow superconducting cylinder with a filling in its central hole. Physica C: Superconductivity, 2013, 485: 58–63. DOI: 10.1016/j.physc.2012.10.003

[35] Feng W J, Han X, Ma P. Flux-pinning-induced stress and magnetostriction in a functionally graded long rectangular superconductor slab. Journal of Applied Physics, 2011, 110(6): 063917. DOI: 10.1063/1.3639302

[36] Yong H D, Zhou Y H. Effect of nonsuperconducting particles on the effective magnetostriction of bulk superconductors. Journal of Applied Physics, 2008, 104(4): 043907. DOI: 10.1063/1.2952042

[37] Yong H D, Yang Y, Zhou Y H. Mechanical behaviours in Bi2223/Ag/Ag alloy composite tape with different volume fractions. Journal of Superconductivity and Novel Magnetism, 2016, 29(2): 329–336. DOI: 10.1007/s10948-015-3358-1

[38] Tomita M, Murakami M. High-temperature superconductor bulk magnets that can trap magnetic fields of over 17 tesla at 29 K. Nature, 2003, 421(6922): 517–520. DOI: 10.1038/nature01350

[39] Zeng J, Yong H D, Zhou Y H. Edge-crack problem in a long cylindrical superconductor. Journal of Applied Physics, 2011, 109(9): 093920. DOI: 10.1063/1.3585830

[40] Zeng J, Zhou Y H, Yong H D. Fracture behaviors induced by electromagnetic force in a long cylindrical superconductor. Journal of Applied Physics, 2010, 108(3): 033901. DOI: 10.1063/1.3456038

[41] Zhou Y H, Yong H D. Crack problem for a long rectangular slab of superconductor under an electromagnetic force. Physical Review B, 2007, 76(9): DOI: 10.1103/PhysRevB.76.094523

[42] Yong H D, Jing Z, Zhou Y H. Crack problem for superconducting strip with finite thickness. International Journal of Solids and Structures, 2014, 51(3): 886–893. DOI: 10.1016/j.ijsolstr.2013.11.013

[43] Yong H D, Xue C, Zhou Y H. Thickness dependence of fracture behaviour in a superconducting strip. Superconductor Science and Technology, 2013, 26(5): 055003. DOI: 10.1088/0953-2048/26/5/055003

[44] Yong H D, Zhou Y H. Crack problem for thin superconducting strip in a perpendicular magnetic field. IEEE Transactions on Applied Superconductivity, 2012, 22(2): 8400905. DOI: 10.1109/TASC.2011.2178093

[45] Gao S W, Feng W J, Liu J X. Fracture problems of a superconducting slab with a central kinked crack. Journal of Applied Physics, 2013, 114(24): 243907. DOI: 10.1063/1.4852495

[46] Gao Z W, Zhou Y H. Crack growth for a long rectangular slab of superconducting trapped-field magnets. Superconductor Science and Technology, 2008, 21(9): 095010. DOI: 10.1088/0953-2048/21/9/095010

[47] Wang X, Yong H D, Xue C, et al. Inclined crack problem in a rectangular slab of superconductor under an electromagnetic force. Journal of Applied Physics, 2013, 114(8): 083901. DOI: 10.1063/1.4818284

[48] Xue C, He A, Zhou Y H. Fracture problem of the thin superconducting strip with transverse crack, In: 13th International Conference on Fracture, Beijing, China, 2013. p. 1–9.

[49] Gao Z W, Zhou Y H, Lee K Y. The interaction of two collinear cracks in a rectangular superconductor slab under an electromagnetic force. Physica C: Superconductivity, 2010, 470(15): 654–658. DOI: 10.1016/j.physc.2010.06.008

[50] Gao Z W, Zhou Y H, Lee K Y. Crack-inclusion problem for a long rectangular slab of superconductor under an electromagnetic force. Computational Materials Science, 2010, 50(2): 279–282. DOI: 10.1016/j.commatsci.2010.08.015

[51] Xue F, Gou X F. Crack problem for a bulk superconductor with nonsuperconducting inclusions under an electromagnetic force. AIP Advances, 2015, 5(4): 047128. DOI: 10.1063/1.4918752

[52] Xue F, Zhang Z X, Gou X F. Fracture behavior of an inclined crack interacting with a circular inclusion in a high-TC superconductor under an electromagnetic force. AIP Advances, 2015, 5(11): 117141. DOI: 10.1063/1.4936422

[53] Feng W J, Zhang R, Ding H M. Crack problem for an inhomogeneous orthotropic superconducting slab under an electromagnetic force. Physica C: Superconductivity, 2012, 477: 32–35. DOI: 10.1016/j.physc.2012.02.027

[54] Feng W J, Liu Q F, Han X. Crack problem for a functionally graded thin superconducting film with field dependent critical currents. Mechanics Research Communications, 2014, 61: 36–40. DOI: 10.1016/j.mechrescom.2014.07.005

[55] Feng W J, Liu Q F, Su R. Fracture behaviors of a functionally graded thin superconducting film with transport currents based on the strain energy density theory. Theoretical and Applied Fracture Mechanics, 2014, 74: 73–78. DOI: 10.1016/j.tafmec.2014.07.002

[56] Yan Z, Gao S W, Feng W J. Fracture problem for an external circumferential crack in a functionally graded superconducting cylinder subjected to a parallel magnetic field. Physica C: Superconductivity and its Applications, 2015, 521–522: 5–12. DOI: 10.1016/j.physc.2015.12.003

[57] Gao Z, Zheng Z Y, Li X Y. Fracture problem of a nonhomogeneous high temperature superconductor slab based on real fundamental solutions. Physica C: Superconductivity and its Applications, 2015, 519: 5–12.

[58] Gao Z W, Zhou Y H. Dynamic stress intensity factors of mode-I crack in high temperature superconductor. Physica C: Superconductivity, 2013, 495: 169–173. DOI: 10.1016/j.physc.2013.09.013

[59] Patel A, Filar K, Nizhankovskii V I, et al. Trapped fields greater than 7 T in a 12 mm square stack of commercial high-temperature superconducting tape. Applied Physics Letters, 2013, 102(10): 102601. DOI: 10.1063/1.4795016

[60] Freund L. Substrate curvature due to thin film mismatch strain in the nonlinear deformation range. Journal of the Mechanics and Physics of Solids, 2000, 48(6): 1159–1174. DOI: 10.1016/S0022-5096(99)00070-8

[61] Olsson E, Gupta A, Thouless M D, et al. Crack formation in epitaxial [1 1 0] thin films of $YBa_2Cu_3O_{7-\delta}$ and $PrBa_2Cu_3O_{7-x}$ on [1 1 0]$SrTiO_3$ substrates. Applied Physics Letters, 1991, 58(15): 1682. DOI: 10.1063/1.105110

[62] Yong H D, Zhou Y H. Interface crack between superconducting film and substrate. Journal of Applied Physics, 2011, 110(6): 063924. DOI: 10.1063/1.3634019

[63] Jing Z, Yong H D, Zhou Y H. Flux-pinning-induced interfacial shearing and transverse normal stress in a superconducting coated conductor long strip. Journal of Applied Physics, 2012, 112(4): 043908. DOI: 10.1063/1.4748338

[64] Jing Z, Yong H D, Zhou Y H. Shear and transverse stress in a thin superconducting layer in simplified coated conductor architecture with a pre-existing detachment. Journal of Applied Physics, 2013, 114(3): 033907. DOI: 10.1063/1.4813869

[65] Valvo P S. A revised virtual crack closure technique for physically consistent fracture mode partitioning. International Journal of Fracture, 2012, 173(1): 1–20. DOI: 10.1007/s10704-011-9658-y

[66] Rybicki E F, Kanninen M F. A finite element calculation of stress intensity factors by a modified crack closure integral. Engineering Fracture Mechanics, 1977, 9(4): 931–938. DOI: 10.1016/0013-7944(77)90013-3

[67] Irwin G R. Onset of fast crack propagation in high strength steel and aluminum, In: Sagamore research conference proceedings, 1956. p. 289–305.

[68] Yu H J. Investigations on fracture mechanics of nonhomogeneous materials with complex interfaces [thesis]. Harbin Institute of technology; 2010.

[69] Yu H J, Wu L Z, Guo L C, et al. Interaction integral method for the interfacial fracture problems of two nonhomogeneous materials. Mechanics of Materials, 2010, 42(4): 435–450. DOI: 10.1016/j.mechmat.2010.01.001

[70] Nagashima T, Omoto Y, Tani S. Stress intensity factor analysis of interface cracks using X-FEM. International Journal for Numerical Methods in Engineering, 2003, 56(8): 1151–1173. DOI: 10.1002/nme.604

[71] Miyazaki N, Ikeda T, Soda T, et al. Stress intensity factor analysis of interface crack using boundary element method. JSME International Journal, 1993, 36(1): 36–42. DOI: 10.1016/0013-7944(93)90266-U

[72] Suo Z G, Hutchinson J W. Interface crack between two elastic layers. International Journal of Fracture, 1990, 43(1): 1–18. DOI: 10.1007/BF00018123

[73] Dundurs J. Mathematical Theory of Dislocations. New York: American Society of Mechanical Engineering, 1969: 70.

[74] Dizon J R C, Gorospe A B, Shin H S. Numerical analysis of stress distribution in Cu-stabilized GdBCO CC tapes during anvil tests for the evaluation of transverse delamination strength. Superconductor Science and Technology, 2014, 27(5): 055023. DOI: 10.1088/0953-2048/27/5/055023

[75] Prigozhin L. The Bean model in superconductivity: Variational formulation and numerical solution. Journal of Computational Physics, 1996, 129(243): 190–200. DOI: 10.1006/jcph.1996.0243

6

Investigation of Mechanical Behaviour of a Bioceramic

Achouak Elghazel, Rym Taktak and Jamel Bouaziz

Abstract

In order to find a convincing method to measure bioceramics fracture toughness, tensile strength and modulus, a novel configuration of the Brazilian test was applied and described in the experimental work. The flattened Brazilian specimens, which are in the shape of discs having parallel flat ends, are subjected to compression for determination of opening mode I fracture toughness K_{IC}. Experiments were done by using tricalcium phosphate-fluorapatite composites, which were tested by compressive loading on the parallel flat ends. The loading angle corresponding to the flat end width is about 2α = 20° in order to guarantee crack initiation at the centre of the specimen according to the Griffith criteria. Fracture toughness was also performed by using semi-circular bend "SCB". Finite-element program, called ABAQUS, is used for numerical modelling for finding stress intensity factors. The effects of fluorapatite additives and fracture toughness were studied. Fracture toughness values of tricalcium phosphate-fluorapatite composites were found to increase with increasing addition of fluorapatite until an appropriate value. It is shown that there is a good agreement among the experimental, analytical and numerical results.

Keywords: composite, toughness, mechanical properties, flattened Brazilian test, semi-circular bend, numerical modelling

1. Introduction

The development of the ceramic material industry poses the necessity for determining the stress intensity factor (SIF) of bioceramic cracking. Researchers are still studying to find a simple and accurate standard method to determine fracture toughness, which is an important parameter to determine the stress required to drive a pre-existing crack which generally exists in materials. However, the International Society for Rock Mechanics (ISRM) has suggested some methods to determine fracture toughness, listed in Ref. [1]; examples include (1) Chevron

Bend (CB) specimens, (2) Short Rod (SR) specimens and (3) Cracked Chevron Notched Brazilian Disc (CCNBD).

Some methods were previously used to find mode I fracture toughness, K_{IC}, such as Modified Ring (MR) test [2, 3], Diametral Compression Method (DCM) [4], Semi-circular Core in three-point Bending (SCB) [5] and finally Brazilian Disc Test (BDT) [6, 7].

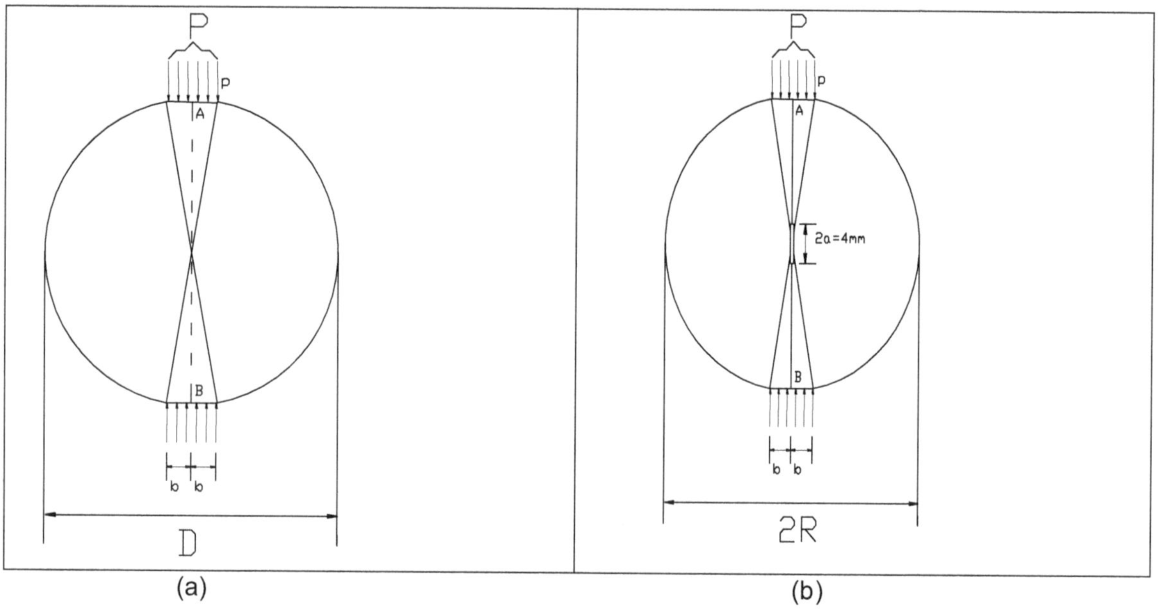

Figure 1. (a) Flattened Brazilian specimen and (b) flattened Brazilian specimen with a central straight-through crack (CSTFBD).

In order to define the mode I elastic modulus, tensile strength and fracture toughness of biomaterials, the central straight through flattened Brazilian disc "CSTFBD" test is ideal for specimens for pure mode I fracture and the well-known test configurations for determining the parameters previously mentioned in just one test. The disc specimen (**Figure 1**) is designed by introducing two equal-width parallel planes in the sample, which are prepared specifically for load application. The loading angle conforming to the flat end width 2α must be greater than a critical value ($2\alpha \geq 20°$) in order to guarantee crack initiation at the centre of the disc [8]. The obtained numerical results are compared with experimental and analytical ones.

Semi-circular bend (SCB) specimens presented in **Figure 2** were used to investigate experimentally the mode I fracture toughness K_{IC} . The SCB specimen was prepared by introducing a straight crack in the semi-disc, which is prepared specifically for measuring K_{IC}.

In this study, we used the commercial tricalcium phosphate (β-CTCP) reinforced with the fluorapatite (Fap) with different amounts of additives (13.26, 19.9 26.52, 33.16 and 40%) sintered at 1300°C. The objective is to determine the stress intensity factors for modified Brazilian test and SCB specimen with analytical formula and numerical simulation. A range of specimen geometries having various crack lengths (a) were modelled and analysed with ABAQUS finite-element program. Fracture toughness values with varying geometric parameters were analysed. The mode I crack growth behaviour of Fap-β-CTCP sample is investigated experimentally and theoretically using both CSTFBD and SCB specimens.

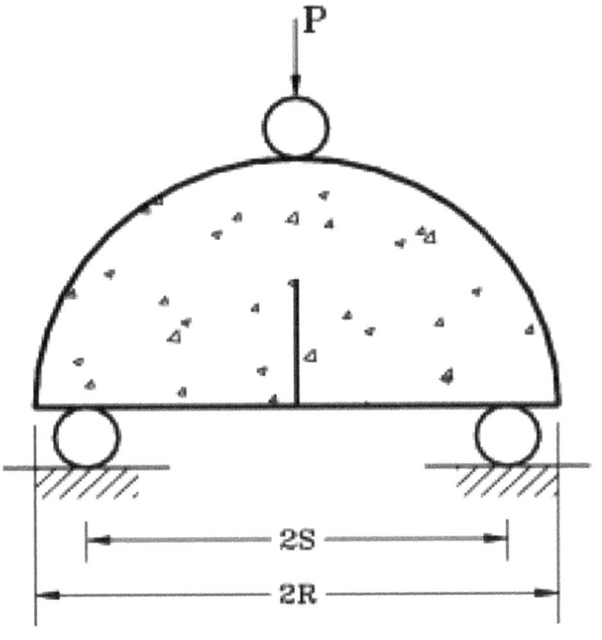

Figure 2. Semi-circular bend specimen (SCB).

2. Materials and methods

In order to elaborate CTCP-Fap, the materials used are the commercial tricalcium phosphate (Fluka) and synthesized fluorapatite. The Fap powder was synthesized by the precipitation method [9]. The approximate representatives Fap-β-CTCP were, respectively [(13.26 wt%, 86.74 wt%), (19.9 wt%, 80.1 wt%), (26.52 wt%, 73.48 wt%), (33.16 wt%, 66.84 wt%) and (40 wt %, 60 wt%)]. Estimated quantities of each powder were milled with absolute ethanol and treated by ultrasound machine for 20 min. The milled powder was dried in a low temperature oven at 80°C to eliminate the ethanol and generate a finely divided powder. Powder mixtures were molded in a metal mould and uniaxially pressed at 67 MPa to form cylindrical compacts with a diameter of 30 mm and a thickness of about 5 mm. The green compacts were sintered in a horizontal resistance furnace (Pyrox 2408) at 1300°C for 1 h 30 min. The heating and cooling rates were 10 and 20°C min^{-1}, respectively.

In this study, we used three different geometries for sample construction: The basic dimensions of the FBD, CSTFBD and SCB specimens were considered to be the same and were as follows:

$D = 30$ mm and $B = 5$ mm, with D is the diameter and t is the thickness.

While the SCB and CSTFBD specimens were being added a crack of 4 mm. A LLOYD model test machine is used for the Brazilian and bending tests for the measurement of the fracture toughness, elastic modulus and tensile strength.

3. Determination of mechanical properties of the ceramics specimen using modified Brazilian tests

The mechanical properties of the compacts were measured by Brazilian test. The maximal rupture strength σ_r is given by the following equation [10, 11]:

$$\sigma_r = \frac{2P}{\pi Dt}. \tag{1}$$

where P is the tensile strength, and D. and t are the diameter and the thickness of the sample, respectively.

Yet, in the flattened Brazilian disc, the previous formula is no longer valid. After choosing an appropriate value of 2α, the rupture strength σ_r can be determined by the following equation [8]:

$$\sigma_r = k\frac{2P_c}{\pi Dt}, \tag{2}$$

where P_c is the critical load (the maximum load during the test) applied on the flat ends and k is the coefficient which is closely related to the loading angle 2α. When $2\alpha = 0°$, we have $k = 1$, and hence the previous formula corresponds to the original Brazilian disc. For the given value of 2α, the value of k can be determined by finite-element analysis. According to the Griffith criterion, at failure, we have $\sigma_G = \sigma_r$ so:

$$k = \frac{\sigma_G}{2P/\pi Dt}, \tag{3}$$

where σ_G is the equivalent stress based on the Griffith strength criteria. To calculate k, we used an approximate formula:

$$k = \left(2\cos^3\alpha + \cos\alpha + \frac{\sin\alpha}{\alpha}\right)^2 8\left(\cos\alpha + \frac{\sin\alpha}{\alpha}\right)\frac{\alpha}{\sin\alpha} \tag{4}$$

For $2\alpha = 20°$, we have $k = 0.9644$.

The elastic modulus E is determined with the modified formula adjusted by Wang et al. [8]:

$$\Delta w = \frac{2P}{\pi E t}\left\{(1-\mu) - \ln\left(1+\frac{4}{\sin^2\alpha}\right)\right\}\frac{\alpha}{\sin\alpha} \tag{5}$$

This formula is inspired from the slope of the load-displacement record before the maximum load, where

- P is the resultant of the uniformly distributed force applied via the flat end (**Figure 1**)
- Δw is the displacement (mm)
- μ is the Poisson's ratio
- $\sin\alpha = \frac{2b}{D}$
- E is the elastic modulus

In this study, as shown in **Figure 1a**, the two parallel flat ends were introduced into the disc for load bearing [8]. The additional flat ends were designed with a special mould for the proposed specimen; thus, the flatness and parallelness of the flat ends are important for a successful test. A crack was adjusted in the precedent geometry to realize the second one as seen in **Figure 1b**.

Research works have proven that only when the load angle satisfies the condition $2\alpha \geq 19.5°$, the centre crack initiation can be guaranteed. This condition should be accomplished for loading the Brazilian disc specimen in the composite fracture toughness test [7]. Then, K_{IC} is determined using the proposed expression in [8], [7] and [12] as:

$$K_{IC} = \frac{P_{min}}{t\sqrt{R}}\varnothing_{max}, \tag{6}$$

where P_{min} is the minimum load, $\varnothing_{max}$ is the maximum stress intensity factor and R is the radius of the disc.

Hence:

$$\varnothing\left(\alpha, \frac{a}{R}\right) = \frac{K_I}{\frac{P}{\sqrt{R}}t}. \tag{7}$$

According to Wang and Xing [12], the SIF $\varnothing$ is calculated by the following formula:

$$\begin{aligned}\varnothing\left(\frac{a}{R}\right) = &-4.2892\left(\frac{a}{R}\right)^7 - 26.6765\left(\frac{a}{R}\right)^6 + 84.9054\left(\frac{a}{R}\right)^5 - 93.087\left(\frac{a}{R}\right)^4 \\ &+ 50.7763\left(\frac{a}{R}\right)^3 - 14.3776\left(\frac{a}{R}\right)^2 + 2.7408\left(\frac{a}{R}\right)\end{aligned} \tag{8}$$

If the record indicates that the test is valid (**Figure 3**), then the fracture toughness K_{IC} can be given using the following formula [8],

$$K_{IC} = 0.789\frac{P_{min}}{B\sqrt{R}} \quad for\ 2\alpha = 20°, \tag{9}$$

where 0.789 is $\emptyset_{max}$ for a loading angle $2\alpha = 20°$ calculated with Formula (8), which corresponds to the critical dimensionless crack length $a_c/R = 0.73$ as illustrated in **Figure 4**.

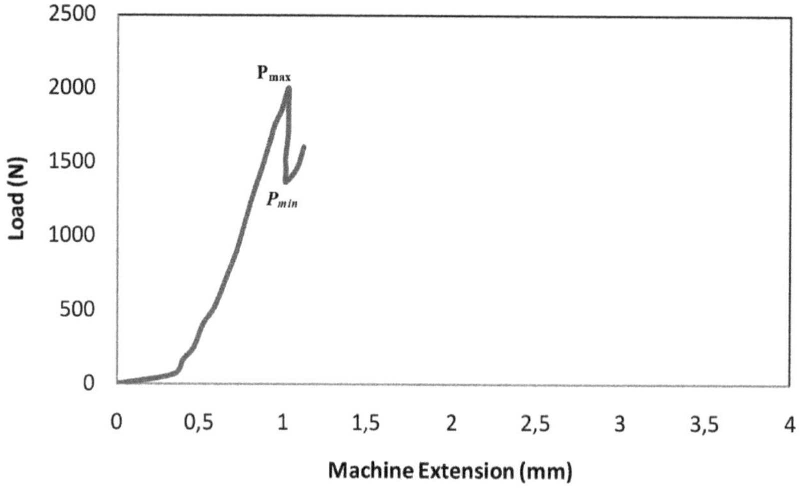

Figure 3. An example for a valid test record for the flattened Brazilian specimen.

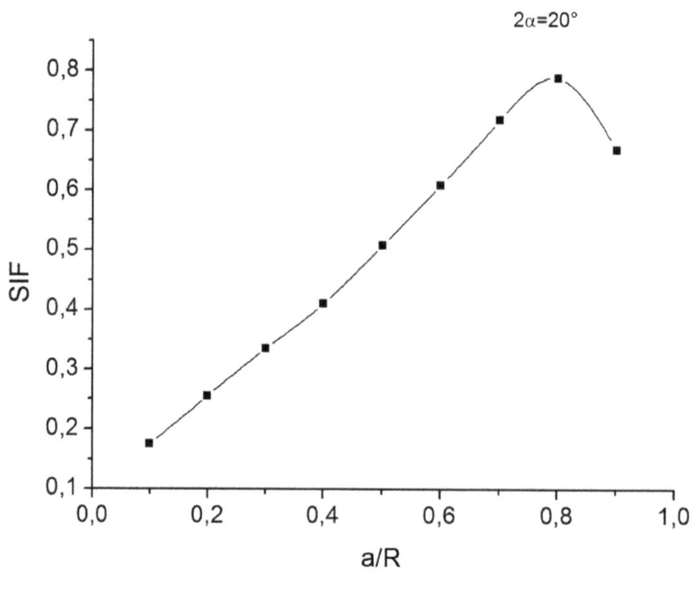

Figure 4. The dimensionless SIF versus dimensionless crack length for the flattened Brazilian disc with a central straight-through crack.

As mentioned previously, when $2\alpha \geq 20°$ the crack can be initiated at the centre of the sample, then the crack expands along the diameter. The value of the SIF gradually rises from zero (crack initiation) to the maximum where $\emptyset_{max}$ is obtained, after that $\emptyset$ decreases until the final rupture of the disc. The critical point corresponds to $\emptyset_{max}$, and the minimum load P_{min} is a turning point between the stable and unstable regions of crack development. This point coincides with the local minimum load immediately succeeding the peak load.

For brittle materials such as bioceramics, the fracture toughness K_{IC} can be considered as a material property. In fact, it requires a valid test that contains the two regions as previously mentioned and characterized with the critical point, which is the unique critical turning point immediately succeeding the top load.

4. Determination of fracture toughness of the ceramics specimen using bending test "SCB"

A new method called the cracked semi-circular bend specimen method, presented in **Figure 2**, is developed for mode I fracture toughness determination using ceramics cores. This method has recently received much attention by researchers and can be used as an alternative to the ISRM standard specimens in determining fracture mode I toughness of brittle materials because of its inherently favourable characteristics, such as simplicity, minimal machining requirements, and easy testability through the application of three-point compressive loading using a standard test frame [13–15].

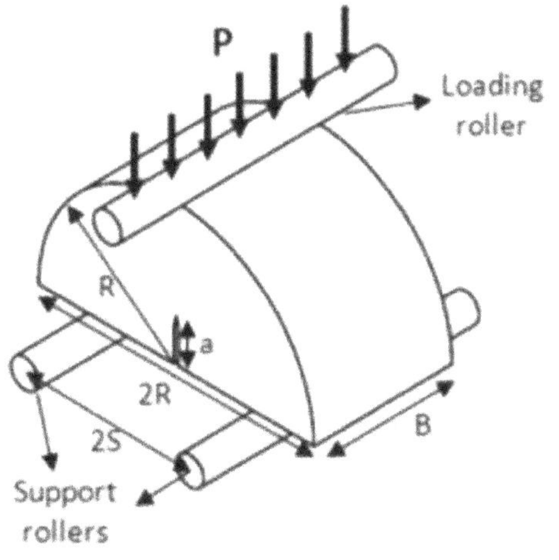

Figure 5. Semi-circular bending (SCB) specimen geometry and loading configuration.

Chong and Kuruppu [16] were among the first who suggested this specimen for conducting fracture tests on brittle materials. Since then, the SCB specimen has been employed frequently to investigate mode I fracture for composite materials. The sample has a simple geometry and can be prepared from typical ceramic cores. Little machining operations and easy test set-up procedure can be considered as major advantages of the SCB specimen.

As shown in **Figure 5**, the sample is a semi-circular disc of radius R with a single edge notch of length a manufactured from the centre of the semi-circle. Fracture test is performed by subjecting the specimen under three-point bending.

The mode I stress intensity factor K_I for the SCB specimen is often written as follows [17]:

$$K_{IC} = \frac{P\sqrt{\pi a}}{2Rt} Y_I\left(\theta, \frac{a}{R}, \frac{S}{R}\right), \tag{10}$$

where P is the compressive applied load and t is the thickness of specimen, the mode I stress intensity factor y_I is the function of crack length ratio a/R, half-span-to-radius ratio S/R and the crack angle θ (the angle between the crack line and the vertical direction). For the pure mode I, θ is zero deg.

Chong et al. [18] also developed a formula for K_I by using both the strain energy release rate method and the elliptical displacement approach.

$$K_I = \frac{P\sqrt{\pi a}}{Dt} Y_k, \tag{11}$$

where Y_k is the dimensionless stress intensity factor as a function of the dimensionless crack length a/D, with D being the disc diameter. Y_k can be calculated by a third-order polynomial as follows:

$$Y_k = 4.47 + 7.40\left(\frac{a}{D}\right) - 106\left(\frac{a}{D}\right)^2 + 433.3\left(\frac{a}{D}\right)^3 \tag{12}$$

The precedent formula (eq. 12) is proven for:

$0.05 \leq \left(\frac{a}{D}\right) \leq 0.4$ and $\frac{S}{D} = 0.25, 0.305, 0.335$ or 0.4 with "s" being the loading span (**Figure 5**).

In our case, the fracture toughness K_{IC} values of SCB with straight crack are calculated with the following formula:

$$K_{IC} = \frac{P_{max}\sqrt{\pi a}}{2Rt} Y_I\left(\frac{a}{R}, \frac{S}{R}\right). \tag{13}$$

where P_{max} is the maximum load, Y_I is the dimensionless stress intensity factor, t, R and a are the thickness, radius of SCB sample and crack length, respectively. Y_I, equally known as a geometry factor, is a function of the ratio of crack length over the semi-disc radius (a/R) and the ratio of half-distance between the two bottom supports "S" over the semi-disc radius (SR), which can be written as the following relation [5]:

$$Y_I = \frac{S}{R}\left(2.91 + 54.39\alpha - 391.4\alpha^2 + 1210.6\alpha^3 - 1650\alpha 4 + 875.9\alpha^5\right), \tag{14}$$

where α is $^a/_R$.

A phosphate calcium-based composite was selected for fracture toughness tests, for these semi-disc specimens of TCP-Fap having a single straight crack were subjected to three-point bending loads (**Figure 6**). Specimens were prepared by a special mould. A straight notch of 4 mm was introduced in each specimen of 15 mm radius and 5 mm thickness, the crack-length-to-diameter ratio was 0.13. The samples were placed on the loading platform, such that the span ratio S/D was 0.36, and then were tested to failure under load-line displacement control and at a loading rate of 0.075 mm/min. The load and load-point displacement (LPD) was recorded as a function of time during each test. A LLOYD machine with a capacity of 5 kN was used for conducting the fracture tests on the SCB specimens.

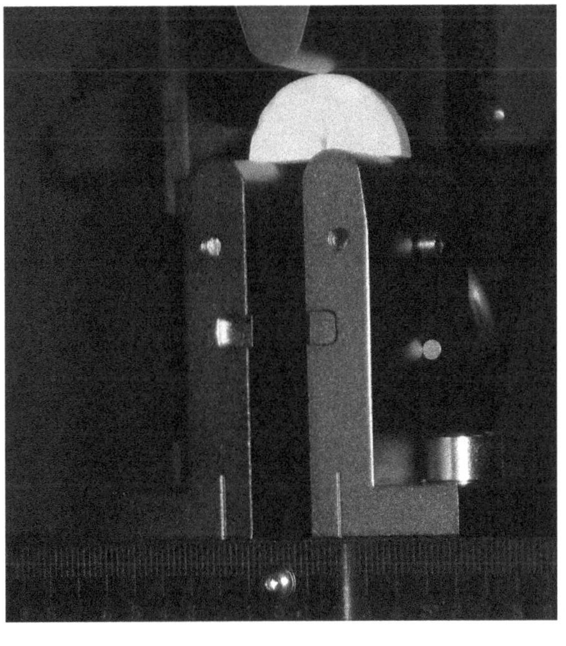

Figure 6. Semi-circular bend (SCB) test specimen.

As shown in the above figure (**Figure 6**), the test procedure in SCB specimens seems easy and cost effective and permits the determination of $K_{IC}3$ for investigating the material behaviour under loading.

5. Numerical computations

Modelling work was done by using ABAQUS finite-element program. In this part, to evaluate the stress intensity factor around the crack tip, a contour integral region is defined and K_I values in this region are computed. In fracture modelling, crack tips are regions of high stress gradients and high stress concentrations, and these concentrations result in theoretically infinite stresses at the crack tip [19]. Hence, to get accurate stresses and strains near the crack tip, finite-element mesh must be refined around the crack tip. The final K_I value at the crack tip is calculated by averaging the K_I values determined for a user-specified number of crack tip concentric mesh rings in the contour integral region.

The mechanical properties were chosen to represent the composite specimens, for which elastic modulus and Poisson's ratio are 31.3, 38.5, 44.5, 60.7 and 66.4 GPa and 0.242, 0.203, 0.286, 0.228 and 0.273, respectively.

5.1. CSTFBD model

Numerical modelling is used for estimating stress intensity factors for Brazilian disc geometry. The flattened Brazilian specimens with a central straight-through crack were used for numerical investigation of mode I fracture in the shaped notches. Since the flattened Brazilian tests do not need 3D modelling the 2D analyses were conducted. To simplify the model, half of it was drawn and the symmetry option was used.

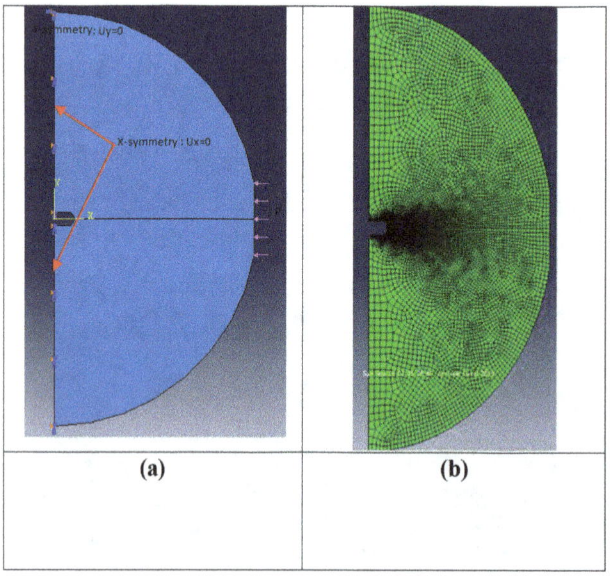

Figure 7. (a) Boundary conditions of the specimen and (b) FE grid generation for the flattened Brazilian specimen with a central straight-through crack.

2DTwo-dimensional-plane strain analyses were performed in this work with a total number of 37,953 Quad 8 elements to simulate the specimen. **Figure 7a** shows the ABAQUS model with different boundary conditions. A large number of elements were used near the crack tip due to its high stress gradient. **Figure 7b** shows a sample FE grid pattern used for the simulation of flattened Brazilian specimen with a central straight-through crack.

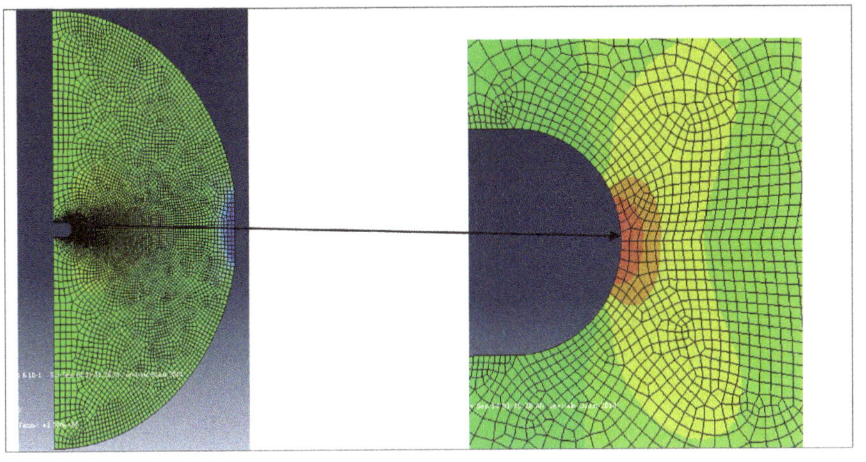

Figure 8. Stress distribution for the flattened Brazilian disc: Distribution of the stress from the crack tip to the parallel flat.

The stress distributions for the flattened Brazilian test and the change of vertical stress while moving away from the crack tip to the parallel flat are revealed in **Figure 8**. These distributions allow the identification of the most stressed area under tensile stresses. The vertical stress $\left(\sigma_{yy}\right)$ is the largest at the crack tip, and it decreases while moving away from the crack tip in the direction of crack propagation. Therefore, the initiation of the crack by tension arises in the disc centre. Actually, many factors contribute to the fact that cracks can be initiated in any place other than the disc centre like material inhomogeneity, which causes local strength variation and the accuracy of specimen preparation, especially the degree of parallelism and loading boundary conditions. All these factors may influence the crack initiation point.

The value of stress intensity factor K_I is given for all compositions for different crack lengths (see **Figure 9**).

The numerical calculation shows that the variation of SIF is similar for all compositions. We have the same trend for every curve. The stress intensity factor K_I increases gradually to reach the maximum. When K_I reaches its maximum value, we have the critical dimensionless crack length ${}^{a}c/_{R}$ corresponding to 0.73. Finally, the K_I decreases until the final breakage of the disc. It is clear that the curve is formed of three regions where K_{Imax}, which corresponds to the fracture toughness for every sample, constitutes the intermediate region. In the first region, when K_I increases progressively, we have also an unstable crack growth because of the evolution of the crack when the load is held constant. In the third region, after the achievement of K_{Imax}, K_I decreases and the crack growth becomes stable [7].

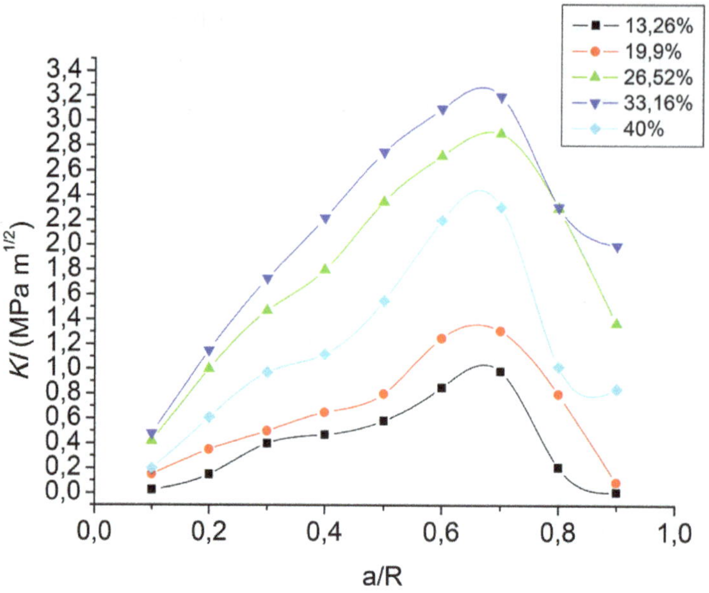

Figure 9. The stress intensity factor versus dimensionless crack length for the flattened Brazilian disc with a central straight-through crack with different percentages of Fap.

5.2. SCB model

In order to compute the stress intensity factor K_I, it is necessary to create an appropriate finite-element model of the specimens that are considered for performing the fracture tests. In this part, the cracked semi-circular (SCB) specimens were performed for numerical calculation of mode I. Two-dimensional modelling of the SCB is used in this work to simulate the specimen and calculate the distribution of stress intensity factor at the crack front, and a large number of elements were used near the crack tip due to its high stress concentration. A typical 2D finite-element modelling of the SCB sample can be seen in **Figure 10a**. Approximately, 5800 Quad 8-node 2D elements were used to mesh this model. As mentioned previously, because of singularity at the crack tip, the elastic singular elements and finer mesh were used around the crack tip as shown in **Figure 10b**.

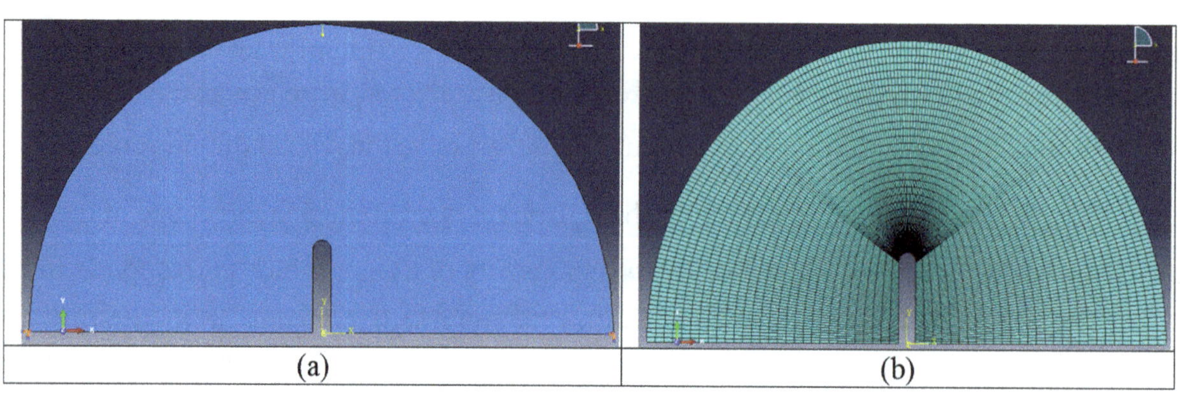

Figure 10. 2D modelling of the SCB specimen.

As the CSTFBD specimen, the vertical stress (σ_{yy}) is the largest at the crack tip and it decreases while moving away from the crack tip in the direction of crack propagation. Therefore, the initiation of the crack by bending test also arises in the centre. After the crack starts from the semi-circular bend centre at the maximum load, the crack propagates symmetrically ahead the loading direction.

The distribution of the stress intensity factor along the crack front for all compositions for different crack lengths is plotted in **Figure 11**.

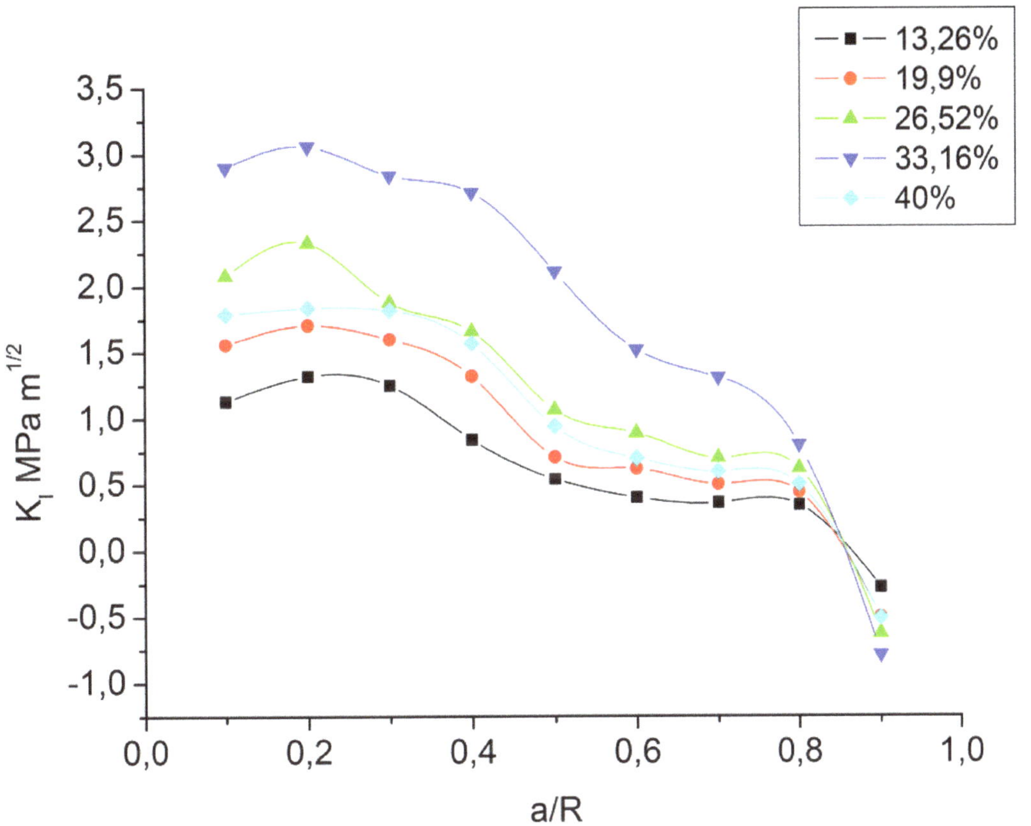

Figure 11. The numerical calculation of the SIF in the SCB specimen versus dimensionless crack length.

As shown in **Figure 11**, the same trend was repeated for different compositions of the ceramic elaborated: The variation of SIF is similar for all compositions. It can also be observed from this figure that the stress intensity factor of the SCB specimen with straight crack first increases until $a/R = 0.2$ and then decreases when the dimensionless crack length exceeds 0.2. In fact, due to the high stress gradient at the crack tip of the specimen, the crack grows sub-critically at first; then after reaching a critical value ($a_c = 0.2$), unstable crack growth occurs rapidly and final failure takes place in the sample. For $\alpha > 0.8$, K_I reaches negative values which affirmed the notice reported by Ayatollah and Aliha [14] that the mode I stress intensity factor in SCB specimens becomes negative for higher values of a/R.

6. Analytical results and discussion

6.1. Determination of elastic modulus E and tensile strength σ_t for valid flattened test

In this part, we used the flattened Brazilian disc without crack. By applying the previous formula Eq. (5), the elastic modulus E is calculated for different percentages of fluorapatite additive.

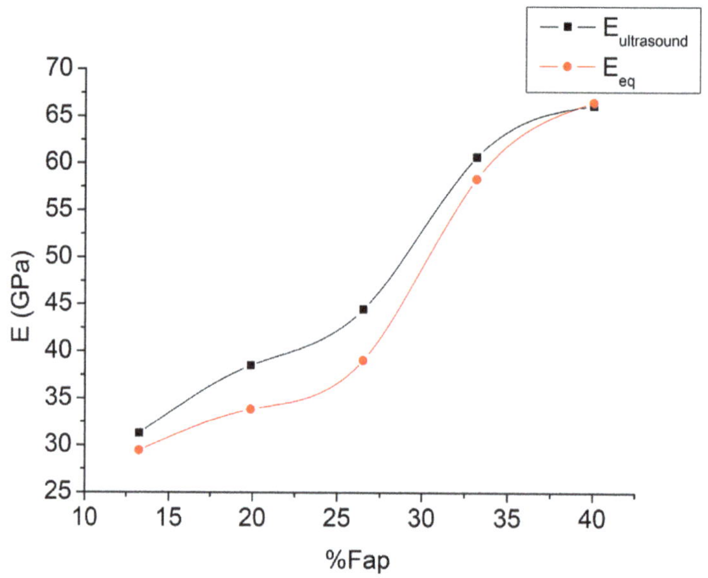

Figure 12. Elastic modulus versus percentage of Fap under optimal conditions.

The calculated values are compared with those found by ultrasound [20], which, as shown in **Figure 12**, was found to be in good compromise.

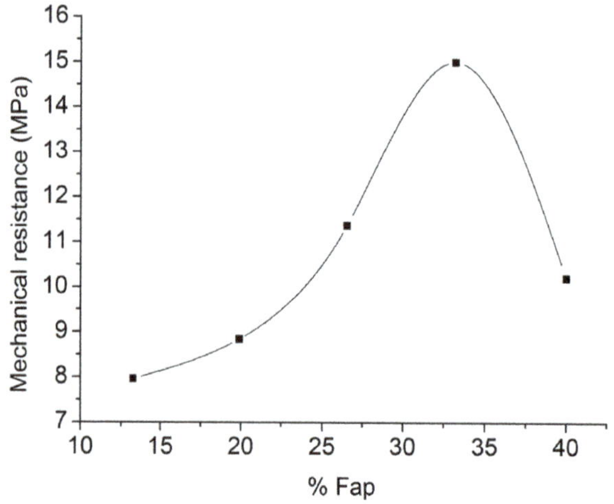

Figure 13. Mechanical resistance versus percentage of Fap of β-CTCP wt% Fap composites sintered at 1300°C for 1 h 30 min.

Figure 13 illustrates the evolution of the mechanical resistance in relation to the percentage of Fap under optimal conditions. According to the work of Bouslama [20], the previously used composite samples reached their optimum at 1300°C, which justifies our choice of sintering temperature. This is attributed to the influence and effect of Fap in the mechanical resistance of the sintered composites. In addition, Fap has good sinterability and mechanical resistance [21]. Ben Ayed et al. (2000a, 2001b, 2006c) have illustrated that the mechanical resistance of Fap increases with temperature and reaches its maximum value at about 14 MPa [9, 21, 22].

In treating the experimental data, the mechanical properties of composites were determined as a function of the sintering temperature. At 1300°C, the rupture strength increases with the percentage of Fap and reaches a maximum value of 33.16% (15 MPa) at 1300°C. Bouslama et al. [23] explained the fall of tensile strength for 40% Fap by the important intergranular porosity existing in the composite's microstructure. It is obvious that the mechanical properties of TCP-Fap composite are also affected by different parameters and operative conditions like temperature, the cycle of sintering, heating time, atmosphere and the presence of micro-crack.

6.2. Determination of the fracture toughness using CSTFBD test

It should be noted that we used different samples sintered under optimal conditions. An experimental–numerical method is proposed to measure bioceramic fracture toughness for flattened Brazilian disc with a central straight-through crack.

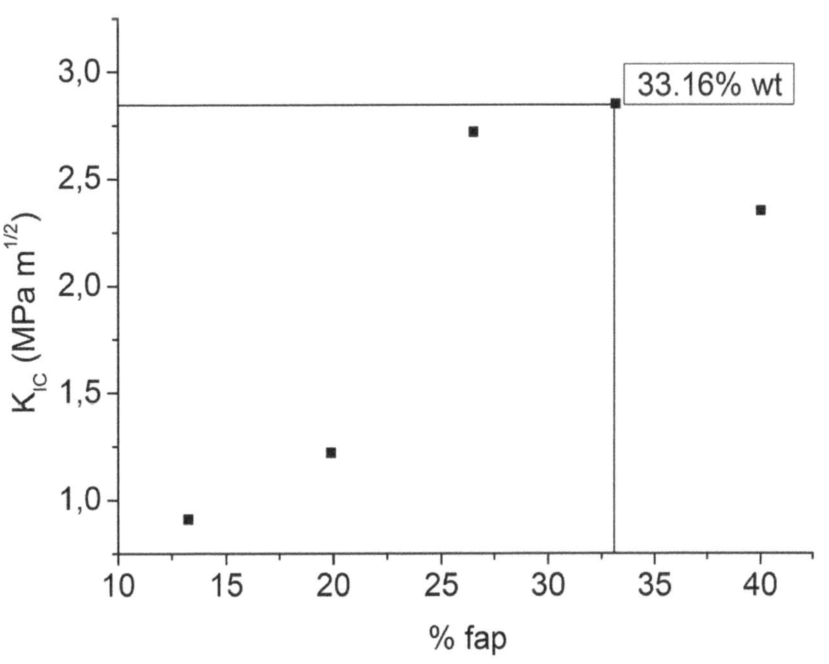

Figure 14. Fracture toughness versus %wt Fap for the CSTFBD specimen.

After the crack starts from the disc centre at the maximum load, the crack propagates symmetrically ahead the loading diameter. Then, the specimen develops into the flattened Brazilian disc with a central straight-through crack, for which there is no stress intensity factor solution in the literature [8]. Thus, we used the finite-element method for the computation of the stress intensity factor. Analytical analysis for this specimen is performed by using Eq. (9). Toughness fracture variation is presented in **Figure 14**.

The fracture toughness values of CTCP-wt% Fap composites range between 0.9 and 2.7 MPam. The lowest toughness (0.9 MPam) is obtained with the 13.26 %wt Fap, while the highest one is approached with the 33.16 %wt fap (2.7 MPam). These results agree well with the mechanical properties evolution of a similar sample sintered under optimal conditions (**Figure 13**), in which the rupture strength reaches maximum when 33.16 wt% Fap are added to the ß-CTCP.

Since Eq. (7) is created, any load P and its corresponding crack length a/R can be placed in this equation to determine the fracture toughness. Luckily, as mentioned above, for the flattened Brazilian disc with a central straight-through crack, the evolution of its stress intensity factor over crack propagation is unique. Indeed, referring to **Figure 4**, $\emptyset$ has a maximum value « $\emptyset_{max}$», which should correspond to a minimum value of load P_{min} (**Figure 3**), which can be easily detected from the load-displacement record.

From the above observation, we come to the conclusion that the assumption of elastic behaviour is determined for the flattened Brazilian test. In fact, the crack initiates from the disc centre and propagates mostly along the loading diameter until the two flat ends. In this way, the disc is broken into two parts as presented in **Figure 15**, while the vacant regions close to the two flat ends involve the existence of crush zones. However, these crush zones developed after the crack initiation at the centre and below the propagation along the diameter. The validity of the flattened Brazilian test is thus further justified experimentally.

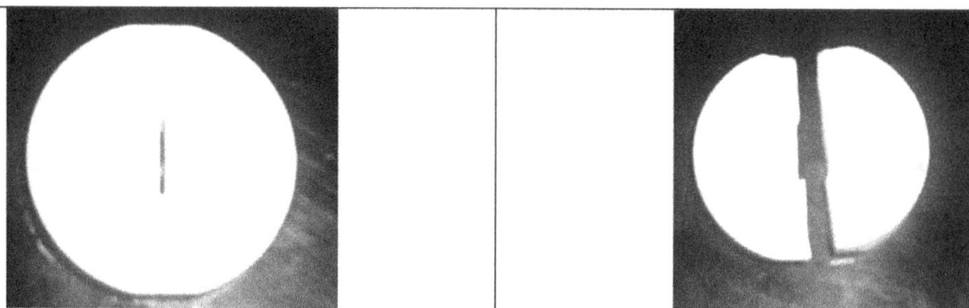

(a)	(b)

Figure 15. (a) CSTFBD before testing and (b) failure mode of CSTFBD specimens.

6.3. Determination of the fracture toughness using bending test

The tests were carried out using the SCB sample configuration shown in **Figure 5**. It should be also noted that bending tests were performed by employing different specimens sintered under optimal conditions. An experimental-analytical method is proposed to investigate the mode I bioceramic fracture toughness for the SCB specimen. Analytical analysis for this geometry is accomplished by using Formula 13. **Figure 16** presents the test results for the calculation of the fracture toughness for the same crack length with different percentages of Fap. Six specimens were performed at each percentage additive. Results of experiments were compared with the results of well-known mode I fracture toughness testing methods.

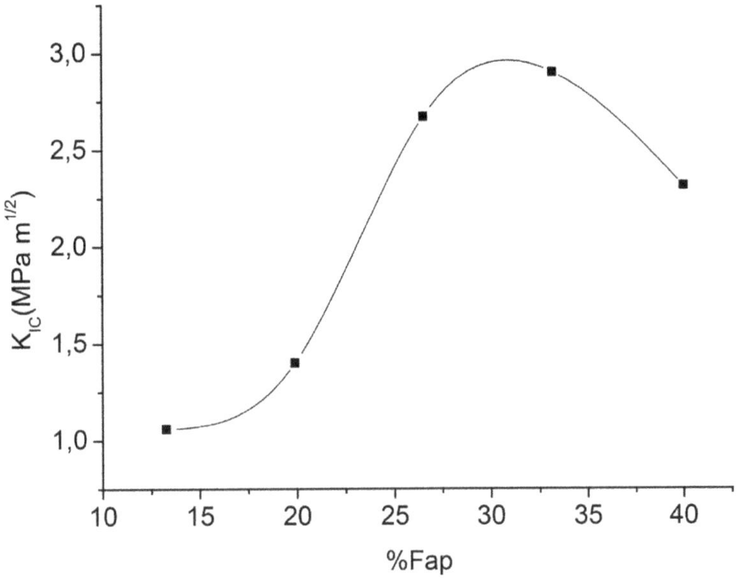

Figure 16. Fracture toughness versus %wt Fap for the SCB specimen.

The fracture toughness values of CTCP-wt% Fap composites range between 1.06 and 2.9 MPam. The lowest toughness (1.06 MPam) is obtained with the 13.26 wt% Fap, while the highest one is approached with the 33.16 wt% Fap (2.9 MPam).These results agree well with the numerical computation effected in the precedent section (**Figure 11**), in which the stress intensity factor reaches maximum when 33.16 wt% Fap are added to the β-CTCP. The mode I fracture toughness measured using SCB specimens is closer to that measured in the precedent section using a flattened Brazilian disc with central straight-through crack (CSTFBD) specimens (2.7 MPam).The variation of the fracture toughness value was due to the differences in the size of the fracture process zone (FPZ) [16]. Aliha et al. [24] reported that the fracture toughness heavily depends on the geometry and loading conditions of the test specimen, for that the fracture toughness of the composite measured using the CSTFBD sample was a little less than that measured using a SCB specimen. Advantages of this new method included easy sample preparation and testing procedure and smaller fracture process zone.

7. Conclusion

The aim of this work is to study the fracture behaviour of the Fap-β-CTCP.

One of the specimens to determine the fracture toughness of bioceramics is the semi-circular bend (SCB) with straight crack. Stress intensity factor at the crack front is an important parameter to find the fracture toughness. On the other hand, a CSTFBD sample is an ideal specimen to be also used for measuring the fracture toughness.

A finite-element modelling study was conducted to evaluate crack propagation in the SCB and CSTFBD specimen during loading. The numerical modelling results are validated by comparing with experimental ones which showed the same outcome for both methods.

Based on the results of both experimental and numerical investigations, the following concluding remarks for the novel configuration Brazilian test can be noticed:

- Three parameters (E,σ and K_{IC}) can be determined in only one test record. E is obtained from the approximate analytical solution for the displacement of the loaded flat end, and when the Poisson's ratio μ is known, the elastic modulus is calculated from the slope of the section of loading-displacement record just before the maximum load. Furthermore, tensile strength is measured from Formula (2) by inserting P_{max} and the coefficient k . Finally, the fracture toughness K_{IC} is determined using the clearly local minimum load P_{min} corresponding to the maximum value of dimensionless stress intensity factor $\emptyset_{max}$ and Eq. (6) is applied.
- The guarantee of the centre crack initiation for the loading angle which satisfies the condition of $2\alpha \geq 20°$: the centre crack initiation being important for test validity.
- The effectiveness and reliability of the new test method for bioceramics fracture test have been demonstrated.

Author details

Achouak Elghazel*, Rym Taktak and Jamel Bouaziz

*Address all correspondence to: elgazelachwak@live.fr

National Engineering School of Sfax, Sfax, Tunisia

References

[1] S.H. Chang, C.I. Lee, S. Jeon. Measurement of rock fracture toughness under modes I and II and mixed mode conditions by using disc type specimens. Eng. Geol. 2002;66:79–97.

[2] ISRM. Commission on testing methods. Suggested method for determining mode I fracture toughness using Cracked Chevron Notched. Int. J. Rock Mech. Min. Sci. Abstr. 1995;32:57–64.

[3] M. Thiercelin, J.C. Roegiers. Fracture toughness determination with the modified ring test. In: Proceedings of the International Symposium on Engineering in Complex Rock Formations (ECRF); Beijing, China, 1986.

[4] R.N. Singh, A.G. Pathan. Fracture toughness of some British rocks by diametral loading of discs. Min. Sci. Technol. 1988;6:179–190.

[5] I.L. Lim, I.W. Johnston, S.K. Choi, J.N. Boland. Fracture testing of a soft rock with semicircular specimens under three point bending. Part I mode I. Int. J. Rock Mech. Min. Sci. Geomech. Abstr. 1994;31:185–197.

[6] H. Guo, N.I. Aziz, L.C. Schmidt. Rock fracture toughness determination by the Brazilian test. Eng. Geol. 1993;33:177–188.

[7] Q.Z. Wang, L.Z. Wu. The flattened Brazilian disc specimen used for determining elastic modulus, tensile strength and fracture toughness of brittle rocks: experimental results. Int. J. Rock Mech. Min. Sci. 2004;41:26–30.

[8] Q.Z. Wang, X.M. Jia, S.Q. Kou, Z.X. Zhang, P.A. Lindqvist. The flattened Brazilian disc specimen used for testing elastic modulus, tensile strength and fracture toughness of brittle rocks: analytical and numerical results. Int. J. Rock Mech. Min. Sci. 2004;41:245–253.

[9] F. Ben Ayed, J. Bouaziz, K. Bouzouita,. Pressureless sintering of fluorapatite under oxygen atmosphere. J. Eur. Ceram. Soc. 2000;20(8):1069–1076.

[10] ISRM. Suggested methods for determining tensile strength of rock materials. Int. J. Rock Mech. Min. Sci. Geomech. Abstr. 1978;15:99–103.

[11] ASTM C496. Standard test method for splitting tensile strength of cylindrical concrete specimens. In: Annual book of ASTM. Standards, 0.042, ASTM, Philadelphia, 1984. p. 336–341.

[12] Q.Z. Wang, L. Xing. Determination of fracture toughness KIC by using the flattened Brazilian disk specimen for rocks. Eng. Fract. Mech. 1999;64:193–201.

[13] M.R. Ayatollahi, M.R. Aliha. Cracked Brazilian disc specimen subjected to mode II deformation. Eng. Frac. Mech. 2005;72:493–503.

[14] M.R. Ayatollahi, M.J. Alborzi. Rock fracture toughness testing using SCB specimen. In: 13th International Conference on Fracture; Beijing, China, 2013.

[15] T. Funatsu, N. Shimizu, M. Kuruppu, K. Matsui. Evaluation of mode I fracture toughness assisted by the numerical determination of K-resistance. Rock. Mech Rock. Eng. 2015;48:143–157.

[16] K. P. Chong, M.D. Kuruppu. New specimen for fracture toughness determination for rock and other materials. Int. J. Fract. 1984;26:R59–R62.

[17] M.R. Ayatollahi, M.R.M. Aliha. On determination of mode II fracture toughness using semicircular bend specimen. Int. J. Sol. Struct. 2006;43:5217–5227.

[18] K.P. Chong, M.D. Kuruppu, J.S. Kuszmaul. Fracture toughness determination of layered materials. Eng. Fract. Mech. 1987;28(1):43–54.

[19] L. Tutluoglu, C. Keles. Mode I fracture toughness determination with straight notched disk bending method. Int. J. Rock Mech. Min. Sci. 2011;48:1248–1261.

[20] N. Bouslama, Y. Chevalier, J. Bouaziz, F. Ben Ayed. Influence of the sintering temperature on Young's modulus and the shear modulus of tricalcium phosphate fluorapatite composites evaluated by ultrasound techniques. Mater. Chem. Phys. 2013;141:289–297.

[21] F. Ben Ayed, J. Bouaziz, K. Bouzouita. Calcination and sintering of fluorapatite under argon atmosphere. J. Alloy Compd. 2001;322(1–2):238–245.

[22] F. Ben Ayed, J. Bouaziz, K. Bouzouita. Résistance mécanique de la fluorapatite frittée. Ann. Chim. Sci. Mater. 2006;31(4):393–406.

[23] N. Bouslama, F. Ben Ayed, J. Bouaziz. Mechanical properties of tricalcium phosphate fluorapatite – alumina composites. Phys. Procedia. 2009;2:1441–1448.

[24] M.R.M. Aliha, M. Sistaninia, D.J. Smith, M.J. Pavier, M.R. Ayatollahi. Geometry effects and statistical analysis of mode I fracture in guiting limestone. Int. J. Rock Mech. Min. Sci. 2012;51:128–135.

7

Fatigue Crack Propagation Rates Prediction Using Probabilistic Strain-Based Models

José António Fonseca De Oliveira Correia, Abílio M.P. De Jesus, Pedro M.G.P. Moreira, Rui A.B. Calçada and Alfonso Fernández-Canteli

Abstract

This chapter proposes an evaluation and extension of the UniGrow model to predict the fatigue crack propagation rate, based on a local strain-based approach to fatigue. The UniGrow model, classified as a residual stress-based crack propagation model, is here applied to derive probabilistic fatigue crack propagation fields (p-da/dN-ΔK-R fields) for P355NL1 pressure vessel steel, covering distinct stress R-ratios. The results are compared with available experimental data. The required strain-life data are experimentally achieved and evaluated. The material representative element size, ϱ^*, a key parameter in the UniGrow model, is assessed by means of a trial-and-error procedure of inverse analysis. Moreover, residual stresses are computed for varying crack lengths and minimum-to-maximum stress ratios. Elastoplastic stress fields around the crack apex are evaluated with analytical relations and compared with elastoplastic finite-element (FE) computations. The deterministic strain-life relations proposed in the original UniGrow model are replaced by the probabilistic strain-life fields (p-ε-N) proposed by Castillo and Canteli. This probabilistic model is also extended by considering a damage parameter to allow for mean stress effects. In particular, a probabilistic Smith-Watson-Topper field (p-SWT-N), alternatively to the conventional p-ε-N field, is proposed and applied to derive the probabilistic fatigue crack propagation fields.

Keywords: fatigue crack propagation, fracture mechanics, local fatigue approaches, probabilistic models, P355NL1 steel

1. Introduction

Research about the fatigue behaviour of materials and structures has engaged both academia and industry. Fatigue has been investigated since the industrial revolution and remains an unsolved research topic due to the lack of reliable predictive models and the continuous emergent materials and applications [1]. The knowledge on fatigue crack propagation is not fully accomplished, despite the great advances achieved in the last decades. In particular, probabilistic approaches are still lacking.

Paris et al. [2] were the first to establish a correlation between the fatigue crack propagation rates and the stress intensity factor, suggesting the well-known and widely accepted Paris' law. Since this contribution, the pioneer Paris' law has been used extensively to model fatigue crack growth under constant amplitude loading but also under random loading. However, Paris' law shows some limitations, as it only models the stable crack propagation regime, excluding near threshold and near unstable fatigue crack propagation phases, and the stress ratio effects are not accounted for. Further, it is a deterministic approach for fatigue crack propagation. Many other fatigue crack propagation relations have been suggested in the literature to solve the shortcomings of the Paris' law including enhanced prediction of fatigue crack growth under variable amplitude loading [3]. The assortment of fatigue crack propagation models available in the literature differs in the number of variables and parameters involved and the application domains. However, their use requires the evaluation of expensive time-consuming fatigue crack propagation tests in order to identify their constants.

Local fatigue strain-based approaches [4–7] represent an alternative to the fracture mechanics fatigue crack propagation once the former being very often applied to model the fatigue crack initiation on notched components [8]. They are usually complemented with fracture mechanics models to allow full fatigue life assessments.

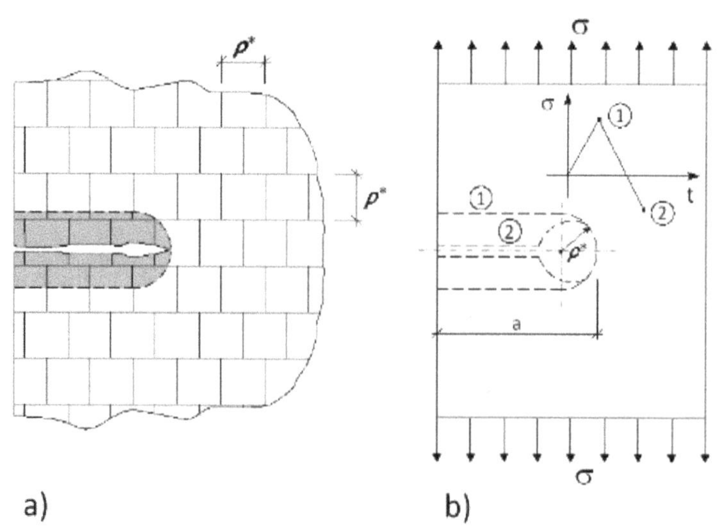

Figure 1. Crack discretization according to the UniGrow model: (a) crack and discrete elementary material blocks and (b) crack shape at the tensile maximum and compressive minimum loads [11].

Some authors, as Glinka [9], Peeker and Niemi [10], Noroozi et al. [11–13] and Hurley and Evans [14], have applied local strain-based fatigue models to represent fatigue crack propagation. Glinka was one of the precursors of modelling fatigue crack propagation using a strain-based fatigue relation [9]. The crack was assumed to have a notch with a tip radius, ρ^*, and the material ahead of the crack tip was assumed to be divided into elemental blocks of finite linear dimension, ρ^* (see **Figure 1**). The crack growth was assumed as the failure of the successive elemental blocks, the fatigue crack growth rate being defined by the following relation:

$$\frac{da}{dN} = \frac{\rho^*}{N_f} \tag{1}$$

where N_f represents the number of cycles to fail the elemental block of dimension, ρ^*. This approach allowed the fatigue crack growth rate to be directly expressed by the strain-life relation constants.

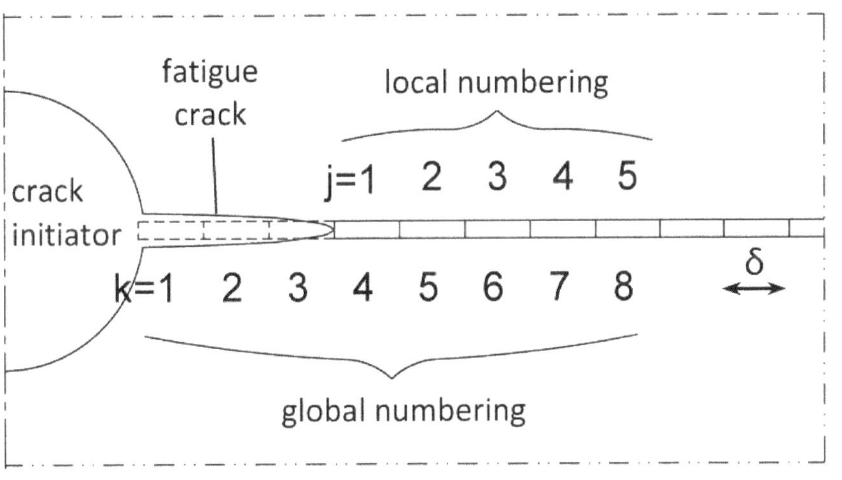

Figure 2. Crack discretization with elements according to the model proposed by Peeker and Niemi [10].

Similarly, the model proposed by Peeker and Niemi [10] allowed the near threshold fatigue crack propagation data and the stable crack growth to be described (see **Figure 2**). For the near threshold fatigue crack propagation, the authors derived the following analytical relation functions of the strain-life constants:

$$\begin{aligned} \frac{da}{dN} &= C_{el} \Delta K^{m_{el}} \\ C_{el} &= 2\delta\left[\left(\sigma_f' - \sigma_m\right)\sqrt{2\pi\delta}\right]^{\frac{1}{b'}} \\ m_{el} &= -\frac{1}{b'} \end{aligned} \tag{2}$$

For the stable crack growth, the above authors derived the following alternative relations that use the cyclic stress-strain constants besides the elastoplastic strain-life constants:

$$\frac{da}{dN} = C_{pl}\Delta K^{m_{pl}}$$

$$C_{pl} = 2\delta\left[\varepsilon'_f\left(\frac{4\pi\delta K'E}{n'+1}\right)^{\frac{1}{n'+1}}\right]^{\frac{1}{c'}} \tag{3}$$

$$m_{pl} = -\frac{2}{c'(n'+1)}$$

The superposition of the two previous relations leads to the following analytical expression for the fatigue crack propagation law covering both fatigue propagation regimes I and II:

$$\frac{da}{dN} = \frac{1}{\frac{1}{C_{pl}\Delta K^{m_{pl}}} + \frac{1}{C_{el}\Delta K^{m_{el}}}} \tag{4}$$

The application of the local approaches to fatigue on fatigue crack growth simulation requires the definition of a crack path discretization. The size of the elements used to discretize the crack path should account for two criteria: (a) the size must be large enough to represent the local material properties by their mean values using continuous variables and (b) its size should be related to micro-structural material parameters, such as the material grain size. For structural steels, and according to reference [10], the criteria result in an average element size of 0.1 mm = 100 μm.

In general, elastoplastic stress analysis at the crack vicinity is required and very often analytical approaches are applied. Nonetheless, Hurley and Evans [14] proposed the use of elastoplastic finite-element analysis. The fatigue life of the process zone was computed using the Walker strain that was correlated directly with the fatigue life from the experimental data using a power relation. This Walker strain is defined according to the following relation:

$$\varepsilon_w = \frac{\sigma_{max}}{E}\left(\frac{\Delta\varepsilon E}{\sigma_{max}}\right)^w \tag{5}$$

where σ_{max} is the maximum stress, E is the Young modulus, $\Delta\varepsilon$ is the strain range and w is a constant varying between 0 and 1. These authors applied the following definition of the cyclic plastic zone that is assumed equal to the fatigue process damage zone under plane strain conditions:

$$\Delta a_i = \frac{1}{3\pi}\left(\frac{\Delta K}{2\sigma_0}\right)^2 \tag{6}$$

where σ_0 is the cyclic yield stress and ΔK is the stress intensity factor range computed from the numerical model. The resulting approach is much simpler than those proposed by previous authors. It is supported by numerical models disregarding analytical aspects that allowed previous authors to verify the relation between these local approaches to fatigue and the fracture mechanics approaches for fatigue crack propagation.

In all previous studies, the fatigue crack growth process has been modelled as a continuous damaging process consisting of successive failure of material elements along the assumed crack path. These crack propagation approaches have provided accurate correlations of crack propagation data from several sources that included the stress ratio or mean stress effects. These crack propagation approaches are local fatigue approaches, where the stress-strain fields ahead of the crack tip are computed supported by elastoplastic stress-strain analyses. The resulting elastoplastic stress-strain fields are used in fatigue damage analysis performed for each material element in the crack path. Analytical methods such as the ones proposed by Neuber [15] and Moftakhar et al. [16] may be applied to perform the elastoplastic analysis taking into account the elastic stress-strain fields computed around the crack tip, using available linear elastic fracture mechanics solutions [11, 16, 17].

This chapter proposes an evaluation and extension of the model proposed by Noroozi et al. [11–13] to predict the fatigue crack propagation rates, based on a local strain-life fatigue approach. This model denoted that the UniGrow model is a residual stress-based crack propagation model [18]. The model is adapted in this work to derive probabilistic fatigue crack propagation fields (p-da/dN-ΔK-R fields) for P355NL1 pressure vessel steel, covering distinct stress R-ratios. The procedure proposed has been already applied to other available experimental data [19–21]. The required strain-life data is experimentally evaluated and can be found in the literature [21]. The material representative element size, ρ^*, a key parameter in the UniGrow model, is assessed by means of a trial and error process. The residual stress field at crack tip vicinity is investigated for various crack lengths and minimum-to-maximum stress ratios. The elastoplastic stress-strain fields evaluated at the vicinity of the crack tip, using analytical solutions, are compared with the stress-strain fields computed using nonlinear elastoplastic finite-element analyses of the tested compact tension (CT) specimens considered in a fatigue crack propagation experimental program.

Deterministic strain-life relations proposed in the UniGrow model are replaced by the probabilistic strain-life fields (p-ε_a-N) proposed by Castillo and Canteli [22]. This probabilistic model is also extended by considering a damage parameter able to account for mean stress effects. In particular, a probabilistic Smith-Watson-Topper field (p-SWT-N) is proposed alternatively to the p-ε_a-N and applied to derive the probabilistic crack propagation fields.

2. Overview of the deterministic UniGrow model

The UniGrow model, originally developed by Noroozi et al. [11], rest on the following premises:

- A continuous material is discretized into elementary finite particles (dimension ρ^*), below which material continuity is no longer assured, (**Figure 1a**).

- The fatigue crack tip geometry exhibits a finite radius equivalent to a round notch of radius ρ^*, (**Figure 1b**).
- The fatigue crack growth is decomposed into successive crack increments, each one of equal size—the material elementary finite particle size. Therefore, this damaging process can be assumed as a continuous crack initiation process occurring at the material element levels characterized by the size ρ^*.
- The fatigue crack growth rate can then be computed using Eq. (1) and the number of cycles required to completely crack the material elementary particles can be computed by a local fatigue relation such as the ones based on a strain-life approach.

Noroozi et al. [11] proposed the application of the Smith, Watson and Topper fatigue damage parameter (SWT damage parameter) [7] in the form of the following fatigue life relation:

$$\sigma_{max} \cdot \Delta\varepsilon/2 = SWT = (\sigma_f')^2 \cdot (2N_f)^{2b}/E + \sigma_f' \cdot \varepsilon_f' \cdot (2N_f)^{b+c} \tag{7}$$

Peeker and Niemi [10] propose, alternatively, the use of the Morrow's equation [6] to compute the failure of the material's representative element:

$$\frac{\Delta\varepsilon}{2} = \frac{\sigma_f' - \sigma_m}{E} \cdot (2N_f)^b + \varepsilon_f' \cdot (2N_f)^c \tag{8}$$

The Morrow's equation is derived from the following Coffin-Manson relation [4, 5] of the material, in order to include mean stress effects:

$$\frac{\Delta\varepsilon}{2} = \frac{\sigma_f'}{E} \cdot (2N_f)^b + \varepsilon_f' \cdot (2N_f)^c \tag{9}$$

In particular, Eq. (7) is derived by the multiplication of the Coffin-Manson Eq. (9) by the Basquin relation [23], for a fully reversal stress ratio, i.e. R $(\sigma_{min}/\sigma_{max}) = -1$, given by

$$\frac{\Delta\sigma}{2} = \sigma_{max} = \sigma_f' \cdot (2N_f)^b \tag{10}$$

In the previous two equations, σ'_f and b represent, respectively, the fatigue strength coefficient and exponent; ε'_f and c represent, respectively, the fatigue ductility coefficient and exponent and E is the Young modulus. The maximum stress, σ_{max}, mean stress, σ_m, and the strain range, $\Delta\varepsilon$, are computed and averaged over the material elementary size ρ^* using an elastoplastic analysis. According to Noroozi et al. [11, 12], the elastoplastic analysis can be performed, following the steps below, in order to estimate the stresses and strains at the elementary material elements placed along the crack path:

i. Application of the Creager-Paris solution [24] to evaluate the elastic stresses in the material region surrounding the crack tip.

ii. Application of analytical elastoplastic formulae such as the Neuber's [15] or Glinka's approaches [25] to transform the elastic stress field into actual elastoplastic stresses and strains ahead of the crack tip. Procedures by Moftakhar et al. [16] and Reinhard et al. [17] may also be applied to perform multi-axial elastoplastic stress-strain analysis.

iii. Determination of the residual stresses along the y-axis direction (normal to crack face) and along the crack path direction (line ahead of the crack tip), using the actual elastoplastic stresses computed at the end of the first load reversal and subsequent cyclic elastoplastic stress range, that is

$$\sigma_r = \sigma_{max} - \Delta\sigma \tag{11}$$

iv. Assuming the residual stress distribution computed ahead of the crack tip to be applied on the crack faces, behind the crack tip, in a symmetric way with respect to the crack tip ($x = 0$), the loading process generates a plastic zone at the crack tip that does not vanish completely during unloading, leading to a cyclic plastic zone, which is controlled by compressive stresses ahead of the crack tip and to some amount of crack opening displacement just behind the crack tip (crack faces does not close completely just behind the crack tip). One way to model crack opening consists in assuming that the compressive residual stress field acting ahead of the crack tip is applied in a symmetrical way, behind the crack tip, directly on crack faces. This compressive stress distribution acting on crack faces is equivalent to a residual stress intensity factor being used to correct the applied stress intensity factor range leading to a total (effective) stress intensity factor range that excludes the effects of the compressive stresses. It is proposed by Noroozi et al. [11, 12] to calculate the residual stress intensity factor, K_r, using the weight function method [26]:

$$K_r = \int_0^a \sigma_r(x) \cdot m(x, a) dx \tag{12}$$

where $m(x, a)$ is the weight function and σ_r is the residual stress field.

v. The applied stress intensity factor values (maximum and range) are corrected using the calculated residual stress intensity, resulting the total $K_{max,tot}$ and ΔK_{tot} values [11, 12]. For positive stress R-ratios, which corresponds to the range covered by the experimental data used in this research, $K_{max,tot}$ and ΔK_{tot} may be computed as follows:

$$\begin{aligned} K_{max,tot} &= K_{max,applied} + K_r \\ \Delta K_{tot} &= \Delta K_{applied} + K_r \end{aligned} \tag{13}$$

where K_r assumes a negative value corresponding to the compressive stress field. For high stress R-ratios, the compressive stresses ahead of the crack tip may be neglected and the applied stress intensity factor range is assumed fully effective and for low stress R-ratios, the compressive stresses increase and the effectiveness of the applied stress intensity factor range decreases.

vi. Using the total values of the stress intensity factors, the first and second steps (steps (i) and (ii)) are repeated to determine the corrected values for the maximum actual stress and actual strain ranges at the material's representative elements. Then, Eq. (7) is applied together with Eq. (1) to compute the fatigue crack growth rates in the original UniGrow model proposition.

The above fatigue crack propagation methodology does not provide close-form expressions for the fatigue crack propagation rates. Nevertheless, considering the material to behave according to dominant elastic or plastic behaviours at the crack tip, close-form solutions for the fatigue crack propagation rates are viable, leading to fatigue crack propagation relations based on two crack-driven parameters with the following form [11, 12]:

$$\frac{da}{dN} = C\left[\left(K_{max,tot}\right)^{p}\left(\Delta K_{tot}\right)^{q}\right]^{\gamma} \tag{14}$$

where C, p, q and γ are constants that can be related with the ones characterizing the cyclic elastoplastic behaviour of the material at the crack tip and the plasticity or elasticity dominance. The dependency of the fatigue crack propagation rates with both the maximum and range of stress intensity factors, $K_{\max}$ and ΔK, allows the mean stress effects to be conveniently accounted on fatigue crack propagation rates. This type of fatigue crack propagation model based on the combination of two parameters crack driving force has recently being proposed by several authors [27, 28].

In this chapter, the methodology proposed by Noroozi et al. [11] is followed with some variations. Material cyclic/fatigue properties are required. In addition, the UniGrow model introduces the material element size, ρ^*, which consists in an additional model parameter that in most of the situations needs to be computed by means of a trial and error iterative approach, in order to result in a good fitting of available experimental fatigue crack propagation data. The analytical elastoplastic analysis, supported by the Creager-Paris [24] elastic solutions and the Neuber's multi-axial approach [15–17], is applied to evaluate the elastoplastic stress-strain field at the first material element at the crack tip. Concerning the residual stresses along the crack path line, it is computed using a nonlinear elastoplastic finite-element analysis overcoming discrepancies verified in the analytical residual stress distributions, as demonstrated later in the chapter.

3. Modelling cyclic plasticity

In the following an elastoplastic constitutive model based on the von Mises yield criterion (J2 plasticity) and associative flow rule is selected to model compact tension specimen's elasto-plastic behaviour and particularly to compute the residual stresses at the fatigue crack region. Multi-linear kinematic hardening is used, which is crucial for the cyclic plasticity simulation and in particular the Bauschinger effect description. The von Mises yield criterion points out a yield function, *f*, defined as follows:

$$f = \left[\frac{3}{2}(S_{ij} - X_{ij})(S_{ij} - X_{ij})\right]^{1/2} - \sigma_k \leq 0 \tag{15}$$

where S_{ij} is the deviatoric stress tensor defined as

$$S_{ij} = \sigma_{ij} - \sigma_h \delta_{ij} \tag{16}$$

X_{ij} is the back-stress tensor that defines the translation of the yield surface, σ_k is the yield stress of the material, σ_{ij} is the stress state of the material, σ_h is the pressure or hydrostatic stress state and δ_{ij} is the Kronecker delta operator. The plastic flow occurs according to the associative flow rule, which means that the plastic strain increment can be computed from the derivatives of the yield function with respect to the stress tensor:

$$\dot{\varepsilon}_{ij}^{P} = \lambda \frac{\partial f}{\partial \sigma_{ij}} \tag{17}$$

where λ is the plastic multiplier related to the amount of plastic deformation. The adopted kinematic hardening rule is based on the Besseling model that is a sub-layer or overlay model [29] with the material behaviour, being composed of various portions (or sub-volumes), all subjected to the same total strain but each sub-volume having a different yield stress. Despite each sub-volume shows a simple stress-strain response, when combined, the model can represent multi-linear stress-strain curves. The following phases are performed in the plasticity calculations [30]:

i. The portion of total volume (the weighting factor) for each sub-volume and its corresponding yield stress are computed.

ii. For each sub-volume, the respective increment of the plastic strain is determined, assuming that each one is subjected to the same total strain.

iii. The individual increments in the plastic strain are summed up using the weighting factors determined in step (i) to result the total or apparent increment in plastic strain.

iv. The plastic strain is updated and the elastic strain is evaluated.

The weighting factor and yield stress for each sub-volume are determined by fitting the material response to the uniaxial stress-strain curve. A perfectly plastic material based on von Mises criterion is assumed. The weighting factor for sub-volume k given by

$$W_k = \frac{E - E_{Tk}}{E - \frac{1-2\nu}{3} E_{Tk}} - \sum_{i=1}^{k-1} W_i \tag{18}$$

where W_k is the weighting factor for sub-volume k and is evaluated sequentially from 1 to the number of sub-volumes N_{sv} and E_{Tk} is the slope of the kth segment of the uniaxial cyclic stress-strain curve. The yield stress for each sub-volume is given by

$$\sigma_{yk} = \frac{1}{2(1-\nu)}[3E\varepsilon_k - (1-2\nu)\sigma_k] \tag{19}$$

where $(\varepsilon_k, \sigma_k)$ is the breakpoint in the uniaxial multi-linear cyclic stress-strain curve. The number of sub-volumes corresponds to the number of breakpoints specified for the definition of the multi-linear stress-strain material relation. Each sub-volume follows the von Mises yield criterion with the associative flow rule and the plastic strain increment for the entire volume is computed according to

$$\Delta\varepsilon_{ij}^{p} = \sum_{k=1}^{N_{sv}} W_k \cdot \Delta\varepsilon_{ij,k}^{p} \tag{20}$$

In the plastic computation algorithm, if the equivalent (von Mises) stress predicted using an elastic trial exceeds the material yield stress given the yield function, then plastic strains will occur. Then a plastic correction will be required, which means that plastic strains reduce the stress state so that it satisfies the yield criterion (return to mapping procedures). The followed scheme for the integration of the elastoplastic constitutive equations consisted of an elastic trial and a return mapping procedure, as proposed by Simo and Taylor [31]. ANSYS commercial code was used in this study with the cyclic plasticity model being already part of the software capabilities [30]. The stabilized cyclic stress-strain curve of the material under consideration in this study is used for the identification of the plasticity constants. In particular, the cyclic stress-strain curve was replaced by a multi-linear representation, the resulting breakpoints being inputted in the model built in the ANSYS commercial code.

4. Probabilistic fatigue damage models

The application of the UniGrow model requires a fatigue damage relation to compute the number of cycles to fail the elementary material blocks. In this section, probabilistic fatigue damage models are proposed rather than the deterministic SWT-N, Coffin-Manson or Morrow models. The probabilistic ε_a-N model proposed by Castillo and Canteli [22] is used. However, since the probabilistic ε_a-N model does not take into account the mean stress effects, a new probabilistic SWT-N field is also proposed, as an extension of the p-ε_a-N field suggested in [22], to include the mean stress influence.

4.1. *p-ε-N* model

Based on physical and statistical considerations, such as the weakest link principle, stability, limit behaviour, range of the variables and compatibility, Castillo and Canteli [22] derived a probabilistic Weibull model to describe the strain-life field using the same formulation as that proposed by the authors for modelling of the stress-life field. Details of each derivation are found in Castillo et al. [32, 33], where the stress version of the model proves to be successfully applied to different cases of lifetime problems. This leads to the following Weibull strain-life model [33]:

$$p = F(N_f^*; \varepsilon_a^*) = 1 - exp\left\{-\left[\frac{log(N_f/N_0)log(\varepsilon_a/\varepsilon_{a0}) - \lambda}{\delta}\right]^{\beta}\right\} \tag{21}$$
$$log(N_f/N_0)log(\varepsilon_a/\varepsilon_{a0}) \geq \lambda$$

where p is the probability of failure, N_0 and ε_{a0} are normalizing values, and λ, δ and β are the non-dimensional Weibull model parameters. Their physical meanings (see **Figure 3**) are as follows:

N_0: threshold value of lifetime,

ε_{a0}: endurance limit of ε_a,

λ: parameter defining the position of the corresponding zero-percentile curve,

δ: scale parameter and

β: shape parameter.

Note that the strain-life model (Eq. (21)) has a dimensionless form and reveals that the probability of failure p depends only on the product $N_f^*\varepsilon_a^*$, where $N_f^* = \log\left(N_f/N_0\right)$ and $\varepsilon_a^* = \log\left(\varepsilon_a/\varepsilon_{a0}\right)$, that is

$$N_f^*\varepsilon_a^* \sim W(\lambda, \delta, \beta) \Leftrightarrow N_f^* \sim W\left(\frac{\lambda}{\varepsilon_a^*}, \frac{\delta}{\varepsilon_a^*}, \beta\right) \tag{22}$$

i.e. $N_f^*\varepsilon_a^*$ follows a Weibull statistical distribution.

This model provides a complete analytical description of the statistical properties of the physical problem being dealt with, including the quantile curves as curves representing the same probability of failure. The model shows, with respect to the conventional strain-life

Coffin-Manson relation, some important advantages. Specifically, it resulted from sound statistical and physical assumptions and not from an empirical arbitrary hypothesis, provides a probabilistic definition of the complete strain-life field without the need of splitting the total strain in its elastic and plastic components but dealing with the total strains directly, includes the run-outs also in the analysis and, finally, also facilitates fatigue damage analysis.

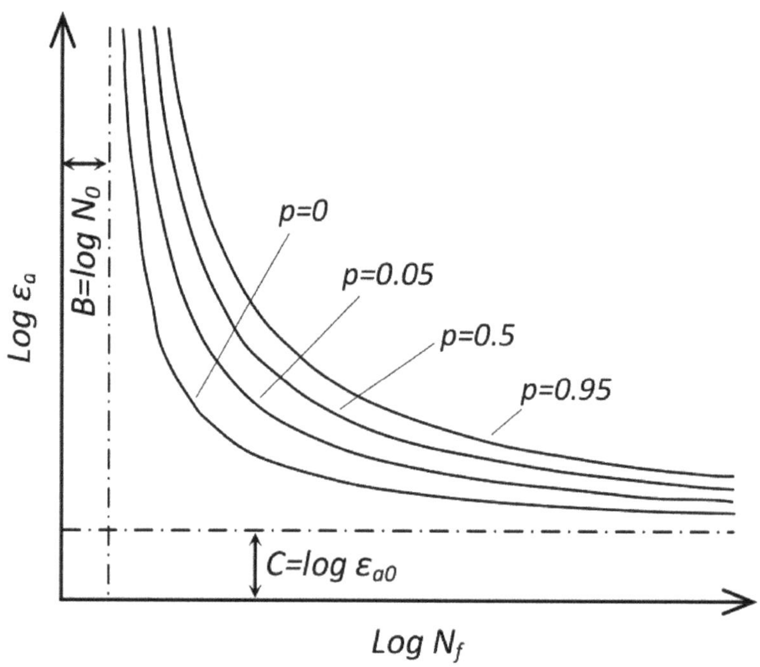

Figure 3. Probabilistic ε_a-N field.

4.2. *p*-SWT-*N* model

The SWT ($=\sigma_{max}.\varepsilon_a$) parameter is proposed by Smith et al. [7] in order to take into account the mean stress effects on fatigue life. Any combination of maximum stress and strain amplitude supplying the same value of the SWT parameter should lead to the same fatigue life. The SWT-*N* and ε_a-*N* fields exhibit similar characteristics. Therefore, the *p*-ε-*N* field proposed by Castillo and Canteli [22] may be extended to represent the *p*-SWT-*N* field as

$$p = F(N_f^*; SWT^*) = 1 - exp\left\{-\left[\frac{log(N_f/N_0)log(SWT/SWT_0) - \lambda}{\delta}\right]^{\beta}\right\}$$
$$log(N_f/N_0)log(SWT/SWT_0) \geq \lambda \qquad (23)$$

where *p* is the probability of failure, N_0 and SWT_0 are normalizing values, and λ, δ and β are the non-dimensional Weibull model parameters. Their physical meanings (see **Figure 4**) are

N_0: threshold value of lifetime and

SWT_0: fatigue limit of SWT,

and the Weibull parameters λ, δ and β with the same meaning as discussed above.

Note that Eq. (23) has a dimensionless form and reveals that the probability of failure p depends only on the product N_f^*SWT*, where $N_f^* = \log\left(N_f/N_0\right)$ and $\text{SWT}^* = \log\left(\text{SWT}/\text{SWT}_0\right)$ that is

$$N_f^* SWT^* \sim W(\lambda, \delta, \beta) \Leftrightarrow N_f^* \sim W\left(\frac{\lambda}{SWT^*}, \frac{\delta}{SWT^*}, \beta\right) \tag{24}$$

i.e. N_f^*SWT* follows a Weibull distribution.

The parameters log N_0 and log ε_{a0} of the p-ε_a-N model, and log N_0 and log SWT_0 of the p-SWT-N model can be estimated by the least square method. The Weibull parameters may be estimated by the maximum likelihood method [32, 33].

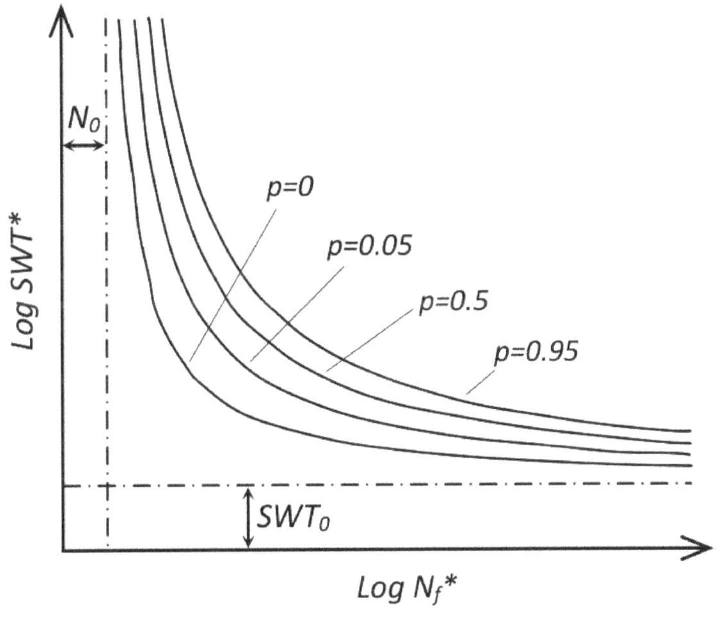

Figure 4. Probabilistic SWT-N_f field.

5. Procedure to generate probabilistic fatigue crack growth rates

A procedure proposed by Correia et al. [34–39] to derive probabilistic fatigue crack propagation fields, based on the UniGrow model, may be summarized into three steps, as follows (see **Figure 5**):

1. Estimation of the Weibull parameters for the p-SWT-N or p-ε_a-N models, described in Section 4, using experimental ε_a-N or SWT-N data from smooth specimens tested under low-cycle fatigue uniaxial tensile loading.

2. Application of the UniGrow model with probabilistic fatigue damage models.
3. Computation of the *p-da/dN-ΔK-R* field.

The UniGrow model is programmed in an Excel workbook, using Visual Basic for Applications (VBA) macros. The problem of the CT specimen geometry was addressed by the code developed [34–39]. The required data are the material properties, loading parameters and geometric dimensions of the CT specimen, including the initial and final crack sizes to be simulated. Also, the elementary material element size, ρ^*, is an input that after the first trial may be iteratively corrected to result in a satisfactory fit between the experimental and numerical data. **Figure 3** gives a general overview of the procedure.

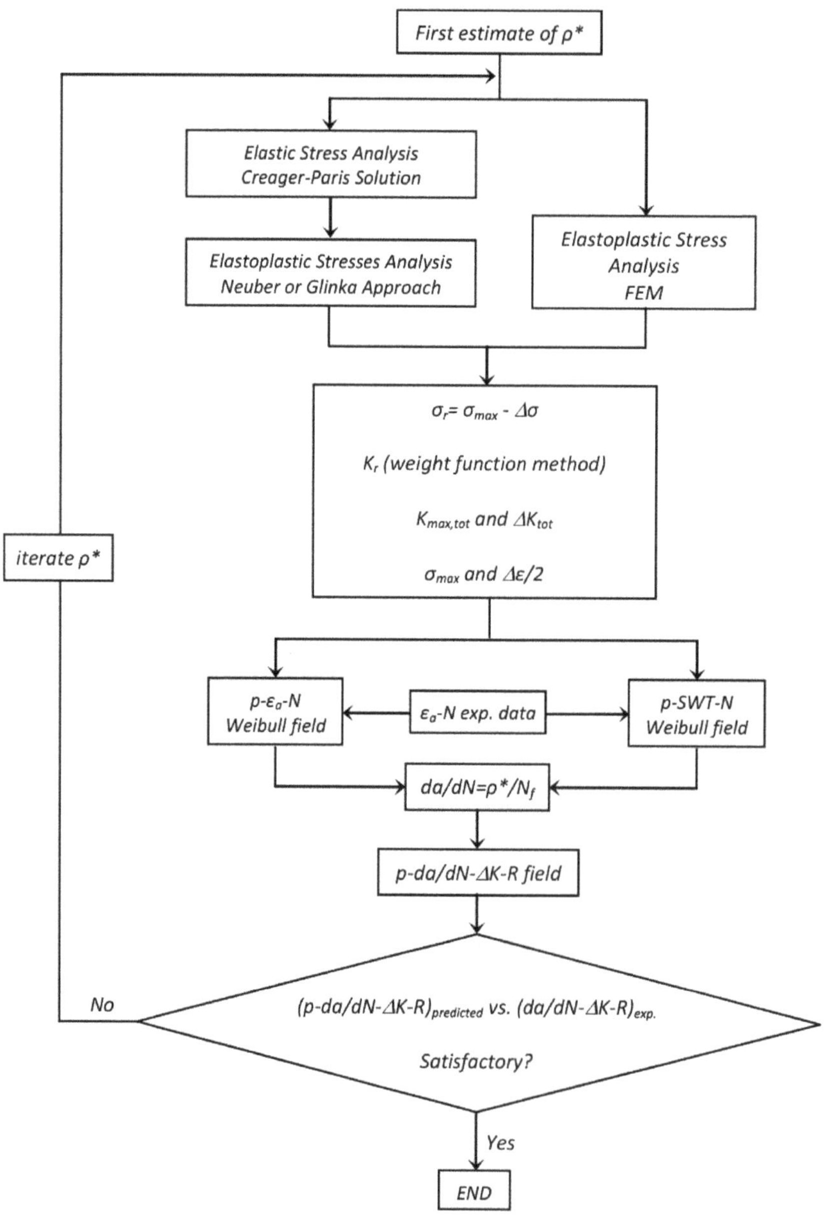

Figure 5. Procedure to generate probabilistic fatigue crack propagation fields.

Neuber and Glinka's approaches for the elastoplastic analysis are both possible to be applied to the cracked CT specimen [15, 25] and a multi-axial elastoplastic calculation as suggested by Moftakhar et al. [16] and Reinhard et al. [17] are programmed. This multi-axial approach is supported by Hencky's total deformation equations that proved to yield reliable predictions for the stress-strain field at notched details under proportional loading. Besides the crack tip material element analysis, the UniGrow model also requires the computation of the elasto-plastic stress-strain response along the crack path, in order to facilitate the residual stress assessment. However, the above referred analytical multi-axial elastoplastic approaches are not able to model the stress redistribution due to yielding, which may lead to inconsistent predictions of the residual stresses. To overcome this limitation of the analytical approaches, in this research the residual stresses are computed preferably using an elastoplastic finite-element approach, the results being verified against those coming from the referred analytical solutions. The residual stresses play a central role in the UniGrow model; therefore, the accuracy of the residual stress evaluation method is vital for attaining consistent fatigue crack growth rate predictions.

The probabilistic fatigue crack propagation fields are alternatively assessed using the probabilistic ε_a-N and SWT-N fields. For each fatigue damage model, independent material element sizes, ρ^*, are identified and compared.

6. Basic fatigue data of the P355NL1 steel

The P355NL1 steel is a pressure vessel steel, the fatigue behaviour of which has been already analysed [21, 35, 39, 40]. This section presents the cyclic elastoplastic fatigue data and the fatigue crack growth data obtained for the P355NL1 steel [21]. The strain-life behaviour of the P355NL1 steel is evaluated through fatigue tests of smooth specimens, carried out under strain-controlled conditions, according to the ASTM E606 standard [41]. Two series of specimens, 19 and 24 specimens, are tested under distinct strain ratios, $R_\varepsilon = 0$ and -1, respectively. The cyclic Ramberg-Osgood [42] and Morrow [6] strain-life parameters of the P355NL1 steel are summarized in **Table 1**, for the conjunction of both strain ratios [21, 35, 39, 40]. The Ramberg-Osgood properties are represented by the cyclic strain hardening coefficient, K' and the cyclic strain exponent, n'. The fatigue ductility coefficient ε_f', the fatigue ductility exponent c, the fatigue strength coefficient σ_f' and the fatigue strength exponent b are parameters of the Morrow strain-life relation. The elastic and monotonic tensile properties of this pressure vessel steel under investigation, as represented by the Young modulus E, Poisson ratio ν, the ultimate tensile strength f_u and upper yield stress f_y, are shown in **Table 2**.

Material	σ'_f	b	ε'_f	c	K'	n'
	MPa	–	–	–	MPa	–
P355NL1	1005.50	−0.1033	0.3678	−0.5475	948.35	0.1533

Table 1. Cyclic elastoplastic and strain-life properties of the P355NL1 steel, $R_\varepsilon = -1$ and $R_\varepsilon = 0$.

Material	E	ν	f_u	f_y
	GPa	–	MPa	MPa
P355NL1	205.20	0.275	568.11	418.06

Table 2. Elastic and tensile properties of the P355NL1 steel.

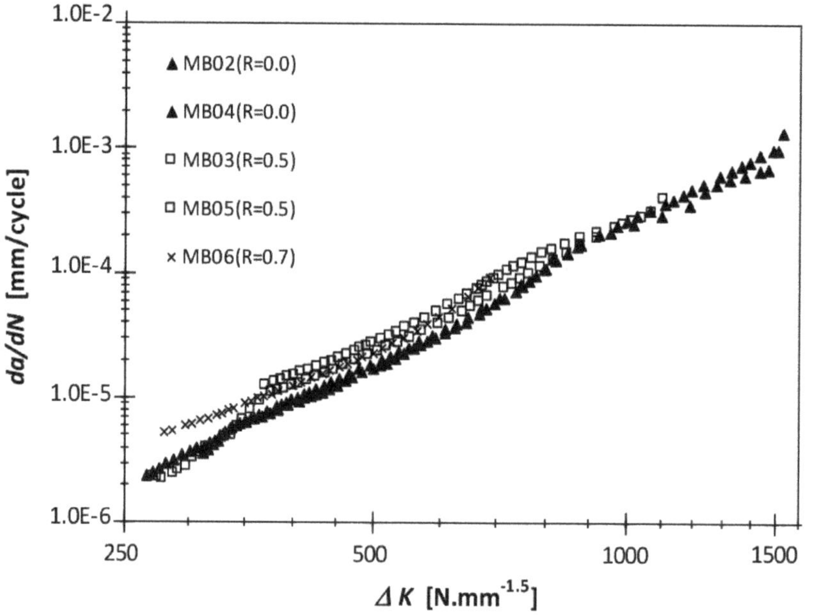

Figure 6. Fatigue crack propagation data obtained for the P355NL1 steel.

Fatigue crack growth rates of the investigated materials are also evaluated for several stress R-ratios, using CT specimens, following the recommendations of the ASTM E647 standard [43]. The CT specimens of P355NL1 steel are defined with a width, $W = 40$ mm and a thickness, $B = 4.5$ mm [21, 35, 39, 40]. All tests are performed in air, at room temperature, under a sinusoidal waveform at a maximum frequency of 20 Hz. The crack growth data derived for the P355NL1 steel, for three tested stress ratios, $R_\sigma = 0.0$, $R_\sigma = 0.5$ and $R\sigma = 0.7$, are illustrated in **Figure 6**. The crack propagation rates are only slightly influenced by the stress ratio. Higher stress ratios provide higher crack growth rates. **Figures 7** and **8** show the p-SWT-N and p-ε_a-N fields, respectively, which are identified for the P355NL1 pressure vessel steel, using the experimental data from the fatigue tests on smooth specimens. The constants of the Weibull fields are also shown in the figures, in particular the threshold constants (B and C) and the Weibull parameters (β, λ and δ). The extrapolations using the Weibull field should be avoided mostly for very low-cycle fatigue regimes. It is verified in this study that the number of cycles to fail the material's representative element, in the fatigue crack propagation regime II, is generally in the low- to very low-cycle fatigue regimes. Thus, it is decided to enhance the Weibull field in this region postulating some fatigue data at the low- to very low-cycle fatigue regimes, using the available deterministic Morrow's equation. The Morrow's equation does not show the vertical asymptote as verified with the Weibull field, being physically more consistent when performing extrapolations for very low number of cycles.

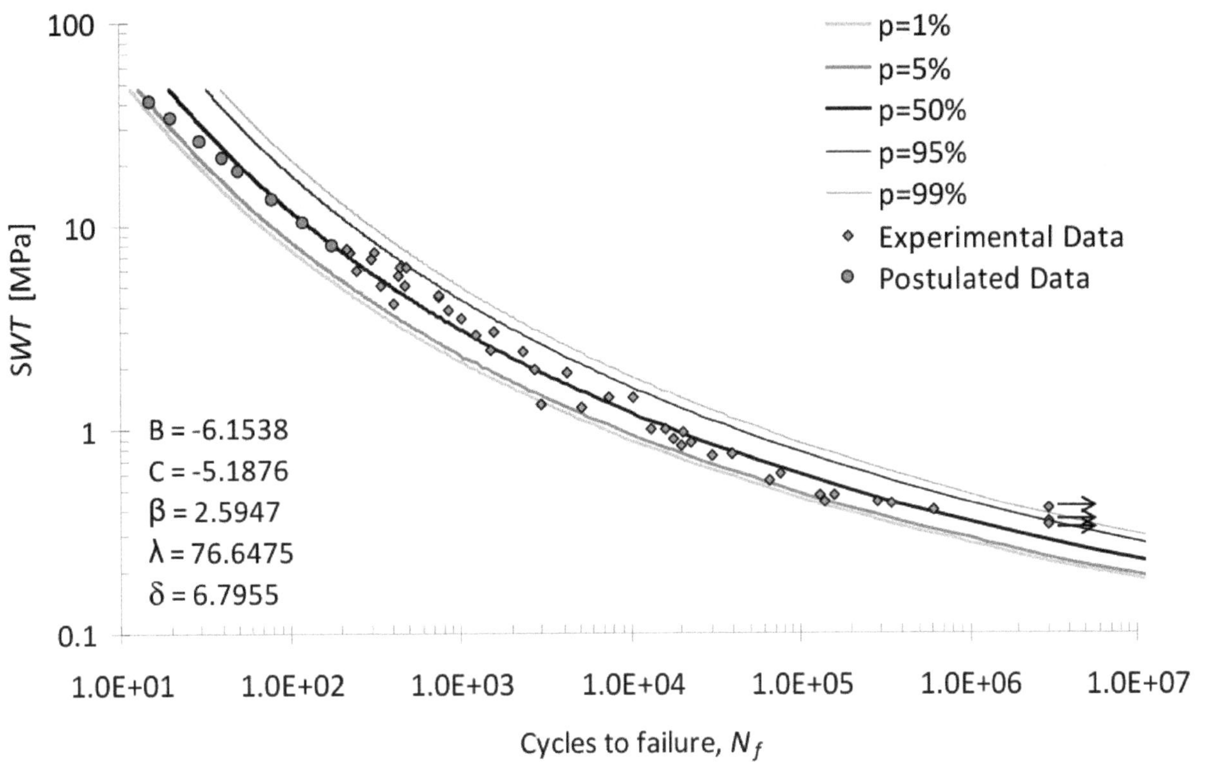

Figure 7. *p*-SWT-*N* field for the P355NL1 steel.

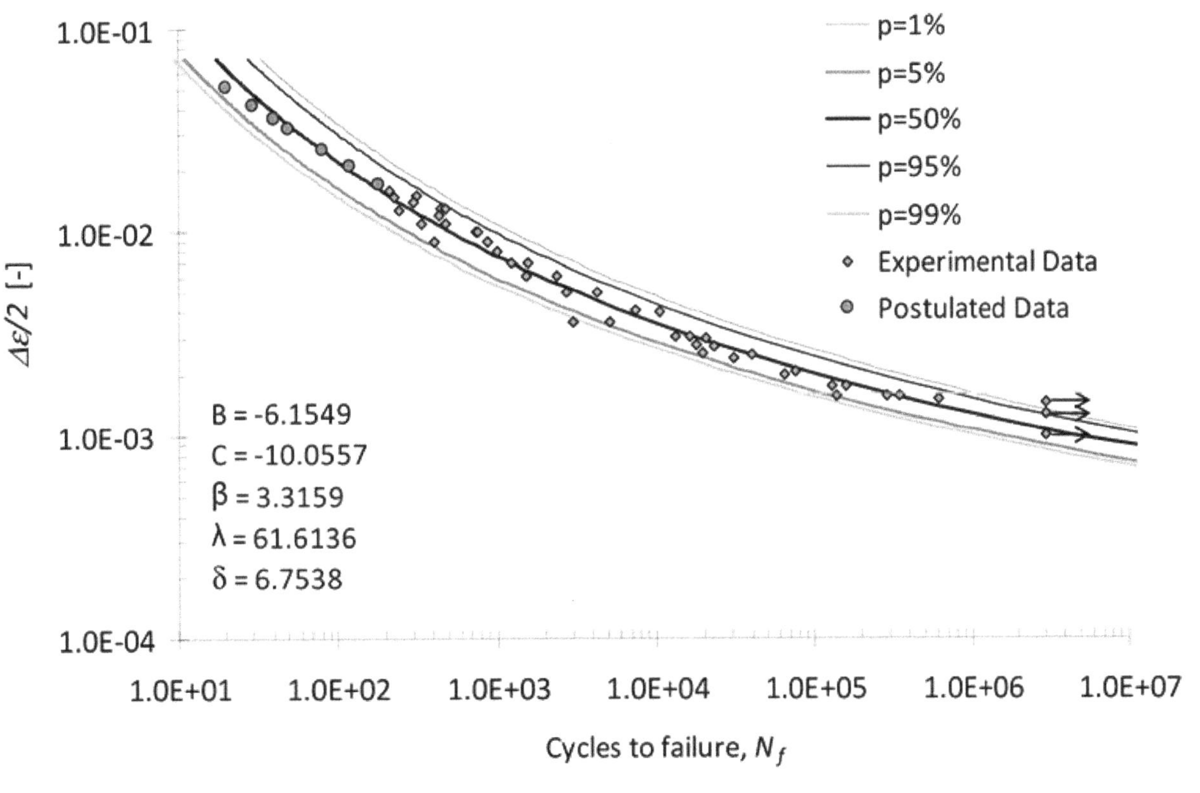

Figure 8. p-ε_a-N field for the P355NL1 steel.

7. Prediction of the probabilistic fatigue crack propagation fields

The prediction of the probabilistic fatigue crack propagation fields is performed through the application of the UniGrow model to CT specimens. Additionally, the elementary material block size, ρ^*, is required and is evaluated by a trial and error procedure in order to achieve good agreement between the predicted and experimental da/dN versus ΔK data, for the P355NL1 steel under consideration. The probabilistic fatigue crack propagation fields are evaluated using, alternatively, the probabilistic p-ε_a-N and p-SWT-N fields (see procedure described in Section 6).

7.1. Numerical FE simulation of the compact tension specimens

Two-dimensional numerical models of the CT specimens are developed using nonlinear elastoplastic finite-element modelling. These models were useful to assess the accuracy of residual stress estimation using the simplified elastoplastic analysis. A very refined finite-element mesh at the crack tip region is used to model conveniently the crack tip geometry with a notch radius, ρ^* (refer to **Figure 1b** for geometric details). **Figure 9** illustrates the finite-element mesh of the CT specimen along with the respective boundary conditions. One-half of the geometry is modelled, taking advantage of the existing symmetry. 2D plane stress elements are used since the specimens' thickness is relatively thin (B = 4.5 mm for the P355NL1 steel).

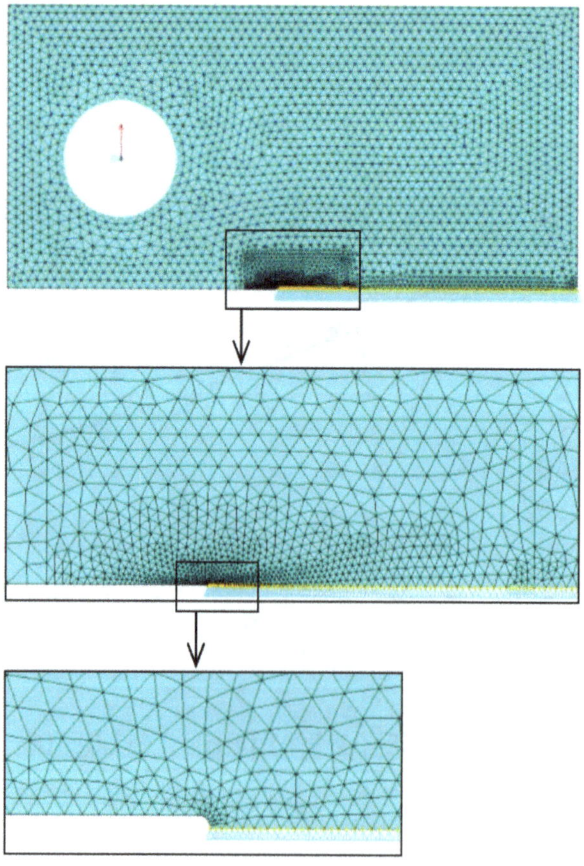

Figure 9. Typical finite-element mesh of the CT specimen.

Quadratic triangular elements (six-noded elements) are selected and applied with a full integration formulation. The pin loading used in the CT testing is simulated with a rigid-to-flexible frictionless contact, the pin being modelled as a rigid circle controlled by a pilot node.

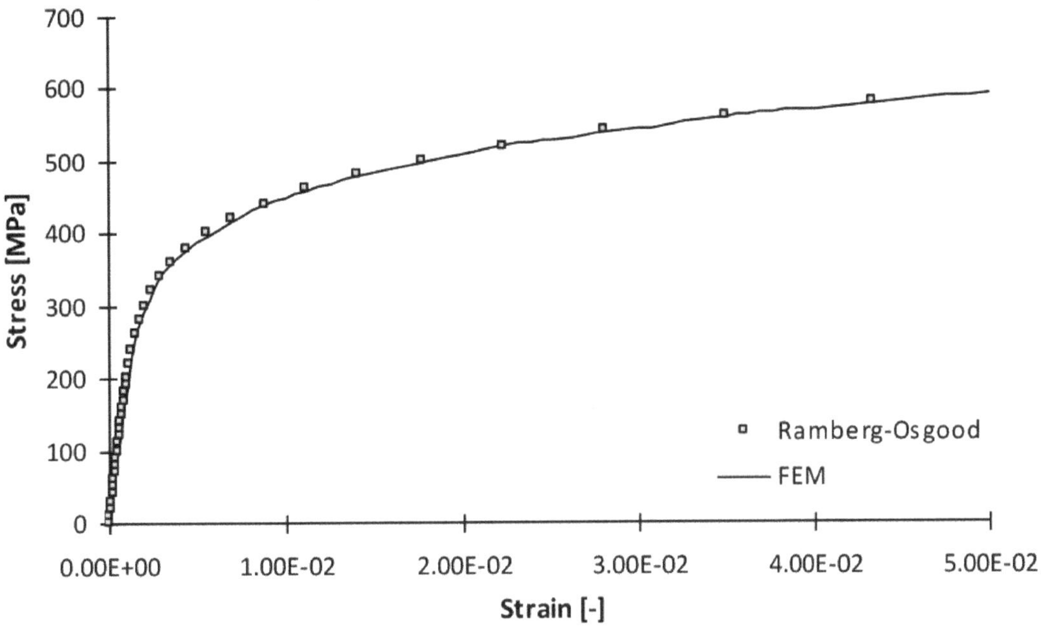

Figure 10. Cyclic stress-strain relation obtained for the P355NL1 steel [39].

Maximum stresses	Mesh 5	Mesh 4	Mesh 1	Mesh 2	Mesh 3
σ_y (MPa)	1367.9	1420.1	1495.8	1605.3	1606.7
Dev. (%)	−14.79	−11.54	−6.82	–	0.09
σ_x (MPa)	347.4	370.3	363.9	354.2	354.1
Dev. (%)	−1.93	4.57	2.74	–	−0.03

Table 3. Maximum elastic stresses for distinct finite-element mesh densities for the P355NL1 steel (F_{max} = 1634.1 N, ρ^* = 30 Âμm).

All numerical simulations are carried out using the ANSYS® 12.0 code [30]. The six-noded plane element adopted in the finite-element analyses is the PLANE181 element available in the ANSYS® 12.0 code library. The contact and target elements used in the pin-loading simulation are, respectively, the CONTA172 and TARGE169 elements available in ANSYS® 12.0 code [30]. A parametric model is built using the APDL language. The surface of the holes is modelled as flexible using CONTA172 elements. The augmented Lagrangian contact algorithm is used. The associative von Mises (J2) yield criterion with multi-linear kinematic hardening is used to model the plastic behaviour. The multi-linear kinematic hardening uses the Besseling model, also called the sub-layer or overlay model, so that the Bauschinger effect is included. The plasticity model was fitted to the stabilized or half-life pseudo-stabilized cyclic curve of the materials. The von Mises yield criterion with multi-linear kinematic hardening is adopted to model the plastic behaviour.

The finite-element model is initially applied to perform elastic and elastoplastic stress analyses in order to allow us the comparison between the elastic and elastoplastic stress distributions as resulting from the Creager-Paris solution [24] and the multi-axial Neuber approach [15], respectively.

One important assumption of the UniGrow model consists in applying the compressive residual stresses that are computed ahead of the crack tip, in the crack faces, in a symmetric way with respect to the normal to crack face that passes through the crack tip. The resulting residual stress intensity factor K_r is computed using the weight function method [26], for each of the stress ratios covered by the testing program.

Figures 9 and **10** show the finite-element mesh of the CT geometry and the cyclic stress-strain curves adopted in the plasticity model of the P355NL1 steel, respectively [39]. The Ramberg-Osgood relation [42] is compared with the response of the finite-element model reproducing a uniaxial stress state in **Figure 10. Table 3** presents the maximum elastic stresses (σ_x and σ_y) ahead of the crack tip, resulting from distinct mesh densities illustrated in **Figure 11**. The mesh 2 is adopted for residual stress computation. Results presented in **Table 3** compare the mesh 2 with the other meshes.

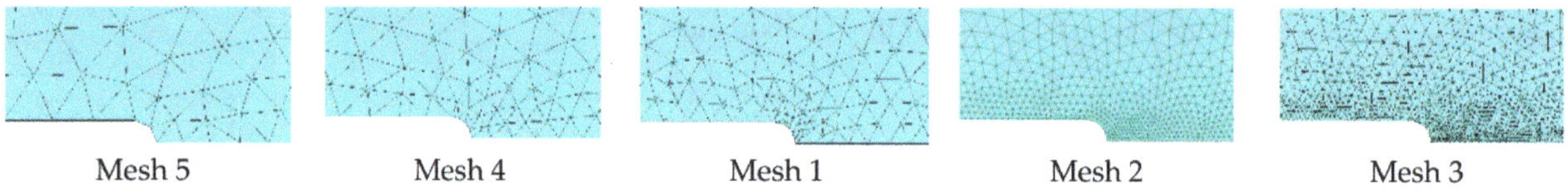

Figure 11. Finite-element meshes used in the convergence study performed for the CT specimen.

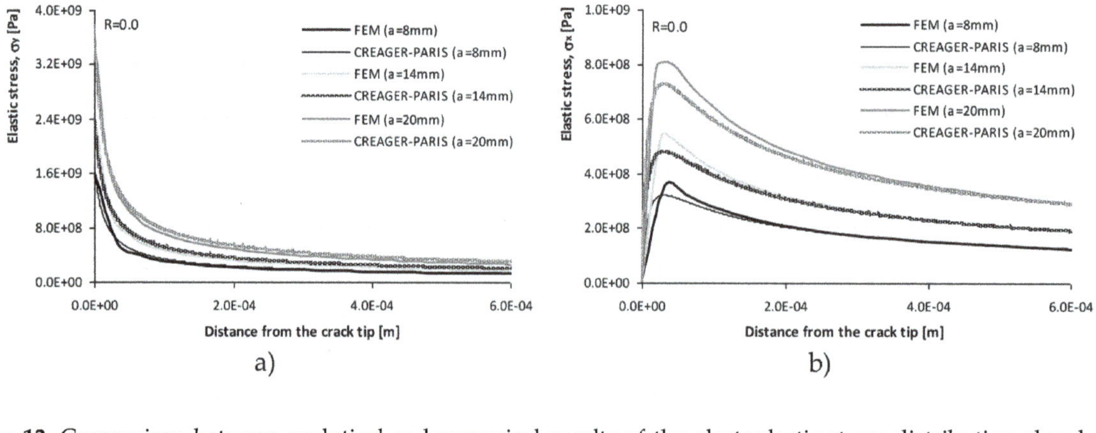

Figure 12. Comparison between analytical and numerical results of the elastoplastic stress distribution ahead of the crack tip and along the crack line (y = 0) for CT specimens made of the P355NL1 steel (F_{max} = 1643.1 N, ρ^* = 30 μm): (a) σ_y stress distribution and (b) σ_x stress distribution.

Figures 12 and **13** illustrate the elastic and elastoplastic stress distributions for the P355NL1 steel, respectively. In these figures, the numerical and analytical solutions for the CT specimens

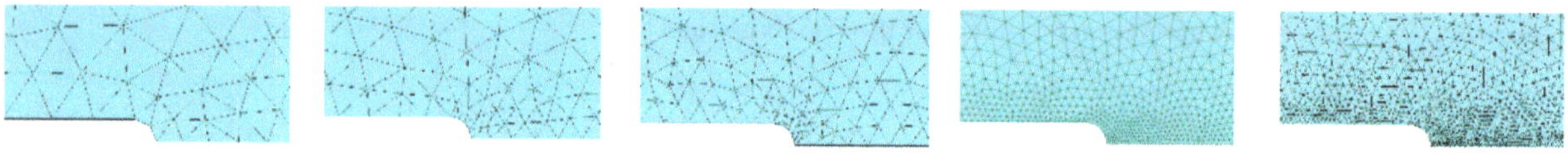

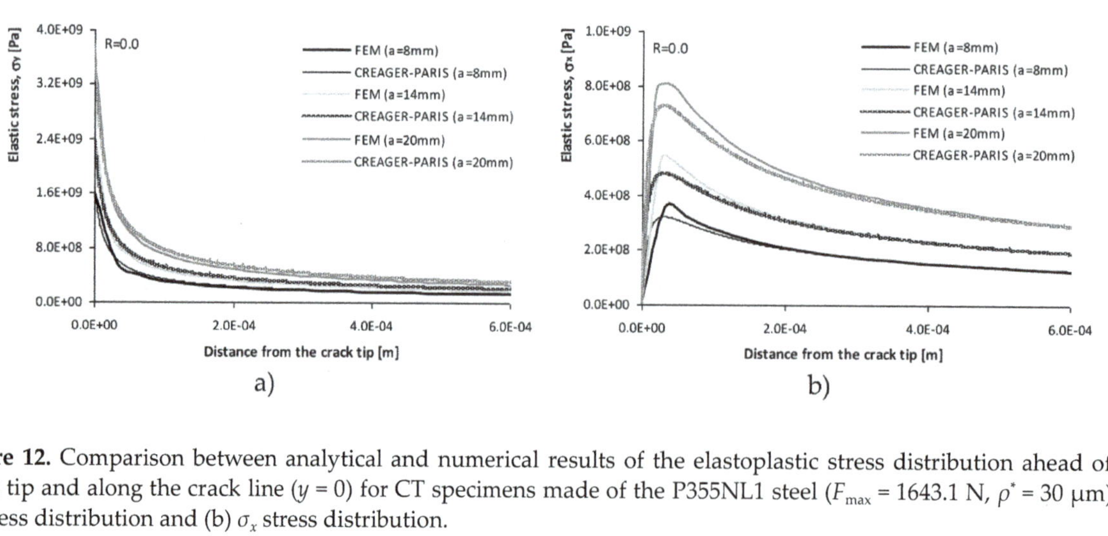

are computed for a crack tip radius, $\rho^* = 30$ μm, which is found to be the best value for the P355NL1 steel. This ρ^* parameter gives the best predictions of the fatigue crack growth rates, using the Morrow relation as referred in [6]. **Figure 14** shows the residual stress distributions for the P355NL1 steel for distinct crack sizes and stress *R*-ratios. The residual stress distributions are computed by means of the analytical and numerical solutions using an elementary material block size, $\rho^* = 30$ μm [39]. **Figure 15** illustrates the stress and strain fields along the *y*- (load) direction obtained for the CT specimens using the elastoplastic finite-element analysis and the properties of the P355NL1 steel, for a material's representative element $\rho^* = 30$ μm, a crack size $a = 14$ mm, a maximum load $F_{max} = 1634.1$ N and a stress *R*-ratio $R_\sigma = 0.0$. The stress and strain fields are shown at the end of the first loading reversal and at the end of the first unloading reversal. **Figure 16** presents the residual stress intensity factor range for an elementary material block size $\rho^* = 30$ μm [39] as a function of the applied stress intensity factor range obtained with the numerical analysis.

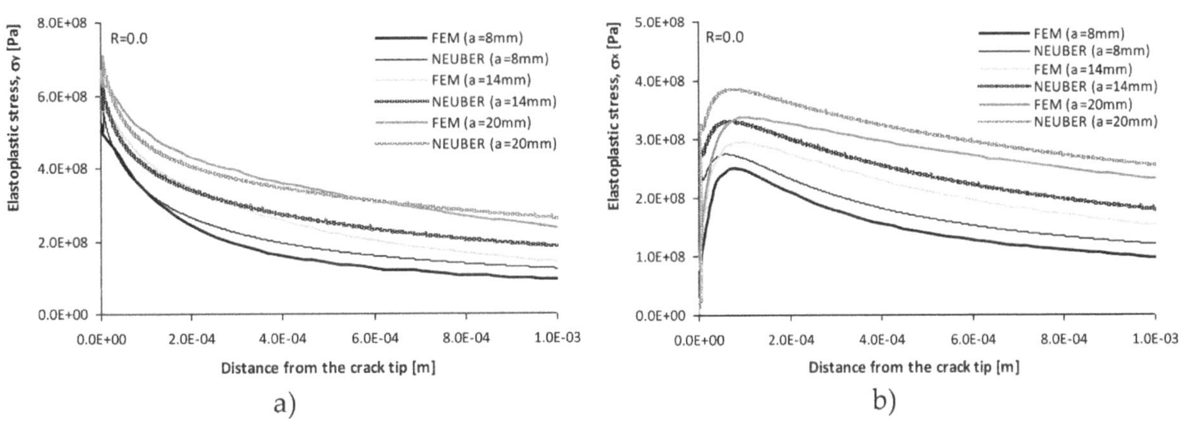

Figure 13. Comparison between analytical and numerical results of the elastoplastic stress distribution ahead of the crack tip and along the crack line ($y = 0$) for CT specimens made of the P355NL1 steel ($F_{max} = 1643.1$ N, $\rho^* = 30$ μm): (a) σ_y stress distribution and (b) σ_x stress distribution.

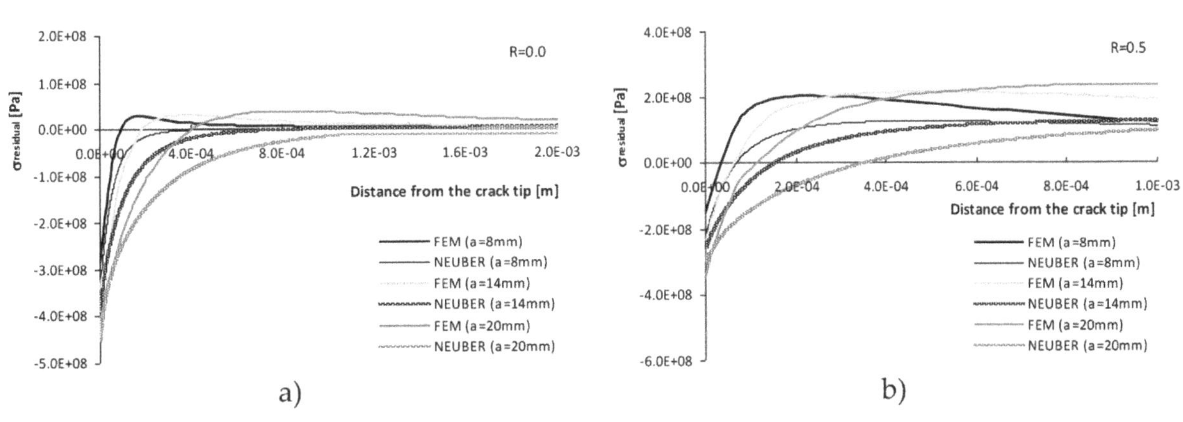

Figure 14. Comparison between analytical and numerical results of the residual stress distribution ahead of the crack tip and along the crack line ($y = 0$) for CT specimens made of the P355NL1 steel ($F_{max} = 1643.1$ N, $\rho^* = 30$ μm).(a) $R_\sigma = 0.0$ and (b) $R_\sigma = 0.5$.

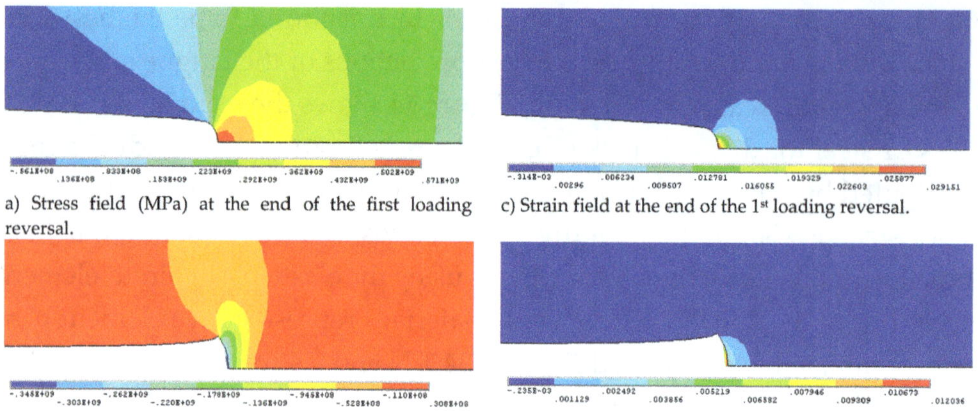

a) Stress field (MPa) at the end of the first loading reversal.

c) Strain field at the end of the 1st loading reversal.

b) Stress field (MPa) at the end of the first unloading reversal.

d) Strain field at the end of the 1st unloading reversal.

Figure 15. Stress and strain fields for the load direction, obtained for the CT geometries made of P355NL1 steel, resulting from elastoplastic finite-element analysis (F_{max} = 1634.1 N, ρ^* = 30 μm, a = 14 mm, R_σ = 0.0). (a) Stress field (MPa) at the end of the first loading reversal. (b) Stress field (MPa) at the end of the first unloading reversal. (c) Strain field at the end of the first loading reversal. (d) Strain field at the end of the first unloading reversal.

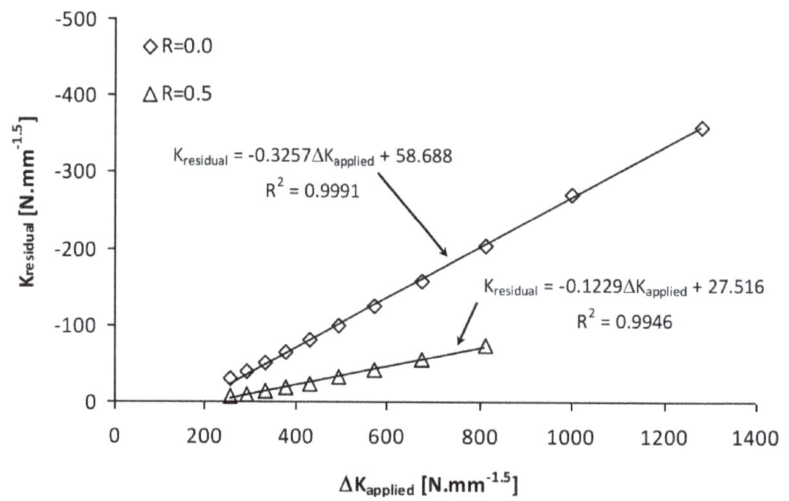

Figure 16. Residual stress intensity factor as a function of the applied stress intensity factor range.

The elastic stress distribution sustains a very good agreement between the analytical and numerical results, for several crack sizes, within a small distance from the crack tip. The analytical solutions lead to higher maximum absolute stresses than the elastoplastic FE analysis. The analytical solution does not show plastic zone behaviour, which is a clear limitation of the analytical approach. The increasing of the stress ratio affects the compressive residual stresses that decrease progressively with increasing stress ratio. This trend is responsible for increasing the effectiveness of the applied stress intensity range as a crack driving force. The extension of the compressive residual stresses increases for higher crack lengths. The numerical compressive stress region is always lower than the one obtained by the analytical analysis. Therefore, the numerical solution, for the residual stresses, is adopted in the crack propagation prediction, based on the UniGrow model. A very high linear correlation between the residual stress intensity factor and the applied stress range is verified for each stress R-ratio. This linear relation agrees with the proposition by Noroozi et al. [12] based on analytical analysis.

7.2. Results and discussion

Finally, in this section, the UniGrow model is applied to compute the fatigue crack propagation rates for the same fatigue crack propagation data described in Section 6. As detailed before, the residual stress intensity factor is calculated using a compressive residual stress distribution computed from numerical analysis and the weight function technique [26]. The strain range and maximum stress, needed by the probabilistic strain-life or SWT-life fields, are evaluated using an analytical calculation applied over the first material element ahead of the crack tip, keeping the original structure of the UniGrow model. Average strain and stress values, along the first elementary material block, are used instead of peak values. The analytical solution produces reliable results at the crack tip notch root as verified in previous section. The original structure of the UniGrow model presents some advantages: (i) provides a direct correspondence with fracture mechanics based analyses, which facilitates the physical understanding of the process, (ii) allows close-form solutions for fatigue crack propagation laws to be achieved in the same format as that of existing fracture mechanics approaches and (iii) requires inexpensive computations.

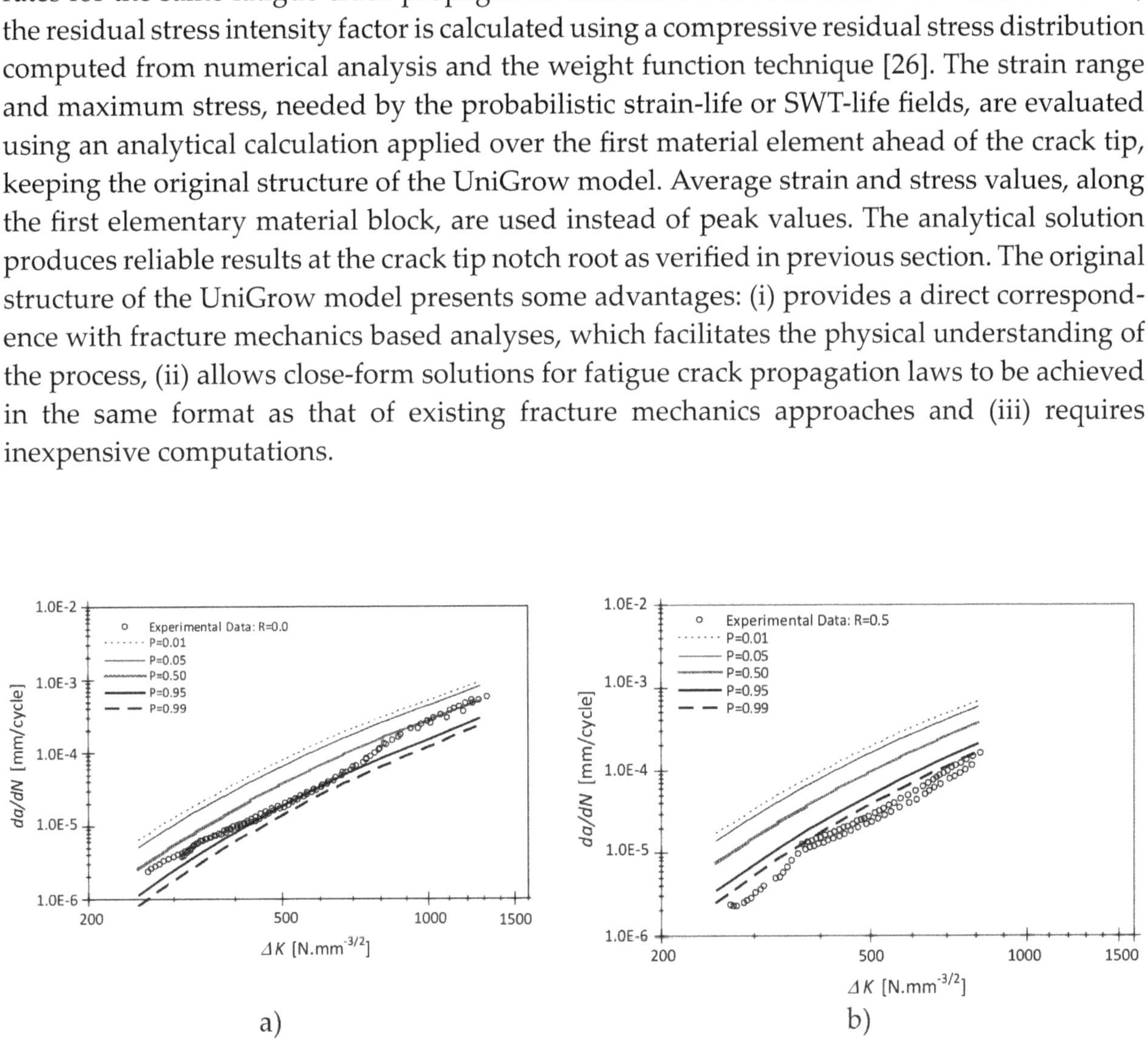

Figure 17. Probabilistic prediction of the fatigue crack propagation based on the p-SWT-N field for the P355NL1 steel ($\rho^* = 3 \times 10^{-5}$ m): (a) $R_\sigma = 0.0$ and (b) $R_\sigma = 0.5$.

The probabilistic ε_a-N and SWT-N fields are used to derive the probabilistic fatigue crack propagation fields (p-da/dN-ΔK-R fields). For each case, an independent identification of the elementary material block size, ρ^*, is performed. **Figure 17** shows the probabilistic fatigue crack propagation fields obtained for the P355NL1 steel, using the p-SWT-N material fields. **Figure 18** illustrates the probabilistic fatigue crack propagation fields predicted for the P355NL1 steel, resulting from the p-ε_a-N material fields.

Elementary material block size of 3×10^{-5} m is found adequate for the P355NL1 steel, resulting from the application of the p-ε-N fields. The elementary material block size found for the P355NL1 steel is higher than those proposed by Noroozi et al. [13] for the 4340 steel ($\rho^* = 2 \times 10^{-6}$ m). So far, there are no conclusive studies relating the micro-structural grain sizes with the elementary material block sizes. Nevertheless, it can be anticipated that ρ^* can be indirectly dependent on the micro-structural features of the analysed material (e.g. grain size) but it cannot be uniquely associated with any specific micro-structural particle size. Therefore, the material grain size may be considered just one of the possible micro-structural features affecting the definition of ρ^*.

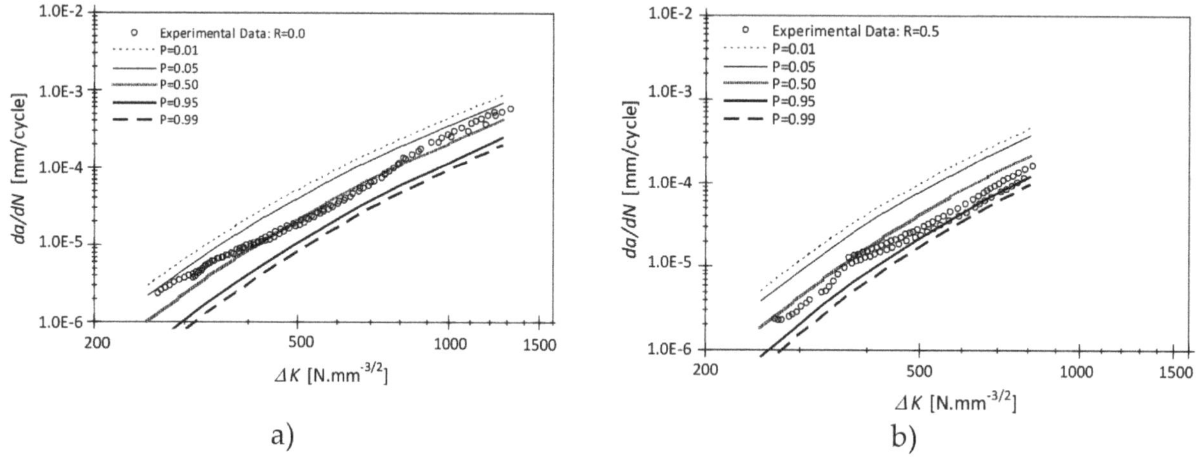

Figure 18. Probabilistic prediction of the fatigue crack propagation based on the p-ε_a-N field for the P355NL1 steel ($\rho^* = 3 \times 10^{-5}$ m): (a) $R_\sigma = 0.0$ and (b) $R_\sigma = 0.5$.

Concerning the probabilistic fatigue crack propagation rate fields of the P355NL1 steel, it is clear that the use of the p-SWT-N model overestimates the effects of the stress R-ratio. Using the probabilistic SWT-N model, the stress ratio effects are accounted twice, through the residual stress intensity factor (compressive residual stresses effects) and through the mean stress of the cycle. Nevertheless, the number of cycles required to crack the material element fall within the very low- to low-cycle fatigue regimes at which a rapid cyclic mean stress relaxation is verified. However, the multi-linear kinematic hardening model used in the investigation is not capable of simulating the cyclic mean stress relaxation. The predictions based on the p-ε_a-N model are satisfactory. First, they define lower and upper bounds for the available experimental data. Secondly, the major influence of the stress ratio is verified when it changes from 0 to 0.5. In the damage computations presented in the chapter, the prior loading history on crack tip elements is not considered. Preliminary calculations performed considering the prior loading history on a set of elements ahead of the crack tip, forming a process zone, show only a marginal influence on da/dN predictions, mainly in the propagation regimes I and II. Nonetheless, the computational costs associated to simultaneous damaged elements increases very significantly.

8. Conclusions

An assessment of the UniGrow model is presented in this chapter, based on available experimental data for the P355NL1 steel under consideration. The UniGrow model is also extended to predict probabilistic fatigue crack propagation fields, replacing the deterministic SWT-N relation proposed in the original UniGrow model by the p-SWT-N or p-ε_a-N fields. In the present chapter, the p-SWT-N field is first proposed as a generalization of the p-ε_a-N field, in order to take into account the mean stress effects. Both p-SWT-N and p-ε_a-N fields led to satisfactory correlations of the experimental data available for the P355NL1 steel under investigation.

The multi-axial analytical Neuber elastoplastic analysis proposed in the UniGrow model to compute the residual stress distribution is assessed using elastoplastic finite-element analysis. Inconsistent compressive residual stress distributions, mainly for $R_\sigma = 0$, are found using this approach. The multi-axial Neuber model does not take into account stress redistribution due to yielding and therefore does not provide the dimensions of the plastic zone. The residual compressive stress intensity factor, computed with the compressive residual stress field from the finite-element analysis, exhibits a linear relation with the applied stress intensity factor range, which confirms the typical trend documented in literature of linear increase of the residual stress intensity factor with the applied stress intensity factor range, for a specific stress R-ratio.

The p-da/dN-ΔK-R fields predicted for the P355NL1 steel, based on material p-ε_a-N field, produce satisfactory results, since this material shows crack propagation rates with relatively small sensitivity to the stress R-ratio. The elementary material block size for the P355NL1 steel is found to be about one order of magnitude higher than the value proposed by Noroozi et al. [14] for the 4340 steel. The ρ^* can be indirectly dependent on the micro-structural features of the analysed material (e.g. grain size) but it cannot be uniquely associated with any specific micro-structural particle size.

Acknowledgements

The authors acknowledge the Portuguese Science Foundation (FCT) for the financial support through the postdoctoral grant SFRH/BPD/107825/2015. The authors gratefully acknowledge the funding of SciTech: Science and Technology for Competitive and Sustainable Industries, R&D project co-financed by Programa Operacional Regional do Norte (NORTE2020), through Fundo Europeu de Desenvolvimento Regional (FEDER). Dr. Moreira acknowledges Programa Operacional Potencial Humano (POPH), Quadro de Referência Estratégico Nacional (QREN)-Tipologia 4.2 promotion of scientific employment funded by the European Social Fund (ESF) and Ministério da Ciência, Tecnologia e Ensino Superior (MCTES).

Author details

José António Fonseca De Oliveira Correia[1*], Abílio M.P. De Jesus[1], Pedro M.G.P. Moreira[1], Rui A.B. Calçada[1] and Alfonso Fernández-Canteli[2]

*Address all correspondence to: jacorreia@inegi.up.pt

1 Faculty of Engineering, University of Porto, Porto, Portugal

2 Department of Construction and Manufacturing Engineering, University of Oviedo, Gijón, Spain

References

[1] Schütz W. A history of fatigue. Eng. Fract. Mech., 1996; 54: 263–300.

[2] Paris PC, Gomez M, Anderson WE. A rational analytic theory of fatigue. Trend Engineering, 1961; 13: 9–14.

[3] Beden SM, Abdullah S, Ariffin AK. Review of fatigue crack propagation models for metallic components. European Journal of Scientific Research, 2009; 28: 364–397.

[4] Coffin LF. A study of the effects of the cyclic thermal stresses on a ductile metal. Trans ASME, 1954;76:931–50.

[5] Manson SS. Behaviour of materials under conditions of thermal stress, NACA TN-1170, National Advisory Committee for Aeronautics, 1954; Report 1170, pp. 591–630. http://naca.central.cranfield.ac.uk/reports/1954/naca-report-1170.pdf

[6] Morrow JD. Cyclic plastic strain energy and fatigue of metals. Int Frict Damp Cyclic Plast ASTM STP., 1965; 378:45–87.

[7] Smith KN, Watson P, Topper TH. A stress-strain function for the fatigue of metals. Journal of Materials, 1970; 5(4): 767–778.

[8] Shang D-G, Wang D-K, Li M, Yao W-X. Local stress–strain field intensity approach to fatigue life prediction under random cyclic loading. International Journal of Fatigue, 2001; 23: 903–910.

[9] Glinka G. A notch stress-strain analysis approach to fatigue crack growth. Engineering Fracture Mechanics, 1985; 21: 245–261.

[10] Peeker E, Niemi E. Fatigue crack propagation model based on a local strain approach. Journal of Constructional Steel Research, 1999; 49: 139–155.

[11] Noroozi AH, Glinka G, Lambert S. A two parameter driving force for fatigue crack growth analysis. International Journal of Fatigue, 2005; 27: 1277–1296.

[12] Noroozi AH, Glinka G, Lambert S. A study of the stress ratio effects on fatigue crack growth using the unified two-parameter fatigue crack growth driving force. International Journal of Fatigue, 2007; 29:1616–1633.

[13] Noroozi AH, Glinka G, Lambert S. Prediction of fatigue crack growth under constant amplitude loading and a single overload based on elasto-plastic crack tip stresses and strains. Engineering Fracture Mechanics, 2008; 75: 188–206.

[14] Hurley PJ, Evans WJ. A methodology for predicting fatigue crack propagation rates in titanium based on damage accumulation. Scripta Materialia, 2007; 56: 681–684.

[15] Neuber H. Theory of stress concentration for shear-strained prismatic bodies with arbitrary nonlinear stress–strain law. Transactions of ASME Journal of Applied Mechanics, 1961; 28: 544–551.

[16] Moftakhar A, Buczynski A, Glinka G. Calculation of elasto-plastic strains and stresses in notches under multiaxial loading. International Journal of Fracture, 1995; 70: 357–373.

[17] Reinhard W, Moftakhar A, Glinka G. An efficient method for calculating multiaxial elasto-plastic notch tip strains and stresses under proportional loading. Fatigue and Fracture Mechanics, ASTM STP 1296, Piascik RS, Newman JC, Dowling NE, Eds., American Society for Testing and Materials, 1997; 27: 613–629.

[18] Mikheevskiy S, Glinka G. Elastic–plastic fatigue crack growth analysis under variable amplitude loading spectra. International Journal of Fatigue, 2009; 31: 1828–1836.

[19] De Jesus AMP, Silva ALL, Figueiredo MV, Correia JAFO, Ribeiro AS, Fernandes AA. Strain-life and crack propagation fatigue data from several Portuguese old metallic riveted bridges. Engineering Failure Analysis, 2010; 17: 1495–1499.

[20] De Jesus AMP, Matos R, Fontoura BFC, Rebelo C, Simões da Silva L, Veljkovic M. A comparison of the fatigue behaviour between S355 and S690 steel grades. Journal of Constructional Steel Research, 2012; 79: 140–150.

[21] De Jesus AMP, Ribeiro AS, Fernandes AA. Influence of the submerged arc welding in the mechanical behaviour of the P355NL1 steel—Part II: Analysis of the Low/High Cycle Fatigue Behaviours. J. Mater. Sci., 2007; 42: 5973–5981.

[22] Castillo E, Fernández-Canteli A. A Unified Statistical Methodology for Modeling Fatigue Damage. Springer, Netherlands, 2009; 232.

[23] Basquin OH. The exponential law of endurance tests. Proc. Annual Meeting American Society for Testing Materials, 1910; 10: 625–630.

[24] Creager M, Paris PC. Elastic field equations for blunt cracks with reference to stress corrosion cracking. International Journal of Fracture Mechanics, 1967; 3: 247–252.

[25] Molski K, Glinka G. A method of elastic-plastic stress and strain calculation at a notch root. Materials Science and Engineering, 1981; 50: 93–100.

[26] Glinka G. Development of weight functions and computer integration procedures for calculating stress intensity factors around cracks subjected to complex stress fields. Progress Report No. 1: Stress and Fatigue-Fracture Design, Petersburg, Ontario, Canada, 1996; 108. http://www.afgrow.net/downloads/documents/SaFFD_1m.pdf

[27] Sadananda K, Vasudevan AK, Kang IW. Effect of superimposed monotonic fracture modes on the ΔK and Kmax parameters of fatigue crack propagation. Acta Materialia, 2003; 51(22): 3399–3414.

[28] Kajawski D. A new (ΔK+Kmax)0.5 driving force parameter for crack growth in aluminium alloys. International Journal of Fatigue, 2001; 23(8): 733–740.

[29] Schiffner K. Overlay models for structural analysis under cyclic loading. Computers and Structures, 1995; 56(2/3): 321–328.

[30] SAS, Version 12.0, ANSYS, Swanson Analysis Systems, Inc., Houston, 2011.

[31] Simo JC, Taylor RL. Consistent tangent operators for rate-independent elastoplasticity. Computer Methods in Applied Mechanics and Engineering, 1985; 48: 101–118.

[32] Castillo E, Galambos J. Lifetime regression models based on a functional equation of physical nature. Journal of Applied Probability, 1987; 24: 160–169.

[33] Castillo E, Fernández-Canteli A, Hadi AS, López-Anelle M. A Fatigue Model with Local Sensitivity Analysis, Fatigue and Fracture of Engineering Material and Structure, 2006; 30: 149–168.

[34] Correia JAFO, De Jesus AMP, Fernández-Canteli A. A procedure to derive probabilistic fatigue crack propagation data. International Journal of Structural Integrity, 2012; 3(2): 158–183.

[35] Correia JAFO, De Jesus AMP, Fernández-Canteli A. Local unified model for fatigue crack initiation and propagation: application to a notched geometry. Engineering Structures, 2013; 52: 394–407.

[36] Correia JAFO. An integral probabilistic approach for fatigue lifetime prediction of mechanical and structural components. Ph.D. Thesis, University of Porto, 2014; p. 382.

[37] Correia JAFO, De Jesus AMP, Fernández-Canteli A, Calçada RAB. Modelling probabilistic fatigue crack propagation rates for a mild structural steel. Frattura ed Integrita Strutturale, 2015; 31: 80–96.

[38] Hafezi MH, Abdullah NN, Correia JAFO, De Jesus AMP. An assessment of a strain-life approach for fatigue crack growth. International Journal of Structural Integrity, 2012; 3(2): 344–376.

[39] De Jesus AMP, Correia JAFO. Critical assessment of a local strain-based fatigue crack growth model using experimental data available for the P355NL1 steel. Journal of Pressure Vessel Technology, 2013; 135(1): 011404-1–0114041-9.

[40] Correia JAFO, De Jesus AMP, Moreira PMGP, Tavares PJ. Crack closure effects on fatigue crack propagation rates: application of a proposed theoretical model. Advances in Materials Science and Engineering, 2016: (2016), Article ID 3026745, 11. http://dx.doi.org/10.1155/2016/3026745

[41] ASTM – American Society for Testing and Materials. ASTM E606–92: Standard practice for strain controlled fatigue testing. In: Annual Book of ASTM Standards, Part 10; 1998; pp. 557–71. http://dx.doi.org/10.1520/E0606-92R04E01

[42] Ramberg W, Osgood WR. Description of the stress-strain curves by the three parameters. NACA TN-902, National Advisory Committee for Aeronautics, Washington, 1943; 29. http://www.apesolutions.com/spd/public/NACA-TN902.pdf

[43] ASTM—American Society for Testing and Materials. ASTM E647: standard test method for measurement of fatigue crack growth rates. In: Annual book of ASTM Standards, vol. 03.01. West Conshohocken, PA: ASTM—American Society for Testing and Materials; 2000; pp. 591–630.

[21] Hanson ME, Anderson GD, Shaffer RJ. Theoretical and Experimental Research on Hydraulic Fracturing. California University, Livermore (USA). Lawrence Livermore Lab. 1978.

[22] Shaffer RJ, Hanson ME, Anderson GD. Theoretical and Experimental Analyses of the Hydraulic Fracturing Process, Lawrence Livermore National Laboratory, Livermore, CA Report UCRL-83419 (1979): also presented at the ASME Energy Technology Conference, New Orleans. 1980.

[23] Daneshy AA. Hydraulic Fracture Propagation in the Presence of Planes of Weakness. SPE 4852. 1974.

[24] Blanton TL. An Experimental Study of Interaction between Hydraulically Induced and Pre-Existing Fractures. SPE 10847. 1982.

[25] Hopkins CW, Frantz Jr JH, Hill DG, Zamora F. Estimating Fracture Geometry in the Naturally Fractured Antrim Shale. In SPE Annual Technical Conference and Exhibition. Society of Petroleum Engineers. 1995.

[26] Beugelsdijk LJ, De Pater CJ, Sato K. Experimental Hydraulic Fracture Propagation in a Multi-Fractured Medium. In SPE Asia Pacific Conference on Integrated Modelling for Asset Management. Society of Petroleum Engineers. 2000.

[27] Zhang X, Jeffrey RG, Bunger AP, Thiercelin M. Initiation and growth of a hydraulic fracture from a circular wellbore. International Journal of Rock Mechanics and Mining Sciences. 2011;48(6):984–95.

[28] Carrier B, Granet S. Numerical modeling of hydraulic fracture problem in permeable medium using cohesive zone model. Engineering Fracture Mechanics. 2012;79:312–28.

[29] Kresse O, Weng X, Chuprakov D, Prioul R, Cohen C. Effect of Flow Rate and Viscosity on Complex Fracture Development in UFM Model, Effective and Sustainable Hydraulic Fracturing, Dr. Rob Jeffrey (Ed.). InTech, 2013;10:56406.

[30] Last NC, Harper TR. Response of fractured rock subject to fluid injection part I. Development of a numerical model. Tectonophysics. 1990;172(1):1–31.

[31] Rahman MK, Hossain MM, Rahman SS. A shear-dilation-based model for evaluation of hydraulically stimulated naturally fractured reservoirs. International Journal for Numerical and Analytical Methods in Geomechanics. 2002;26(5):469–497.

[32] Mohammadi S. Extended Finite Element Method: For Fracture Analysis of Structures. 2008.

[33] Renshaw CE, Pollard DD. Are large differential stresses required for straight fracture propagation paths? Journal of Structural Geology. 1994;16(6):817–822.

[34] Pirayehgar A, Dusseault MB. The Stress Ratio Effect on Hydraulic Fracturing in the Presence of Natural Fractures. The 48th US Rock Mechanics/Geomechanics Symposium, Minneapolis. 2014.

[35] Bandis S, Lumsden AC, Barton NR. An experimental studies of scale effects on the shear behaviour of rock joints. International Journal of Rock Mechanics and Mining Sciences & Geomechanics Abstracts. 1981;18(1):1–21.

[36] Barton N, Bandis S, Bakhtar K. Strength, deformation and conductivity coupling of rock joints. International Journal of Rock Mechanics and Mining Sciences & Geomechanics Abstracts. 1985;22(3):121–140.

8

Toughening of Low-Alloy Steel by Ultrafine-Grained Structure (Development of Fracture Control from Microstructure Design)

Tadanobu Inoue

Abstract

Microstructural design for improving the strength–toughness balance was studied in low-alloy steel. Medium-carbon steel bars with microstructures of two types such as an ultrafine elongated grain (UFEG) structure and an ultrafine equiaxed grain (UFG) structure were fabricated by multi-pass warm caliber rolling and subsequent annealing. Conventionally, quenched and tempered steel with a martensitic structure and low-carbon steel with a ferrite–pearlite structure were also prepared. The tensile and three-point bending tests were conducted for all samples. In particular, the fracture behavior after the bending test was observed in detail and investigated, including the effect of microstructure features. As a result, the strength–toughness balance of the UFEG steel was excellent compared with that of all other steels. The present results provide useful guidelines for designing microstructure to improve the strength–toughness balance in metallic materials.

Keywords: iron and steel, strength–toughness balance, grain refinement, anisotropic properties, crack propagation

1. Introduction

Strong and tough materials are always needed in order to improve safety and performance and to reduce weight in transportation and heavy machinery. It is not difficult to achieve high strength with the help of the current materials science and technology. However, as shown in **Figure 1**, strength and toughness in materials have a strong relation, and toughness decreases with increasing strength, that is, strengthening of materials deteriorate toughness [1]. Few structural metallic materials are limited by their strength; rather, they are limited by their fracture toughness. Unfortunately, most of materials research is still focused on the quest for high strength.

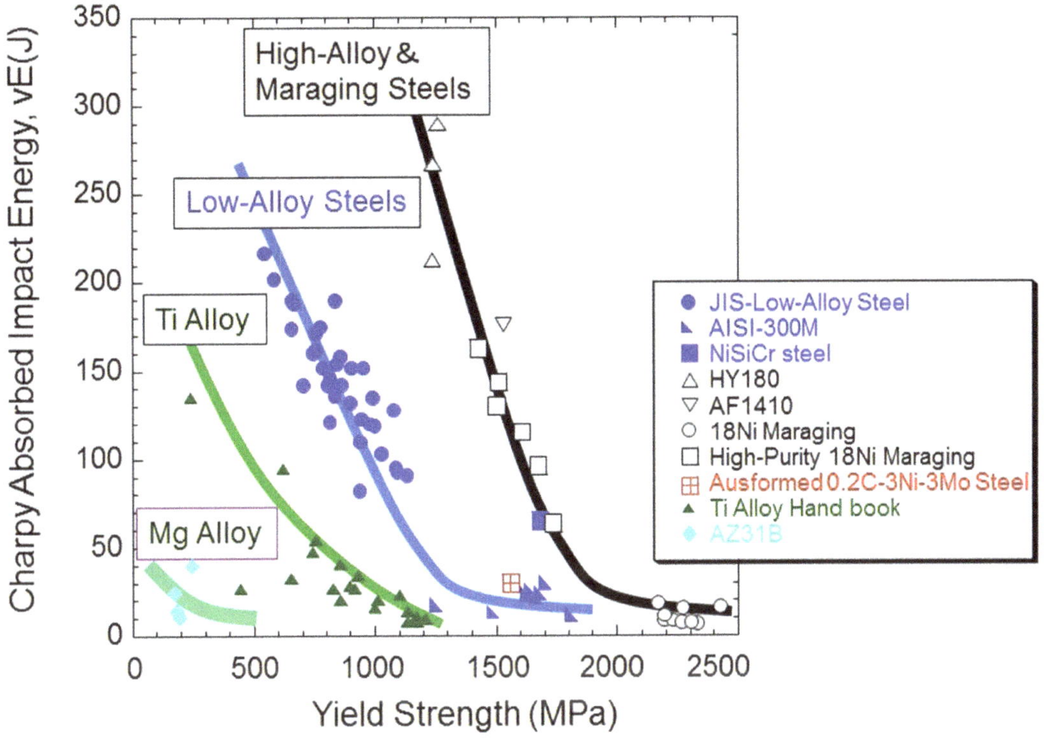

Figure 1. Correlation between yield strength and Charpy full-size V-notch impact energy at ambient temperature in structural metallic materials.

It was believed that refinement of crystal grains is an effective method for developing strength and toughness in metallic materials without the addition of alloying elements; hence, ultrafine-grained (UFG) materials are very attractive in materials science [2–4]. However, although a refinement of crystal grain until submicronsized or nanosized grains are obtained normally leads to higher strength on the basis of the Hall–Petch relation [4, 5], it does not always lead to the improvement of toughness. Many results [5–9] indicated that the ductile-to-brittle transition (DBT) temperature of bcc steels accompanied by a change in the fracture mechanism from void coalescence to cleavage was improved by grain refinement, although the upper-shelf energy became lower due to a decrease in ductility by strength enhancement. However, for such UFG steels produced through severe plastic deformation processes, the relaxation of the triaxial tension toward a state of biaxial tension resulting from the presence of separations of the crack-divider type (see **Figure 2a**) appears to be one of the reasons [10, 11]. The fracture of this type has often been observed elsewhere, such as in rolled steel pipes and plates and Al–Li alloys, and it leads to the improvement of toughness at low temperatures because of relaxation of the triaxial tension stresses generated by the localized plastic constraint at the crack tip. Bourell [12] showed that for low-carbon steel, the separations markedly appeared with increasing strain and decreasing temperature due to the development of {1 0 0}<1 1 0>

texture in a rolled sheet. As a result, the upper shelf energy in the sheet decreased and the lower shelf energy increased. The microstructure in Al–Li alloys used for the main fuel tank of the space shuttle was pancake-shaped grains elongated in the rolling direction, and the alloys had high toughness at liquid helium temperatures [13, 14]. This is due to delamination toughening of crack-divider type, which is caused by the weak sites, such as the segregation to grain boundaries and coarse Fe–Cu- or Mg–Cu-rich constituent particles.

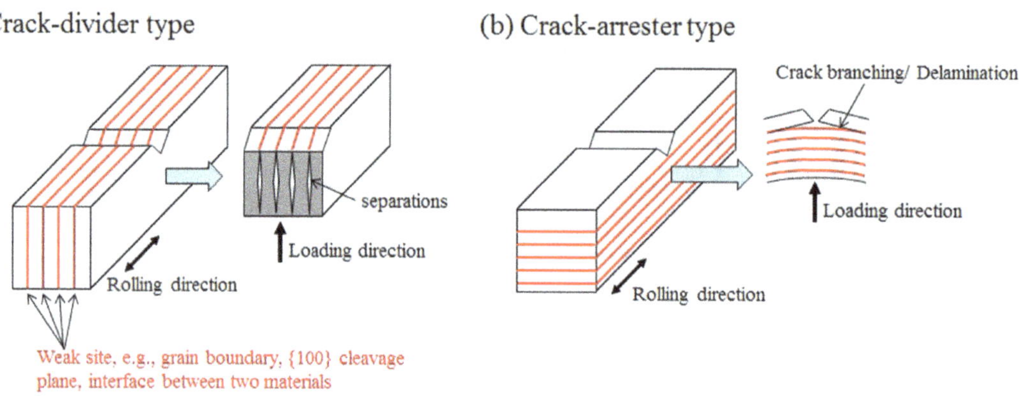

Figure 2. Fracture types of delamination/splitting in anisotropic material containing specific planes of weakness in one direction.

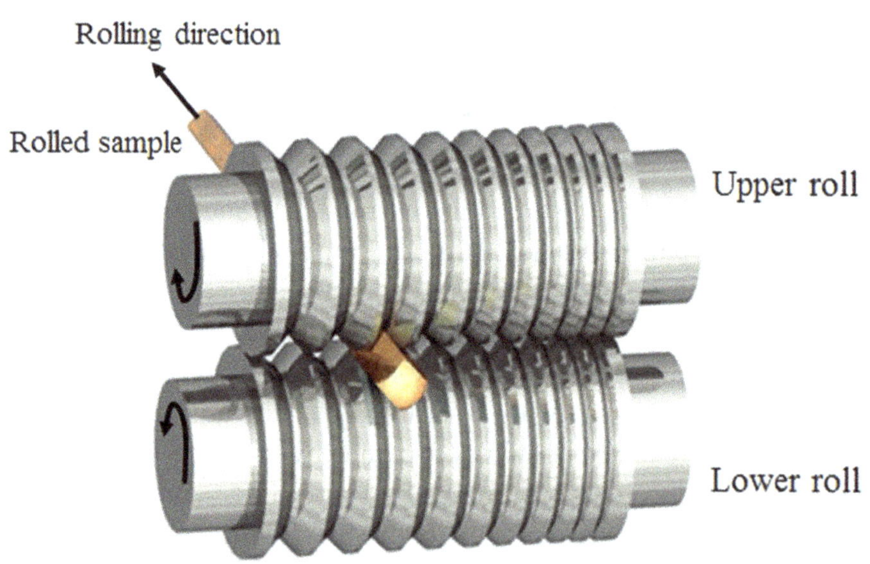

Figure 3. Schematic drawing of the caliber rolling with square grooves.

In our studies, warm caliber rolling (WCR) shown in **Figure 3** has been proposed as a deformation process to fabricate bulk UFG bars with a length of over 1000 mm, and this process has already been reported to be effective for producing various metallic materials with UFG structures. In these studies, a medium-carbon low-alloy steel of 1800 MPa tensile strength level with ultrafine elongated grain (UFEG) structures showed an inverse temperature dependence of toughness in the Charpy V-notch impact tests [2]. In addition, in a low-carbon low-alloy steel of 800 MPa strength with UFEG structures, the strength increased with the decrease in thickness of the elongated ferrite grain structures, and the upper-shelf energy was maintained until a low temperature was reached [15]. These toughening are attributed to mechanisms of the crack arrester type (see **Figure 2b**), which arrests the propagation of main cracks. However, in case of impact toughness such as the Charpy test, it is difficult to clarify the fracture behavior because the occurrence, propagation, and linkage of microcracks are caused almost simultaneously during the impact loads. And, the property of toughness is sensitive to not only the size of crystal grains but also their orientation and shape. It is of interest to systematically study the strength–toughness balance on the static fracture toughness in UFG steel, the effect of microstructure features with respect to its balance and fracture behaviors including crack propagation.

In this chapter, medium-carbon steel bar with UFEG structure is fabricated by multi-pass WCR, and steel bar with UFG structure is created by subsequent annealing. Also, two conventional steels, medium-carbon steel with a martensitic structure and low-carbon steel with a ferrite–pearlite structure are prepared. First, a three-point bending test is conducted at ambient temperature, the fracture behaviors of each sample are compared, including bending load–displacement curves and microstructure observation of crack propagation from the initial notch. Next, the three-point bending test is conducted at a temperature range from 200 to −196°C, and the effect of the heterogeneous microstructure is examined based on the strength–toughness balance as well as fracture behavior.

2. Experimental procedure

2.1. Specimen preparation

A chemical composition of medium-carbon steel used in this study is Fe–0.39C–2.01Si–1.02Cr–1.0Mo–0.21Mn–0.004Al–0.0022N–0.001O–<0.001P–<0.001S (all in mass pct). A 100-kg ingot was prepared by vacuum melting and casting, homogenized at 1200°C, and then hot rolled to a 40-mm-thick plate. A block of 40 × 120 L mm^3 was cut out of the plate, solution-treated at 1200°C for 3600 s and then hot rolled into a bar of about 31 mm^2, followed by water quenching. The quenched bar was soaked at 500°C for 1 h, subjected to a caliber rolling simulator [16] without any lubricant, and then air cooled. Eventually, a 14.3 × 930 L mm^3 rolled bar was fabricated. Hereafter, this sample is designated as *the TF sample*. To clarify the shape effect of crystal grains on the strength–toughness balance, the TF sample was annealed at 700°C for 1 h [2]. Hereafter, this sample is designated as *the TFA sample*. For comparison, conventional steels of *the QT sample* and *the SM490 sample* were prepared. To fabricate the QT sample with a martensitic structure, a bar was solution-treated at 950°C for 0.5 h, followed by oil quenching, tempered at 500°C for 1 h and then water cooled. The SM490 sample with a ferrite–pearlite structure in a chemical composition of 0.15C–0.3Si–1.5Mn was heated to 900°C and held for 1 h, followed by air cooling [15]. The principal axes of the rolled bar in this study are defined as shown in **Figure 4a**.

2.2. Microstructure and mechanical properties

The appearance of specimens after the three-point bending test was observed through a digital camera, and fracture surfaces were observed through a scanning electron microscope (SEM) operated at 15 kV. The microstructures at the central parts in the rolled bars were observed using the electron backscattered diffraction (EBSD) method in an SEM equipped with a field emission gun.

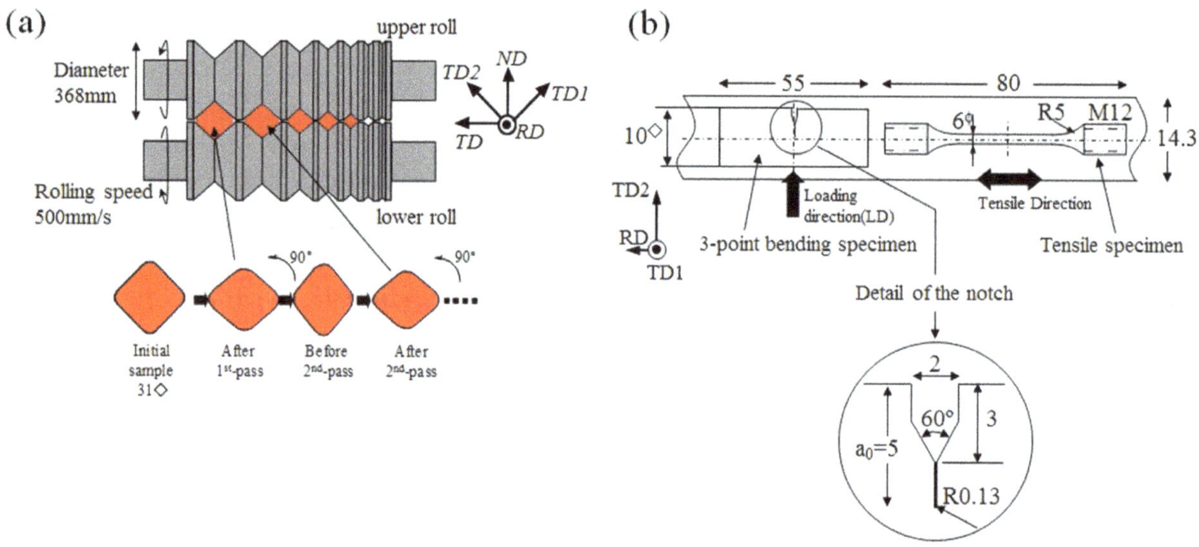

Figure 4. (a) Schematic drawing of the caliber rolling used in the present study and (b) position relation between a rolled bar in a 14.3 mm², a three-point bending specimen in a 10 mm², and tensile specimen in round type.

All mechanical test samples were taken from the center in the rolled bars, as shown in **Figure 4b**. Tensile tests were conducted with a crosshead speed of 0.85 mm/min (initial strain rate 5×10^{-4}/s) using specimens with a round cross section of 6 mm and a gage length of 30 mm. To prepare single-edge bend specimens in a three-point bending test, rectangular bars of 10W × 10B ×55L mm³ were first machined along the RD, and then a notch with a depth of a_0 = 5 mm and a root radius of 0.13 μm as illustrated in **Figure 4b** was introduced using electro-discharge machining with fine wire 0.2 mm in diameter. A three-point bending test with a support distance of 40 mm at a crosshead speed of 0.5 mm/min was conducted at a temperature range of 23 and −196°C in the QT, TFA, and SM490 samples and a range of 200 and −196°C in the TF sample. The test was terminated when the specimen fractured completely or the displacement reached 10 mm, which corresponds to the specimen thickness. Nonlinear fracture mechanics methods on the basis of ASTM Standard E1820-01 were used to evaluate the fracture toughness. The apparent fracture energy, *J* (KJ/m²), was calculated through the following formula:

$$J = 2\frac{A_{pl}}{Bb} = \frac{2}{Bb}\int_{0}^{u_a} P\,du$$

where A_{pl} is the area under the *P-u* curve, *B* is the specimen thickness, *b* is the ligament length (W-a_0), and u_a denotes the displacement at which the test was terminated. All the values presented are an average of two to four measurements.

3. Results

3.1. Microstructure evolution

The orientation maps for the QT, TF, and TFA samples and an SEM image for the SM490 sample are shown in **Figure 5**. The QT sample has a martensitic structure of a random crystallographic orientation, as shown in **Figure 5a**. On the other hand, the TF and TFA samples have a strong α-fiber texture parallel to the RD, that is, RD//<1 1 0>, as shown in **Figure 5b** and **c**. There are no significant differences in texture between the TF and TFA samples despite the annealing treatment. In the TF sample, the average of transverse linear interceptions for the elongated grains with misorientation angles of more than 10° was 310 nm. Furthermore, spheroidal nanosized carbide particles of 50 nm and below were dispersed in the elongated grain matrix [17]. In the TFA sample which resulted in the development of a granular grain structure, the transverse size of the ferrite grains increased to 660 nm. Relatively large carbide particles (200–300 nm) appeared to exist on the grain boundaries, while finer carbide particles were dispersed in the grain matrix. On the other hand, the distributions of Kernel average misorientation, KAM, which has a strong correlation with the dislocation densities within the cell interior of the deformed structures, were analyzed from electron backscatter patterns (EBSP) maps shown in **Figure 5b** and **c**. The average value of KAM for the TF and TFA samples was 0.65° and 0.38°,

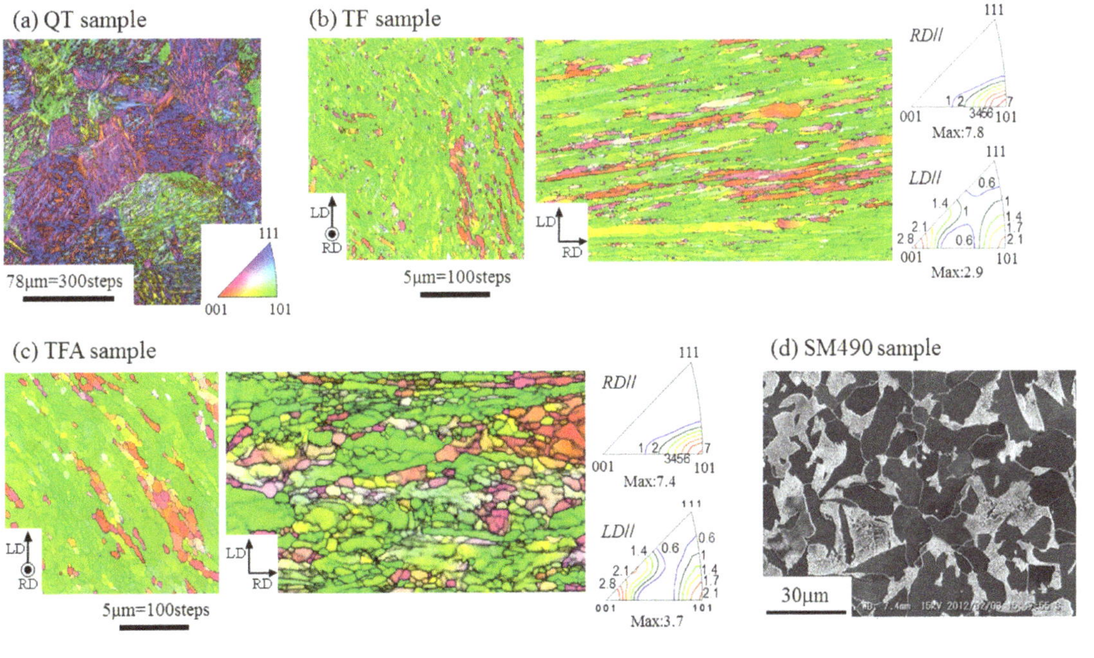

Figure 5. Orientation maps on cross-sectional plane for (a) QT, (b) TF, and (c) TFA samples by EBSD analysis and (d) an SEM image of SM490 sample.

respectively. This indicated that the dislocation density for the TF sample was somewhat larger than that for the TFA one. In the SM490 sample, the average size of the ferrite grain in **Figure 5d** was approximately 20 μm [18].

3.2. Mechanical properties

3.2.1. Mechanical properties at ambient temperature

The static mechanical properties, including fracture toughness, *J*, at ambient temperature, are summarized in **Table 1**. The TF sample with UFEG structure exhibited superior ductility, despite the high yield strength. Generally, the plastic instability or necking during tensile tests occurred immediately after the tensile stress reached the yield point by grain refinement. However, the TF and TFA samples had superior uniform elongation, as well as reduction in area. Such superior tensile–ductility balance has been seen for other warm rolled steels with different compositions [15, 19, 20]. Similarly, steel wires with UFEG structures, dominated by a strong α-fiber texture, exhibit superior tensile ductility despite their high strength [21]. The steel with UFG structure created by the WCR exhibited superior reduction in area despite showing an increase in strength and a decrease in uniform elongation [15, 22]. On the other hand, the presence of finer carbide particles, homogeneously dispersed in the ferrite matrix, improves the uniform elongation characteristics of UFG steels [23, 24]. These results explain the superior ductility in the TF and TFA samples.

	Strength (GPa)		**Ductility (%)**			**Toughness (kJ/m²)**	**Strength–toughness balance (GPa × kJ/m²)**
	0.2% yield strength σ_{ys}	**Tensile strength** σ_B	**Uniform elongation** ε_u	**Total elongation** ε_t	**Reduction in area** δ	**Fracture energy** J	$\sigma_{ys} \times J$
QT	1.51	1.82	4.60	9.20	28.3	134	202
TF	1.86	1.86	7.00	14.8	40.2	5433	10105
TFA	0.99	1.06	9.80	22.0	51.3	1376	1362
SM490	0.36	0.53	18.4	30.0	79.0	2157	766

Table 1. Static mechanical properties at ambient temperature.

Figure 6 shows the *P*–*u* curves and the appearance of the samples after the bending test at ambient temperature. The cracks in the QT sample propagated directly across the center portion of the test bar, and the sample fractured with a peak loading (P_{max}) of 12.3 kN, and showed typical brittle fracture behavior. The fracture surface exhibited a quasi-cleavage of a martensite structure, as shown in **Figure 7a**. In the TF sample, the crack branched parallel to the longitudinal direction of the test bar, that is, the crack propagated vertically to the LD. The steel was not broken as shown in **Figure 6c**. The fracture surface for the crack branching planes normal to the LD was characterized by a quasi-cleavage, and that for the planes roughly parallel to the LD was characterized by a fine dimple structure (**Figure 7b**). Namely, the fracture surface consisted of the delamination structure (⊥ LD) and the fine dimple structure (// LD). The TFA and SM490 samples exhibited fully ductile fracture as shown in **Figure 6a**. In those samples, shear lip was observed (**Figure 6d** and **e**), and the fracture surface consisted of a dimple structure. In particular, in the TFA sample, a very fine dimple structure was observed (**Figure 7c**).

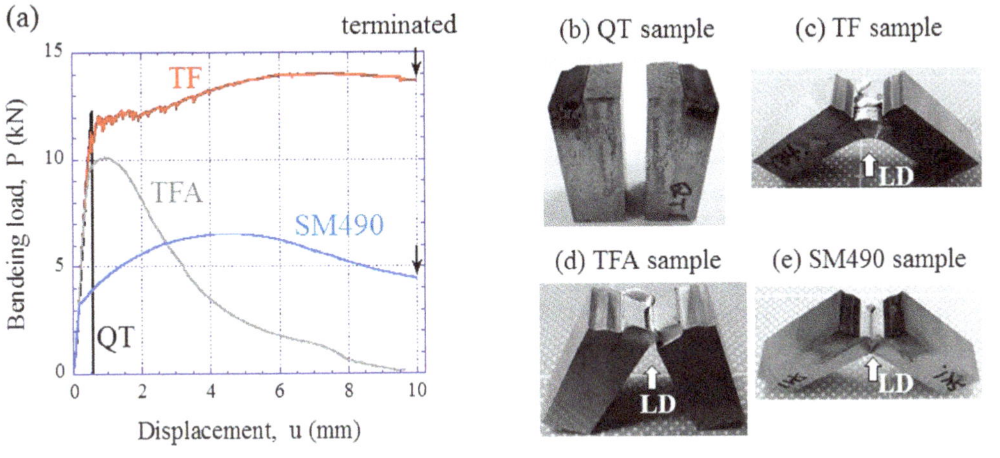

Figure 6. (a) Bending load-displacement curves at ambient temperature. (b–e) Appearance of samples after the bending test.

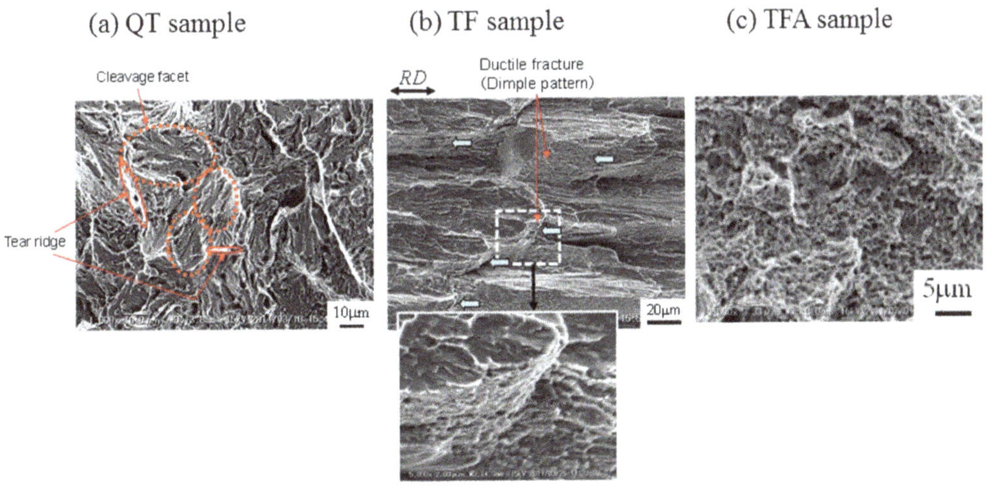

Figure 7. SEM micrographs of fracture surfaces of (a) QT, (b) TF, and (c) TFA samples.

3.2.2. *Temperature dependence on strength and toughness*

Figure 8 shows the yield strength, σ_{ys}, and the fracture energy, J, as a function of the test temperature for all samples. In **Figure 8a**, as is well known, the σ_{ys} increases with decreasing temperature. In **Figure 8b**, it can be seen that the J of the QT sample is very low due to brittle fracture, regardless of temperature. The TFA and SM490 samples had a typical energy transition curve, in which the J decreases with decreasing temperature. Furthermore, the energy transition curve of the TFA sample is very similar to that of the SM490 sample exhibiting a DBT from −40 to −100°C, although the J of the TFA sample is lower than that of the SM490 sample due to a difference in strength. The result suggests that it is difficult to improve toughness by grain refinement only. On the other hand, in the TF sample, the J increased remarkably as the temperature decreased from 200°C, reached a maximum near ambient temperature and then decreased. Namely, the steel showed inverse temperature dependence of the toughness, such as the Charpy impact toughness reported by [2].

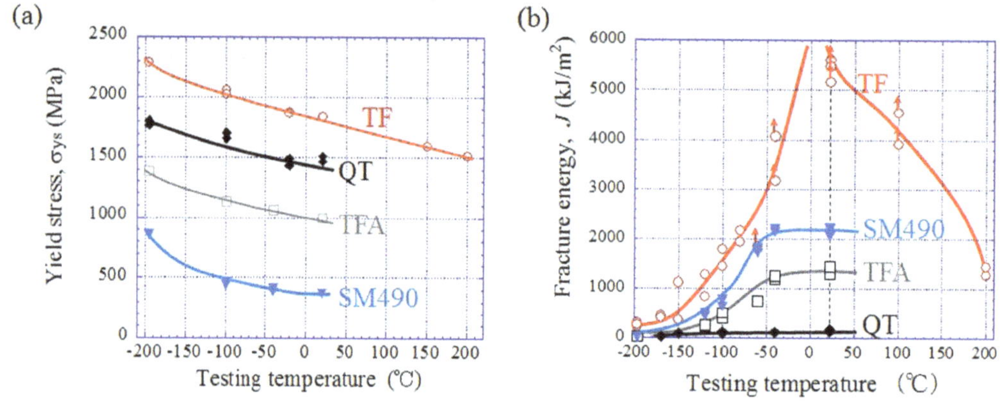

Figure 8. Variations of (a) yield strength, that is, 0.2% proof stress and (b) fracture energy with temperature. In (b), data points with upward-pointing arrows indicate that the specimens did not separate into two pieces.

4. Discussion

4.1. Crack branching and crack propagation behavior

As shown in **Table 1**, although the σ_{ys} of the TF sample improved to 1.86 GPa, compared with the QT sample, its J improved remarkably to 5433 kJ/m^2 and was about 40 times higher than that of the QT sample. On the contrary, the σ_{ys} of the TFA sample decreased to 0.99 GPa, and, similarly, its J decreased to 1376 kJ/m^2 by annealing, although it improved about 10 times more than that of the QT sample. Reduction in strength from TF to TFA samples is mainly attributed to an increase in the sizes of transverse ferrite grains and carbide particles and a decrease in dislocation density [17]. These samples had superior strength–toughness balance compared with the QT sample. In particular, the TF sample had excellent strength–toughness balance in all samples despite its highest yield strength.

Although the first load drop occurred at P_1 = 10.1 kN in the TF sample, which was smaller than the P_{max} = 12.3 kN of the QT sample, the steel exhibited a noncatastrophic fracture behavior with the evidence that the stepwise load increased beyond the P_1 and yielded a maximum load of 14 kN (**Figure 6a**). Finally, the test was terminated at u = 10 mm. In order to obtain a better insight into the mechanism responsible for the zigzag crack propagations, an interrupted bend test was carried out. The test was stopped at u = 0.74 and 1.58 mm. **Figure 9a** presents the P–u curves in the test interrupted at u = 0.74 and 1.58 mm and also, for comparison, the curves from **Figure 6a**. Three curves for the TF sample showed the same features. The optical microscope images near the initial notch at mid-thickness for the interrupted tests at u = 0.74 and 1.58 mm and for the terminated test at u = 10 mm are shown in **Figure 9b–f**. And **Figure 9g** and **h** shows SEM images of the delaminating cracks in **Figure 9e**. It is found from **Figure 9b** that a crack started to propagate vertical to the LD, from near the initial notch root, and then the cracks propagated in a zigzag pattern along the longitudinal direction (**Figure 9c**). At u = 10 mm, many zigzag microcracks branching from the main zigzag crack, starting from the notch root were observable in the test bars (**Figure 9d**). Furthermore, some microcracks (⊥ LD and ∠45°LD) were seen ahead of and near the zigzag cracks (**Figure 9e** and **f**) [25].

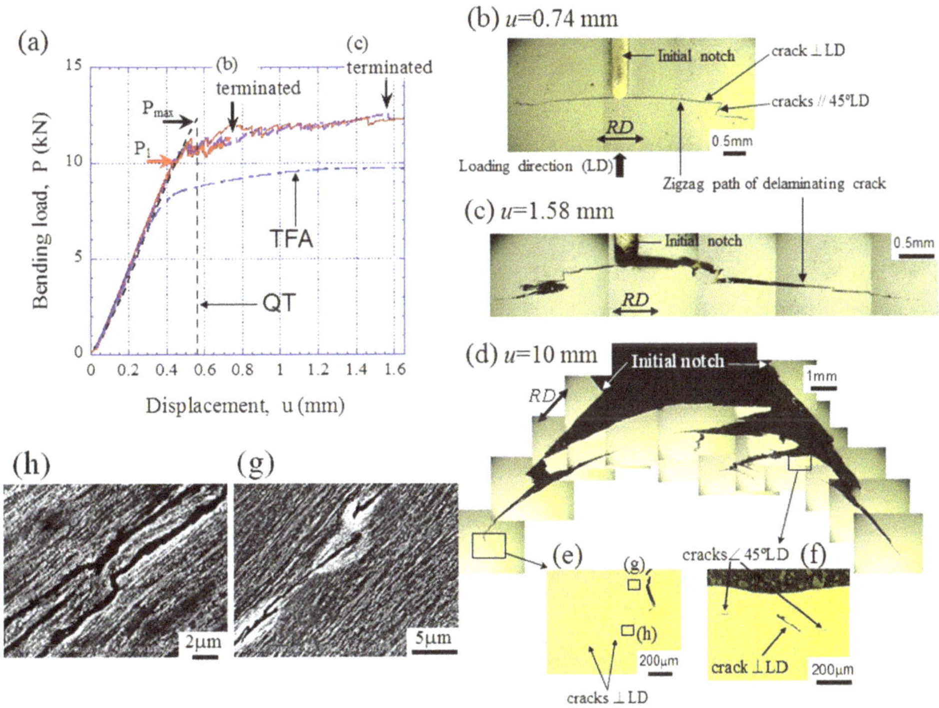

Figure 9. (a) Bending load variations in a range of u = 1.6 mm or less. OM images near the initial notch at (b) u = 0.74 mm, (c) u = 1.58 mm, and (d–f) u = 10 mm of the TF sample at mid-thickness. (g) and (h) SEM images of the delaminating crack.

In the case of laminate composites [26] having a weak interface normal to the LD (**Figure 2b**), the number of delaminations is dependent on the interfaces between layers. The ideal situation for high toughness is that all interfaces will delaminate during the applied load. As a general result, the load drops sharply after it attains the maximum value ((1)→(2) in **Figure 10**). Subsequently, a plateau region (3) in which the load becomes constant appears, and the load drops again. These load drops are due to crack propagation through the block of layers until the crack is arrested at the interfaces, and delamination along the interfaces appears as the plateau region. The extension of delamination is characterized by the ductility of the next layer, where a new crack is renucleated. This pattern is repeated until the sample is fully fractured. Therefore, for delamination toughening, it is important to have not only a weak interface but also a layer with plastic deformation abilities. As seen in **Figures 6a** and **9a**, the load drops in the TF sample are very small during bending tests, and many small load drops are seen until $u = 5$ mm by increasing the load. Lath martensite, the initial microstructure in the TF sample, shows a complicated hierarchical microstructure consisting of prior austenite grains, packets, blocks, sub-blocks, and laths [17]. This characteristic is significantly different from laminate composites. The composites bonded two or more materials through fabrication have relatively straight and long interface. In the TF sample, the weak site that causes delamination is predicted to be located in an elongated {1 0 0} cleavage plane and grain boundaries [2, 15, 27]. Furthermore, a UFEG structure with RD//<1 1 0> texture has superior plastic deformation abilities. The extensive delaminating crack plane was produced with a fine dimple structure, as observed in **Figure 7b**. Namely, not only a cleavage plane (crack∠45°LD) related to {1 1 0} <1 1 0> but also a fracture plane associated with a plastic deformation was induced, together with crack ⊥ LD related to {1 0 0} <1 1 0> and grain boundaries, during the bending test. Even if many micro-cracks during the applied load occur by the presence of many cleavage planes, the propagation of micro-cracks is arrested by many grain boundaries because the UFEG structures are three-dimensionally intertwined. Under such a phenomenon, the microstructural damage is not localized, however, rather, is widely distributed over very large dimensions, such as nacre [28, 29] or bio-inspired ceramic composites [30]. In the *P*–*u* curve, many load drops, as shown in **Figure 6a**, appeared, and the *P* did not decrease with an increase in the *u* by two effects, that is, the stress shielding associated with the interference of multiple cracks and the improved plastic deformation associated with grain refinement and texture. As a result, high strength with excellent toughness was achieved in the TF sample.

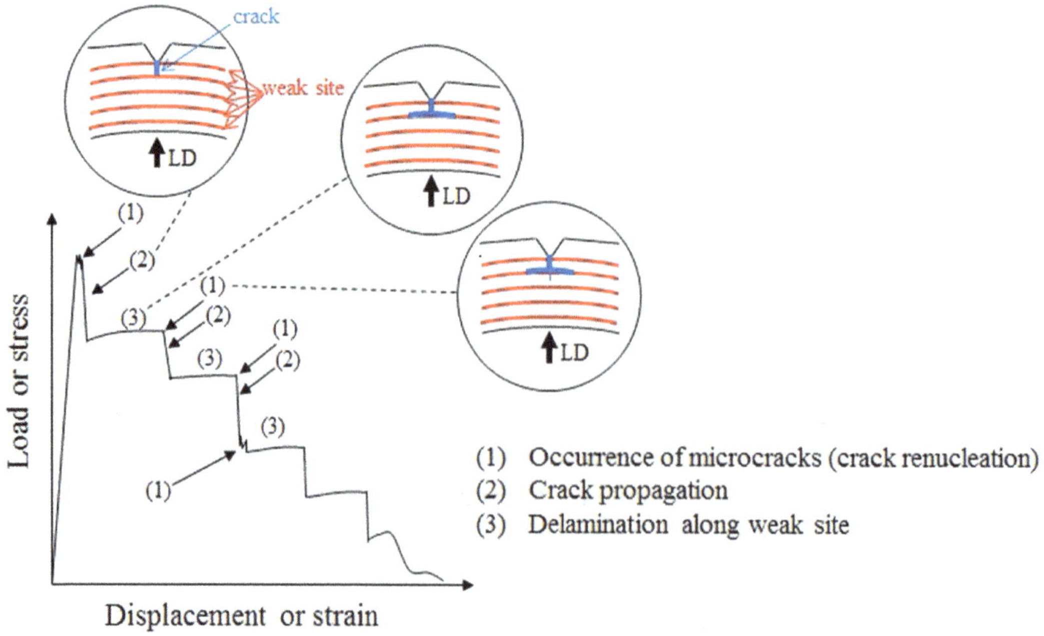

Figure 10. Typical load–displacement curve of laminate composite during three-point bend test.

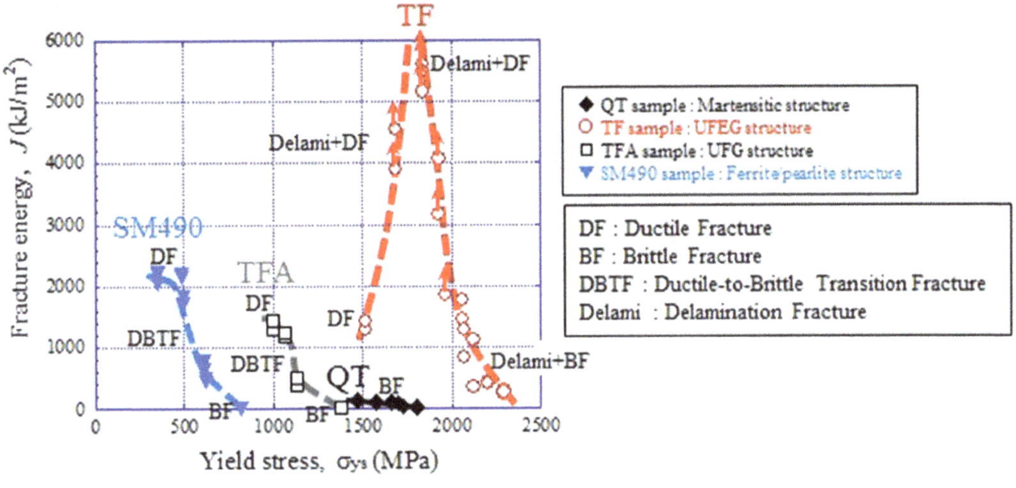

Figure 11. Strength–toughness balance for all samples. Data points with upward-pointing arrows indicate that specimens did not separate into two pieces during the bending load.

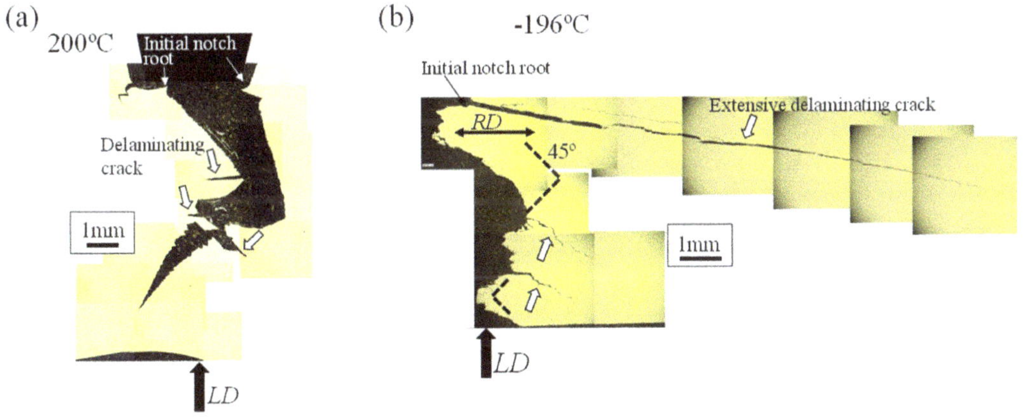

Figure 12. Cross-sectional OM images of the TF sample at mid-thickness after the bending test at (a) 200°C and (b) −196°C.

4.2. Strength–toughness balance

From the results in **Figure 8**, variations of toughness as a function of strength are obtained. **Figure 11** shows correlation between J and σ_{ys} for all samples. The results of the QT, TFA, and SM490 samples exhibited a typical strength–toughness balance. In the QT sample with strength of more than 1510 MPa, the J showed a low value of below 134 kJ/m^2 due to its brittleness at a temperature range from 23°C to −196°C. In the TF sample, first, the J increases as the σ_{ys} increases from 1500 MPa, and then it yields a maximum at near 1860 MPa which corresponds to yield strength at ambient temperature. Finally, it decreases with increasing σ_{ys}. This variation is attributed to unusual fracture behavior related to delaminating crack. The OM images of the TF sample after the bending test at 200°C and −196°C are shown in **Figure 12**. At 200°C, fracture manner exhibited ductile fracture (DF), and the crack propagated across the center portions of the test bars, although some delaminating micro-cracks (arrows in **Figure 12a**) were observed. At ambient temperature (**Figure 9d**), a significant delaminating crack propagated in a zigzag pattern along the longitudinal direction (Delami + DF). This zigzag crack occurred because crack ⊥ LD and crack∠45°LD were linked, and this behavior was repeated on the basis of the cleavage delamination mechanism. At −196°C, extensive delamination starting from the initial notch root and zigzag delaminating cracks is observable in **Figure 12b**. Macroscopically, the zigzag fracture paths appeared to have an angle of ±45° to the LD (Delami + BF). As a result, the TF sample exhibited the excellent strength–toughness balance in all samples.

In **Figure 8b**, the *J* for the TFA sample with UFG structure was lower than that for the SM490 sample with coarse grain structure, despite the same ductile fracture. **Figure 13** shows the OM images of the TFA and SM490 samples after the bending test at ambient temperature and SEM micrographs of fracture surfaces. It is found that the crack path in the TFA sample is relatively smooth compared with the crack path in the SM490 sample. As seen from SEM micrographs in **Figures 7c** and **13a**, the fracture surface in the TFA sample consists of a very fine dimple structure. On the contrary, the dimple structure of the SM490 sample observed in **Figure 13b** is larger and deeper than that of the TFA sample. Hence, low toughness of the TFA sample is attributed to difference in morphology of dimple structures related to grain size. As crystal grain becomes smaller, decrease in ductile fracture toughness is considered. The plot (DF) at σ_{ys} = 1500 MPa for the TF sample shown in **Figure 11** leads its prediction.

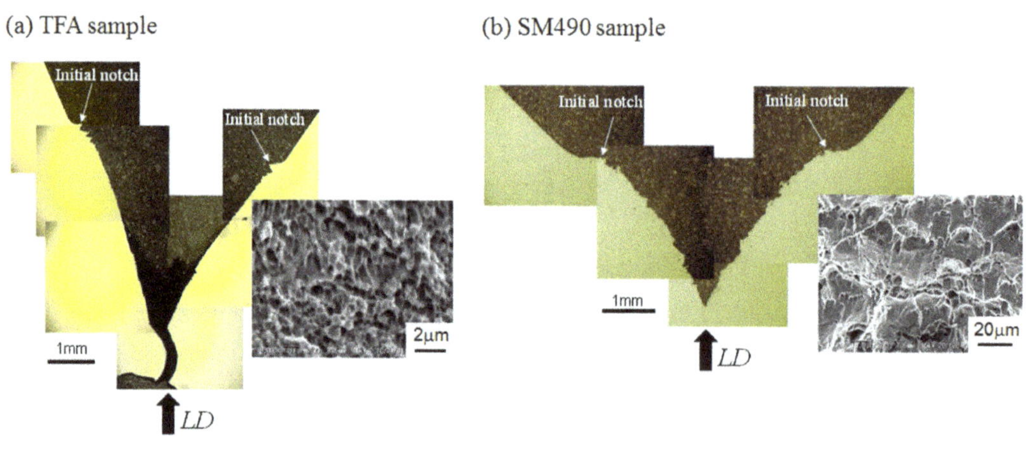

Figure 13. Cross-sectional OM images of (a) the TFA sample and (b) the SM490 sample at mid-thickness after the bending test at ambient temperature, and SEM images of fracture surfaces.

In **Figure 11**, the TF sample exhibited a ductile fracture at σ_{ys} = 1500 MPa. On the contrary, the QT sample exhibited a brittle fracture at the same strength. As a result, the *J* in the TF sample is about 10 times higher than that of the QT sample. This is attributed to a difference in the brittle fracture stresses in each direction associated with microstructural features. The brittle fracture stress, σ_F, related to weak sites such as the cleavage planes or the grain boundaries is a function of the crack size (it is replaced by the effective grain size, d_{eff}), and it is generally independent of temperature [31, 32]. Hence, the σ_F increases with the decreasing grain size. The grain size dependence of a fracture stress is higher than that of a yield stress [33]. The brittle fracture occurs when the tensile stress near the crack tip exceeds the σ_F. Normaly, the d_{eff} of the steel with a martensitic structure such as the QT sample is reported to be packet or prior austenite grain. On the contrary, the d_{eff} of microstructures such as the TF, TFA, and SM490 samples is ferrite grain size. The tensile direction of the maximum stress near the notch induced by a bending test corresponds to the RD, as shown in **Figure 14** [34]; hence, the crack generally propagates parallel to the LD. However, in the TF sample with the UFEG structure of $d_t << d_L$, the $\sigma_{F//RD}$ is very higher than the $\sigma_{F//LD}$, due to $\sigma_{F//RD} \alpha d_t^{(-1/2)}$ and $\sigma_{F//LD} \alpha d_L^{(-1/2)}$. Hence, the fracture is dominated by brittle fracture stress, $\sigma_{F//LD}$, parallel to the LD. If a ductile fracture occurred before the delaminating crack, the fracture manner exhibits the DF in **Figure 11**, and the J becomes larger than that of the QT sample exhibiting the brittle fracture. On the contrary, even if the delaminating crack of σ_{LD} (the tensile stress near the notch parallel to LD) $\leq \sigma_{F//LD}$ occurred before a ductile fracture, the delamination relaxes the triaxial tensile stress near the notch and arrests the propagation crack. As a result, the J remarkably increases by the fracture manner of the Delami + DF.

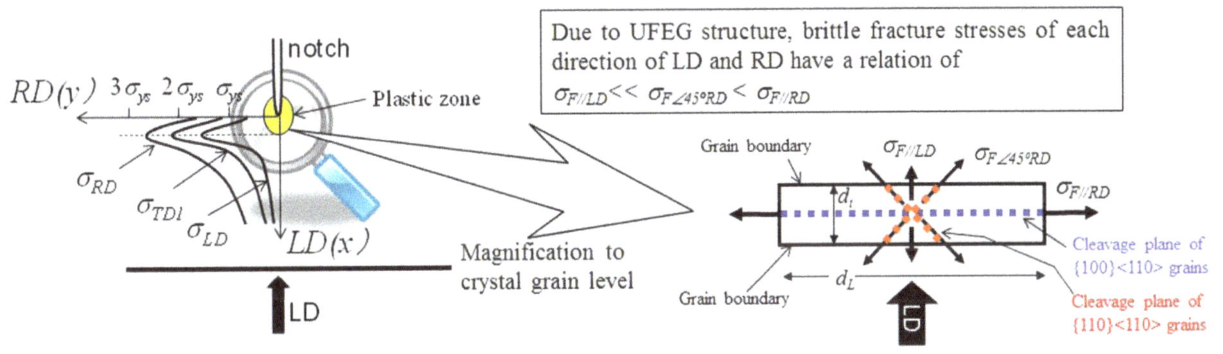

Figure 14. Schematic illustrations of (a) tensile stress triaxiality near the notch under the plane strain condition and of fracture stress for the elongated grains with a strong RD//<110>.

5. Conclusions

Low-alloy steel with an UFEG structure with an average transverse grain size of 300 nm was fabricated by multi-pass caliber rolling at 500°C, and the steel with an UFG structure with a grain size of 700 nm was fabricated by subsequent annealing at 700°C. Theses steels have yield strength, YS, of 1.86 and 0.99 GPa at ambient temperature. For comparison, two conventional steels, medium-carbon steel (YS = 1.51 GPa) with a martensitic structure and low-carbon steel (YS = 0.36 GPa) with a ferrite (20 μm)–pearlite structure, were also prepared. They were studied for the strength–toughness balance, the effect of microstructure features with respect to its balance and fracture behavior including crack propagation. The main results are as follows:

1. The fracture toughness of the UFEG steel was about 40 times higher than that of the steel with a martensitic structure, which exhibited a brittle fracture. This result is attributed to delamination toughening with many zigzag cracks. The UFG steel exhibited a fully ductile

fracture, and its toughness was about 10 times higher than that of the steel with a martensitic structure.

2. The strength and the toughness in the UFEG steel exhibited anisotropic properties. In particular, the toughness anisotropy was remarkable compared to the strength. The notch orientation dependence on toughness is due to differences in the spatial distribution of weak sites, such as {100} cleavage planes and boundaries of elongated grains.

3. The UFG steel and the ferrite–pearlite steel exhibited a typical energy transition curve, and the *J* of the UFG steel was lower than that of the ferrite–pearlite steel at a temperature range from 23 to −196°C. This result suggests that it is difficult to improve toughness by grain refinement only.

4. The UFEG steel showed an unusual energy curve, that is, inverse temperature dependence of the toughness, and the steel was not broken into two pieces at a temperature range from −40 to 100°C due to the fracture manner with delamination and ductile structures. The strength–toughness balance of the UFEG steel was excellent compared with that of all other steels.

Acknowledgements

We thank S. Kuroda, Y. Taniuchi, and K. Nakazato for materials processing; T. Kanno and G. Aragane for sample preparation of three-point bending test including electronic beam welding; and Ms. Yasuda for her experimental assistance in the microstructural observations. This study was supported in part by grants from the KAKENHI A (No.26249107) and the IKETANI Foundation. These grants are gratefully appreciated.

Author details

Tadanobu Inoue

Address all correspondence to: INOUE.Tadanobu@nims.go.jp

National Institute for Materials Science, Tsukuba, Japan

References

[1] Kimura Y, Inoue T, Tsuzaki K: Tempforming in medium-carbon low-alloy steel. Journal of Alloys Compounds. 2013; 577:S538–S542. DOI: 10.1016/j.jallcom.2011.12.123

[2] Kimura Y, Inoue T, Yin F, Tsuzaki, K: Inverse temperature dependence of toughness in an ultrafine grain-structure steel. Science. 2008; 320-5879: 1057–1060. DOI: 10.1126/science.1156084.

[3] Inoue T, Yanagida A, Yanagimoto J: Finite element simulation of accumulative roll-bonding process. Materials Letters. 2013; 106: 37–40. DOI: 10.1016/j.matlet.2013.04.093

[4] Inoue T, Horita Z, Somekawa H, Ogawa K: Effect of initial grain sizes on hardness variation and strain distribution of pure aluminum severely deformed by compression tests. Acta Materialia. 2008; 56(20): 6291–6303. DOI: 10.1016/j.actamat.2008.08.042

[5] Takaki S, Kawasaki K, Kimura Y: Mechanical properties of ultra fine grained steels. Journal of Materials Processing Technology. 2001; 117(3): 359–363. DOI: 10.1016/S0924-0136(01)00797-X

[6] Nagai K: Ultrafine-grained ferrite steel with dispersed cementite particles. Journal of Materials Processing Technology. 2001; 117(3): 329–332. DOI: 10.1016/S0924-0136(01)00789-0

[7] Song R, Ponge D, Raabe D: Mechanical properties of an ultrafine grained C–Mn steel processed by warm deformation and annealing. Acta Materialia. 2005; 53(18): 4881–4892. DOI: 10.1016/j.actamat.2005.07.009

[8] Tsuji N, Okuno S, Koizumi Y, Minamino Y: Toughness of ultrafine grained ferritic steels fabricated by ARB and annealing process. Material Transactions. 2004; 45(7): 2272–2281.

[9] Fujioka M, Abe Y, Hagiwara Y: Refinning of ferrite grain by using of transformation or recrystallization induced by heavy deformation. CAMP–ISIJ. 2000; 13: 1136–1139.

[10] Heiser F A, Hertzberg R W: Structural control and fracture anisotropy of banded steel. Journal of Iron Steel Institute. 1971; 209: 975–980

[11] McNicol R C: Correlation of Charpy test results for standard and nonstandard size specimens. Welding Journal. 1965; 44: 385–393.

[12] Bourell D L: Cleavage delamination in impact tested warm-rolled steel. Metallurgical and Materials Transactions A. 1983; 14: 2487–2496. DOI: 10.1007/BF02668890

[13] Rao K T V, Ritchie R O: Mechanical properties of Al–Li alloys. Part 1. Fracture toughness and microstructure. Materials Science and Technology. 1989; 5(9): 882–895. DOI: 10.1179/mst.1989.5.9.882

[14] Launey M E, Ritchie R O: On the fracture toughness of advanced materials. Advanced Materials. 2009; 21(20): 2103–2110. DOI: 10.1002/adma.200803322

[15] Inoue T, Yin F, Kimura Y, Tsuzaki K, Ochiai S: Delamination effect on impact properties of ultrafine-grained low carbon steel processed by warm caliber rolling. Metallurgical and Materials Transactions A. 2010; 41: 341–355. DOI: 10.1007/s11661-009-0093-x

[16] Inoue T, Yin F, Kimura Y: Strain distribution and microstructural evolution in multi-pass warm caliber rolling. Materials Science and Engineering A. 2007; 466: 114–122. DOI: 10.1016/j.msea.2007.02.098

[17] Kimura Y, Inoue T, Yin F, Tsuzaki K. Supporting online material for [2]. www.sciencemag.org/cgi/content/full/320/5879/1057/DC1

[18] Inoue T, Kimura Y: Toughening of low-carbon steel by ultrafine-grained structure. Transactions of the Japan Society of Mechanical Engineers, Series A. 2013; 79(804): 1226–1238 (in Japanese).

[19] Tsuchida N, Inoue T, Enami K: Estimations of the true stress and true strain until just before fracture by the stepwise tensile test and Bridgman equation for various metals and alloys. Materials Transactions. 2012; 53: 133–139. DOI: 10.2320/matertrans.MD201112

[20] Lee T, Park C H, Lee D L, Lee C S: Enhancing tensile properties of ultrafine-grained medium-carbon steel utilizing fine carbides. Materials Science and Engineering A. 2011; 528: 6558. DOI: 10.1016/j.msea.2011.05.007

[21] Yutori T, Katsumata M, Kanetsuki Y: Bull. JIM. 1989; 28: 313 (in Japanese)

[22] Torizuka S, Muramatsu E, Murty S V S N, Nagai K: Microstructure evolution and strength-reduction in area balance of ultrafine-grained steels processed by warm caliber rolling. Scripta Materialia. 2006; 55: 751–754. DOI: 10.1016/j.scriptamat.2006.03.067

[23] Oh Y S, Son I H, Jung K H, Kim D K, Lee D L, Im Y T: Effect of initial microstructure on mechanical properties in warm caliber rolling of high carbon steel. Materials Science and Engineering A. 2011; 528: 5833–5839. DOI: 10.1016/j.msea.2011.04.016

[24] Ohmori A, Torizuka S, Nagai K: Strain-hardening due to dispersed cementite for low carbon ultrafine-grained steels. ISIJ International. 2004; 44(6): 1063–1071. DOI: 10.2355/isijinternational.44.1063

[25] Inoue T, Kimura Y, Ochiai S: Shape effect of ultrafine-grained structure on static fracture toughness in low-alloy steel. Science and Technology of Advanced Materials. 2012; 13(3): 035005. DOI: 10.1088/1468-6996/13/3/035005

[26] Pozuelo M, Carreno F, Ruano O A: Delamination effect on the impact toughness of an ultrahigh carbon–mild steel laminate composite. Composite Science and Technology. 2006; 66: 2671. DOI: 10.1016/j.compscitech.2006.03.018

[27] Kimura Y, Inoue T, Yin F, Tsuzaki K: Delamination toughening of ultrafine grain structure steels processed through tempforming at elevated temperatures. ISIJ International. 2010; 50: 152–161. DOI: 10.2355/isijinternational.50.152

[28] Meyers M A, Chen P Y, Lin A Y, Seki Y: Biological materials: structure and mechanical properties. Progress Materials Science. 2008; 53: 1–260. DOI: 10.1016/j.pmatsci.2007.05.002

[29] Kakisawa H, Sumitomo T: The toughening mechanism of nacre and structural materials inspired by nacre. Science and Technology of Advanced Materials. 2011; 12: 064710. DOI: 10.1088/1468-6996/12/6/064710

[30] Munch E, Launey M E, Alsem D H, Saiz E, Tomsia A P, Ritchie R O: Tough, bio-inspired hybrid materials. Science. 2008; 322: 1516–1520. DOI: 10.1126/science.1164865

[31] Ma H: The effect of stress triaxiality on the local cleavage fracture stress in a granular bainitic weld metal. International Journal of Fracture. 1998; 89:143. DOI: 10.1023/A:1007484026645

[32] Kameda J, Nishiyama Y: Combined effects of phosphorus segregation and partial intergranular fracture on the ductile–brittle transition temperature in structural alloy steels. Materials Science and Engineering A. 2011; 528: 3705–3713. DOI: 10.1016/j.msea.2011.01.018

[33] Gillbert A, Hahn G T, Reid C N, Wizcox R A: Twin-induced grain boundary cracking in b.c.c. metals. Acta Metallurgica. 1964; 12(6): 754–755. DOI: 10.1016/0001-6160(64)90230-5

[34] Inoue T, Kimura Y: Effect of initial notch orientation on fracture toughness in fail-safe steel. Journal of Materials Science. 2013; 48(13): 4766–4772. DOI: 10.1007/s10853-012-6874-4

9

On the Fractography of Impact-Tested Samples of Al-Si Alloys for Automotive Alloys

Zheyuan Ma, Agnes M. Samuel,
Herbert W. Doty and Fawzy H. Samuel

Abstract

Castings were prepared from both industrial and experimental 319.2, B319.2 and A356.2 alloy melts, containing Fe levels of 0.2–1.0 wt%. Stontium-modified (~200 ppm) melts were also prepared for each alloy/Fe level. Impact testing of heat-treated samples was carried out using an instrumented Charpy impact testing machine. At low Fe levels and high cooling rates (0.4% Fe, dendrite arm spacing (DAS) of 23 μm), crack initiation and propagation in unmodified 319 alloys occur through the cleavage of β-Al_5FeSi platelets (rather than by their decohesion from the matrix). The morphology of the platelets (individual or branched) is important in determining the direction of crack propagation. Cracks also propagate through the fracture of undissolved $CuAl_2$ or other Cu intermetallics, as well as through fragmented Si particles. In Sr-modified 319 alloys, cracks are mostly initiated by the fragmentation or cleavage of perforated β-phase platelets, in addition to that of coarse Si particles and undissolved Cu-intermetallics. In A356.2 alloys, cracks initiate mainly through the fracture of Si particles or their debonding from the Al matrix, while crack propagation occurs through the coalescence of fractured Si particles, except when β-Al_5FeSi intermetallics are present, in which case the latter takes precedence. In the Sr-modified case, cracks propagate through the linkage of fractured/debonded Si particles, as well as fragmented β-iron intermetallics. In samples exhibiting low-impact energies, crack initiation and propagation occur mainly through cleavage of the β-iron intermetallics.

Keywords: fracture mechanism, Al-Si alloys, impact testing, SEM, metallography

1. Introduction

Fractography is defined as the study and documentation of fracture surfaces [1]. The purpose of fractography is to analyze fracture features and attempt to relate the topography of the fracture surface to the causes and/or basic mechanisms of fracture. The knowledge of fracture behavior is important in upgrading material specifications, improving product design, and analyzing failures for improved reliability [2]. A study of the characteristics of fracture surfaces is often carried out using optical microscopy (viz., light-microscope fractography), particularly when a low magnification of the fracture surface is adequate. The magnification is usually selected such that a good resolution is obtained and can range from "macroscopic" or low magnification fields (up to 50 diameters) to "microscopic" or high magnification fields (50 diameters and above).

When very fine details of the fracture surface, e.g., dimples and microvoids, are required to be observed, a scanning electron microscope (viz., fractography) is used instead, where the magnifications can go from about 5 to 240,000 diameters. With a resolution limit of ~100 Å and a depth of field about 300 times that of an optical microscope, the SEM is generally preferred over the latter for the analysis of fracture surfaces.

Four major types of fracture (or failure) modes have been discussed in the literature. Briefly, these modes and the sources of their occurrence are as follows [3]:

1. Dimpled rupture (or microvoid coalescence) caused by ductile fracture or rapid overload fracture;
2. Cleavage or quasi-cleavage caused by brittle fracture or premature or overload failure by catastrophic rapid fracture;
3. Intergranular fracture caused by stress corrosion cracking, hydrogen embrittlement, or subcritical growth under sustained load; and
4. Ductile striations caused by fatigue cracking or subcritical growth under cyclic load.

The first three types are microstructurally dependent. Complications arise in determining the failure type when complex microstructures are involved. For example, in aluminum alloys containing extensive intermetallic compounds, the failure may occur by a combination of ductile rupture of the matrix and brittle cleavage of the intermetallic particles. Such intermetallic compounds (>1 μm) also reduce the alloy toughness. Often, transgranular fracture of the brittle compounds may also occur. Multiple microstructural phases can lead to numerous combinations of microscopic fracture modes.

Al-Si alloys are essentially a combination of the high strength, brittle silicon phase, and low-strength, ductile aluminum matrix containing, in addition, other microstructural features such as iron intermetallic phases, porosity, and inclusions. While it is natural to assume that the Si particles play an important role in the fracture behavior, the contribution of these other microstructural aspects must also be considered, even if their relative importance is affected by the amount and morphology of the constituents and the interactive effects between them.

Consequently, the alloying and melt-processing parameters (e.g., strontium modification, grain refining) and the solidification rates [different dendrite arm spacings (DASs)] which directly affect the microstructure also come into play.

Figure 1 shows two main types of fracture observed in the Al-Si alloy under tensile overloading. **Figure 1(a)** illustrates an example of the simple rupture mode of fracture, where the particles are observed at the bottom of the dimples. These particles are seen to have decohered from the surrounding matrix. In such cases, where decohesion of the particles takes place, the resultant fracture surface attains the dimpled appearance as seen in the figure. The alloy is said to have failed by the process of microvoid coalescence. The microvoids nucleate at regions of localized strain discontinuity, such as that associated with second-phase constituents, inclusions, grain boundaries, or dislocation pileups [4].

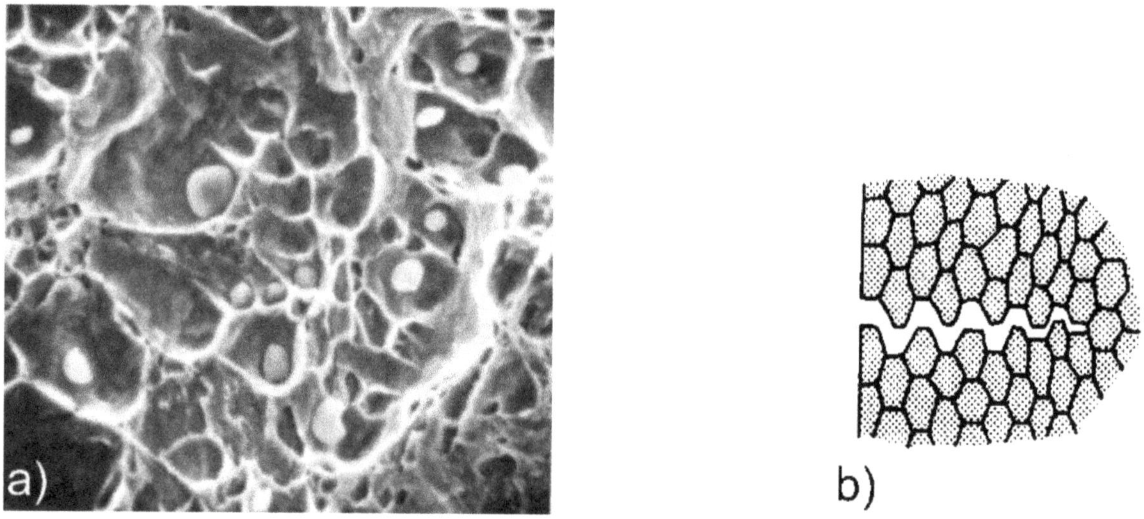

Figure 1. Two main types of fracture observed in Al-Si alloys under tensile overloading: (a) dimple rupture and (b) cleavage fracture of intermetallics [4].

Figure 1(b) represents an example of the cleavage fracture of intermetallics. Cleavage is a low-energy fracture that propagates along well-defined low-index crystallographic planes known as cleavage planes. Theoretically, a cleavage fracture should have perfectly matching faces and a completely flat and featureless surface. However, as the Al-Si alloys used are polycrystalline and contain grain boundaries, inclusions, and dislocations, these imperfections affect the propagating cleavage and alter the perfectly flat characteristics to a certain extent. Sometimes, the microconstituents present in the path of the propagating crack can result in the alloy sample exhibiting a mixed fracture mode, as was observed in the present study.

In this chapter, the fracture behavior of Al-Si-Cu and Al-Si-Mg cast alloys as reported by other researchers in the literature will be reviewed, to obtain a basic idea about the subject. Based on this, the fracture behavior of selected impact samples tested in the present study will be discussed. The focus is on alloy sample conditions corresponding to optimum or minimum impact energy conditions.

1.1. Si particles

Wang and Caceres [5] studied the fracture behavior of Al-Si-Mg (A356) alloys using tensile samples of different section sizes and under different conditions (as-cast/heat-treated, unmodified/Sr-modified), where *in situ* observations of the surface crack initiation and propagation using scanning electron microscopy (SEM) were presented. Cracks were found to initiate the fracture of Si particles at relatively low strain values. Crack propagation also proceeded by the same mechanism, preferably through the eutectic region of the microstructure, avoiding the primary Al dendrite cores wherever possible. Continued strain resulted in localized plastic deformation around the fractured Si particles, which eventually led to the formation of microcracks in the eutectic region. Further straining caused selected microcracks to link up with other neighboring microcracks in the same eutectic region to initiate the primary crack, or to link up with the propagating crack front. The overall fracture paths tended to follow the eutectic regions in the microstructure.

Ammar et al. [6] described the microstructure of A356 alloys consisting of primary Al-1% Si dendrites and a eutectic with silicon particles (12 pct volume) embedded in the Al-1% Si matrix. Room temperature constant amplitude fatigue tests revealed that for a low crack-tip driving force, the surface fatigue crack propagated primarily through the Al-1% Si dendrite cells. The silicon particles in the eutectic remained intact and served as barriers at low-fatigue crack propagation rates. When the fatigue crack crossed the three-dimensional Al-Si eutectic network, it propagated mainly along the interface between the silicon particles and the Al-1% Si matrix. Therefore, nearly all of the silicon particles were progressively debonded by fatigue cracks propagating at low rates, with the exception of elongated particles with a major axis perpendicular to the crack plane, which were fractured. For a high crack tip driving force, silicon particles ahead of the crack tip were fractured, and the crack subsequently propagated through the weakest distribution of prefractured particles in the Al-Si eutectic. Only small rounded silicon particles were observed to debond at high crack propagation rates. The crack process in the two cases is schematically depicted in **Figure 2**.

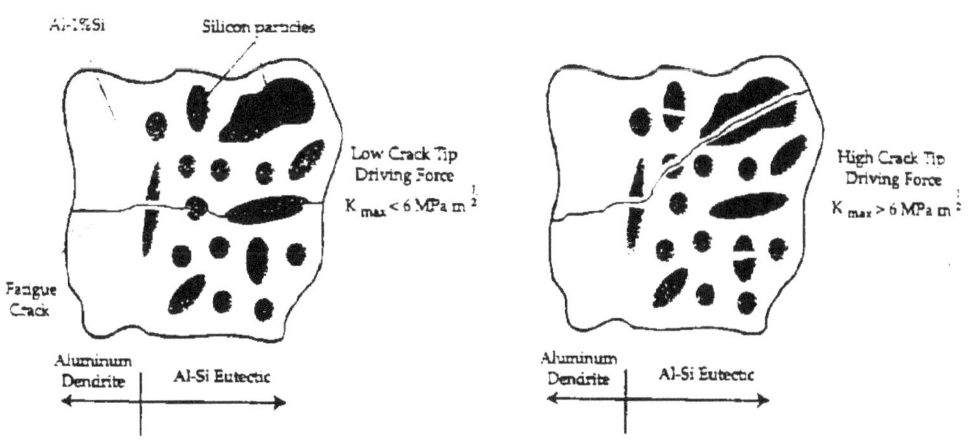

Figure 2. Schematic of the fatigue crack process through the silicon particle-rich eutectic regions at low and high crack-tip driving forces [7].

Hafiz and Kobayashi [8] conducted a study on the microstructure-fracture behavior relations in Al-Si casting alloys using tensile testing. The main focus was on Si particles, where different morphologies of the Si particles were obtained using a variety of solidification rates and different Sr levels. Fracture paths in the mid-sections were examined using optical microscopy, while the features of the fracture surface were examined using SEM. In general, the voids were found to initiate at silicon particles. The individual voids then grew and coalesced, creating microcracks in the eutectic region. These microcracks linked up to form the main crack, resulting in the final fracture.

Lee et al. [9] observed that the fatigue crack growth in Al-Si-Mg casting alloys occurs either by the nucleation and linkage of microcracks or voids formed as a result of the decohesion of non-fractured Si particles from the surrounding matrix (such as those encountered in modified alloys), or by cleavage or the cracking of silicon particles (such as those occurring in unmodified alloys), or both. In alloys containing intermediate Si particle size ranges, a mixed fracture morphology is observed, with both the cleavage fracture of Si particles and particle decohesion taking place.

Thus, the fracture mechanism proceeds as follows:

1. Cracking of the Si particles or their decohesion from the Al matrix at low strain;
2. Localized plastic deformation with increase in strain, leading to the formation of micro-cracks in eutectic regions;
3. Microcrack coalescence followed by crack propagation, leading to final fracture [3, 10].

Figure 3 shows a schematic representation of the process.

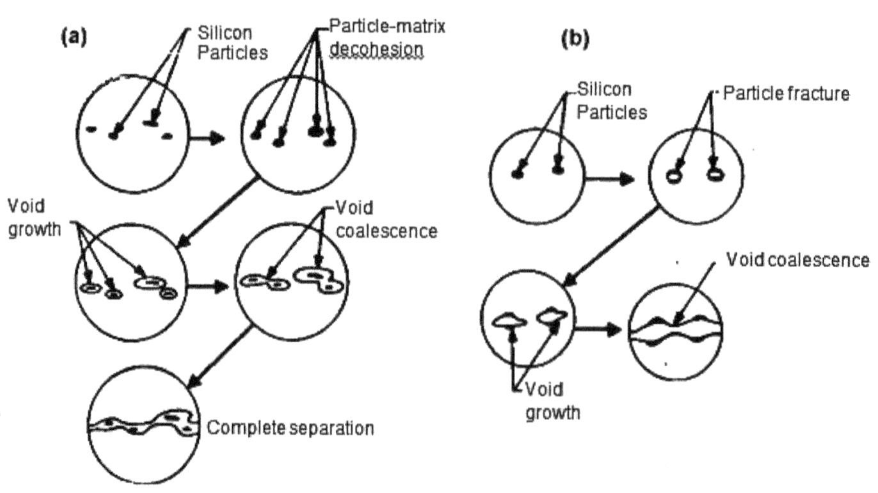

Figure 3. Schematic illustration of fracture mechanisms based on Si particles. (a) Initiation by Si particle-Al matrix interface decohesion and (b) initiation by Si particle fracture [11].

According to Ma et al. [3] in commercial A356 alloy, the fracture path of tensile-tested specimens proceeds mainly through the largest silicon particles, which constitute less than 1% of the overall population of silicon particles in the bulk microstructure. Li et al. [10] and Tiryakiolu [12] pointed out that the larger and longer silicon particles are more prone to cracking. In coarser structures, silicon particle cracking occurs at low strains, while in finer structures, the progression of damage is more gradual. They also found that in the former case, broken particles were observed at both cell and grain boundaries, with no evident preference for either, whereas in the case of finer structures, cracking is initiated on the grain boundaries. Lee et al. [9] also reported that the fatigue behavior of Al-Si-Mg casting alloys depends on the size, orientation, and local distribution of the Si particles.

1.2. Strontium modification

As strontium modification is commonly used to alter the morphology of the eutectic Si particles in Al-Si alloys, the process is expected to affect the fracture behavior of the alloy. Elsebaie et al. [13, 14] studied the role of microstructure in relation to the toughness of hypoeutectic Al-Si casting alloys, using unnotched Charpy impact test samples. Additions of 0.017 or 0.03 wt % Sr were used to modify the alloys. In the unmodified alloy, the fracture followed a path marked by the eutectic Si, circumventing the primary Al dendrites, where Si particles could be seen adhering to the sides of the crack path. The detection of submicron cracks showed that the fracturing of the Si particles corresponded to an early stage of the crack propagation process, i.e., the Si particles tended to fracture ahead of the main crack. In the modified alloy, the fracture propagated through both the eutectic region and the Al matrix, with sheared aluminum cells being detected in the fracture path, and the submicron cracks observed in the eutectic regions clearly indicated that the interdendritic structure had failed first.

Hafiz et al. [15] observed that in unmodified Al-8% Si alloy, the fracture surface mainly consists of a complicated array, resembling the Si array in the eutectic region containing cleaved Si particles. In addition, broken Si particles could also be detected. In the Sr-modified condition, two main features were observed: dimple colonies which represented the majority of the fracture surface and a smooth ripple pattern surrounding the dimple colonies, typical of ductile fracture.

According to several researchers [16–19], even though the fracture details are influenced by the Si particle morphology and the Al matrix, the basic fracture sequence is independent of changes in the microstructural characteristics due to the strontium modification. However, the addition of strontium does increase the amount of strain required to initiate and propagate the microcracks, particularly with respect to crack initiation.

1.3. Iron intermetallics

Their brittle nature, lateral or faceted growth mode (which makes them poorly bonded to the Al matrix) and the multiple (001) growth twins (parallel to the growth direction) of the plate-like β-Al_5FeSi iron intermetallic particles make them pro-crack sites. Hrong et al. [20] studied the fracture behavior of A356 alloys with different iron contents under resonant vibration. According to them, the cracks were found to initiate and grow along the eutectic Si- and Fe-rich intermetallic phase particles. The crack paths are propagated predominantly through the Si particles and occasionally through the Fe-rich intermetallics. When the iron content was below 0.57 wt%, the iron intermetallics were mainly α-$Al_8Mg_3FeSi_6$ or α-$Al_5Fe_3Si_2$, whereas above that, the Fe-intermetallics occurred mostly as β-Al_5FeSi, which was found to be detrimental to the vibration fracture resistance.

According to Taylor [21], directly after the onset of plastic deformation, Al-7Si-0.3Mg sand-cast alloys display microcracks which are associated with brittle grain boundary cracking and cleavage of plate-shaped iron intermetallics, although the iron content is as low as 0.1 wt%. Intermetallics can be seen only in part of the grain boundaries that failed by brittle fracture. In die-cast alloys, microcracks occur due to fracture of small intermetallic particles located at interdendritic boundaries. With the increase in strain, Si particles at grain boundaries begin to fracture, followed by decohesion from the matrix.

Dinnis and Taylor [22] examined the microstructure of Al-7.0% Si-1.0% Fe alloys during tensile deformation using optical microscopy and *in situ* SEM. He found that the needle-like β-iron intermetallics are cracked easily in the earlier stages of deformation, whereas the α-iron Chinese script intermetallics can withstand a higher stress. When the form of the iron intermetallics was changed from Chinese script to needle-like, crack initiation did not occur at the silicon particles but at the iron intermetallics.

Villeneuve et al. [23] found that in Al-13% Si-Fe alloy castings, cracks appeared within the β-Al_5FeSi platelets rather than at the β/Al interface. This is due to the brittle nature of the β-phase, whereby the platelets are easily split into two halves.

1.4. Solidification rate

As the solidification rate directly affects the microstructure of an alloy casting, it is expected to affect the fracture behavior as well. The work of Alkahtani et al. [24] on the tensile testing of Al-7Si-0.3Mg alloy showed that, in sand-cast alloys, microcracks are associated with brittle grain boundary cracking and the cleavage of plate-shaped intermetallics and can be observed only in those parts of the grain boundaries that failed by brittle fracture, indicating an intergranular fracture mode. In die-cast alloys, all microcracks observed are due to the fracture of small intermetallic particles and located at interdendritic boundaries, indicating the fracture mode to be the transgranular type.

The work of Cáceres et al. [25] on fracture behavior of an Al-7Si-0.4Mg (A356.0) casting alloy shows that under low cooling rate conditions, when coarse structures are observed, the cell boundaries are distinct, the cracked particles are located with equal probability in the cell and grain boundaries, and the fracture mode is of the transgranular type. At high cooling rates (viz., in fine structures), the cracked particles are preferentially located in the grain boundaries, and the fracture mode is of the intergranular type.

In a similar study, Wang et al. [26] observed that in large cell-size materials, the fracture tends to occur along the dendritic cell boundaries (transgranular fracture), while for smaller cell-size materials, the fracture path runs along grain boundaries in an increasing proportion (intergranular fracture). The proportion of intergranular fracture increases from zero at large DAS values to about 60% for small DASs.

The work of Samuel and Samuel [27] on 319.2 alloys using end-chill castings showed that, in general, fracture surfaces of alloys obtained at 5-mm distance from the chill (DAS ~15 μm) displayed a transgranular mode of fracture, typical of ductile materials. At the highest distance from the chill (100 mm, DAS ~95 μm), however, the fracture was of the intergranular type, customarily associated with brittle, low-ductile materials.

1.5. Porosity

According to Dinnis et al. [28], after the onset of plastic deformation which generates micro-cracks, intense shear bands commonly form. They are oriented in directions of about 45 degrees with respect to the tensile axis and propagate preferentially along grain boundaries. Microcracks grow and link along these shear bands. The presence of porosity can support shear band formation but does not initiate shear bands in the absence of the intermetallic particles. In certain cases, pores can even act as crack stoppers if located at the tips of sharp cracks.

Estensoro et al. [29] studied the fatigue behavior of two cast Al-7% Si alloys, with magnesium contents of 0.3 and 0.45%, respectively, and observed a very strong effect of defects (porosity) on the fatigue life. Fractographic examination showed that the strong scatter in fatigue life was associated with the amount of porosity in the specimen and determined by the number, size, and location of the pores in the fatigue specimens. The dispersion was more significant when the porosity emerged at the specimen surface. They also found that porosity exerts more influence on fatigue life than chemical composition or heat treatment.

In their investigation of the high-cycle fatigue testing of AlSi11 cast aluminum alloy, Lados and Apelian [30] also showed that casting voids reduce the fatigue strength and lead to early fracturing. Crack initiation occurs at these voids, particularly if the porosity acts as a stress concentrator. The crack initiation times are strongly influenced by the position of the porosity, viz., whether it is close to the surface or to the center of the specimen. Savelli et al. [31] also reported that, in cast AlSi7Mg0.3 aluminum alloys, internal pores are responsible for the crack initiation of fatigue cracks. Based on this, a mathematical model was proposed by them to predict the fatigue life using pore dimensions as the main parameter.

2. Experimental procedure

The primary A356.2 and experimental 319 alloys used in this study were supplied in the form of 12.5-kg ingots. The chemical compositions of the as-received alloys are listed in **Table 1**. The two 319 alloys used differ from each other in their Mg contents, where the B319.2 alloy contained 0.4 wt% Mg compared to 0.002 wt% in the 319.2 alloy. The Mg level of 319 alloys used in automotive applications is generally of the order of ~0.4 wt%.

Study code	AA alloy	Element (wt%)									
		Si	Cu	Mg	Fe	Mn	Zn	Ti	Sr	Pb	Al
EA	319.2	6.06	2.98	0.002	0.12	0.023	0.006	0.007	0.0003	–	bal.
IA	319.2	6.12	3.57	0.0719	0.405	0.0939	0.0483	0.138	0.00024	–	bal.
IB*	B319.2	6.22	3.21	0.40	0.15	0.026	0.0034	0.0075	–	–	bal.
IC	A356.2	6.78	0.02	0.33	0.11	0.04	0.04	0.08	–	0.03	bal.

*IB alloy was prepared from EA alloy where the Mg level was increased to 0.4 wt%.

Table 1. Chemical compositions of 319 and 356 alloys.

Fracture surfaces of impact-tested samples were examined in order to determine the crack initiation and propagation behavior with respect to the alloy compositions and melt treatment conditions. Samples were cut approximately a quarter inch from the fracture surface where the upper part was used for examination. Extra care was taken during the cutting to avoid contamination of the fracture surface. The bottom of the sample was ground to flatten it and then mounted on a standard base with a special glue. Several samples were kept together in a box and classified systematically according to their chemical compositions and melt treatment conditions before they were placed into the specimen chamber of a scanning electron microscope. The SEM work was carried out using a scanning electron microscope operating at 15 kV.

3. Fractography results

Optical microscopy and SEM were used to examine the facture surfaces of selected impact-tested samples. In each alloy, samples with the lowest and highest iron levels were taken, obtained at the highest and lowest cooling rates (i.e., at DASs of ~23 and ~85 μm). In each case, at least two SEM micrographs were taken, one from the edge and one from the central region of the fracture surface, to observe the crack initiation and crack propagation characteristics, respectively. In addition, interesting features appearing on the fracture surface were also examined. Optical micrographs from longitudinal sections (perpendicular to the fracture surface) of these samples, particularly those of A356.2 alloy, were also obtained, to further interpret the fracture process.

3.1. SEM fractography

The popular use of a scanning electron microscope for fractography studies is on account of the fact that it allows for a direct examination of the fracture surface without the need for a replica, as in the case of transmission electron microscopy. One of the great advantages of the SEM is its ability to examine specimens at low magnifications of about 50 diameters and then enlarge regions of special interest to very high magnifications.

Both backscattered electron and secondary electron modes of operation are normally used for fractographic purposes. The secondary electron image (SEI) offers a better resolution, whereas the backscattered electron image (BEI) offers an improved image contrast that is necessary in the case of smooth specimens and at low magnifications. The high image contrast is, however, accompanied by a loss of resolution, inherent with the backscattered mode of operation.

Backscattered electrons are electrons with high energy (equivalent to that provided by the accelerating voltage, which is usually in the range of 20–25 kV). The electrons escape from as deep as 300 Å in the specimen. Because of scatter within the specimen, the resultant source is much larger in diameter than the incident-beam diameter. On the contrary, the secondary electrons that are generated possess energies of about 20–50 eV (maximum), which is no more than is required for the electrons to escape from the top 100 Å layer of the specimen. A zone within 100 Å of the surface, then, constitutes the source of secondary electrons that can be captured by the electron collectors. Secondary electrons generated further within the specimen do not have enough energy to escape. The difference in the volume of the source for the two types of electrons accounts for the difference in resolution [32]. Most of the fractographs presented here are BEIs, to bring out the contrast in the various features observed on the fracture surfaces of the samples studied.

3.2. Alloys 319.2 and B319.2 (alloys A and B)

The SEM fractographs of the 319.2 alloy sample corresponding to the 0.4% iron level and highest cooling rate are presented in **Figures 4** and **5**, taken respectively from the sample edge where the sample was hit during impact testing and the center of the sample surface.

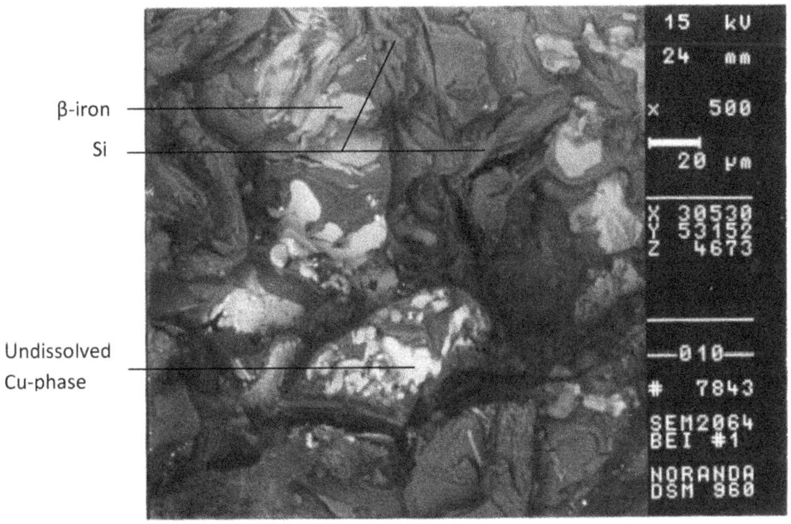

Figure 4. SEM micrograph showing fracture surface of alloy 319.2 (0.4% Fe, DAS 23 μm, unmodified, edge).

The features exhibited in **Figure 4** are typical of cleavage fracture, as is expected to be the case at high deformation rates during impact loading. The fracture plane changes orientation from grain to grain. A certain amount of intergranular cleavage is also observed in the form of secondary cracks. The fracture surface reveals the presence of β-Al_5FeSi iron intermetallic platelets, undissolved $CuAl_2$ or Al_7Cu_2Fe (insoluble phase) particles (since all alloy samples were T6-heat treated), and acicular Si particles surrounding them. The fragmentation of all three particle types indicates that possibly all three could have acted as crack initiation sites. The microstructural constituents of interest in **Figure 4** and those that follow have been marked in each case.

Figure 5, taken from the central region of the fracture surface of the same sample, shows a cracked α-iron script particle, surrounded by brighter Cu-containing phase particles on either side. The presence of cracks within these particles confirms how crack formation is facilitated when such intermetallics are present in the microstructure.

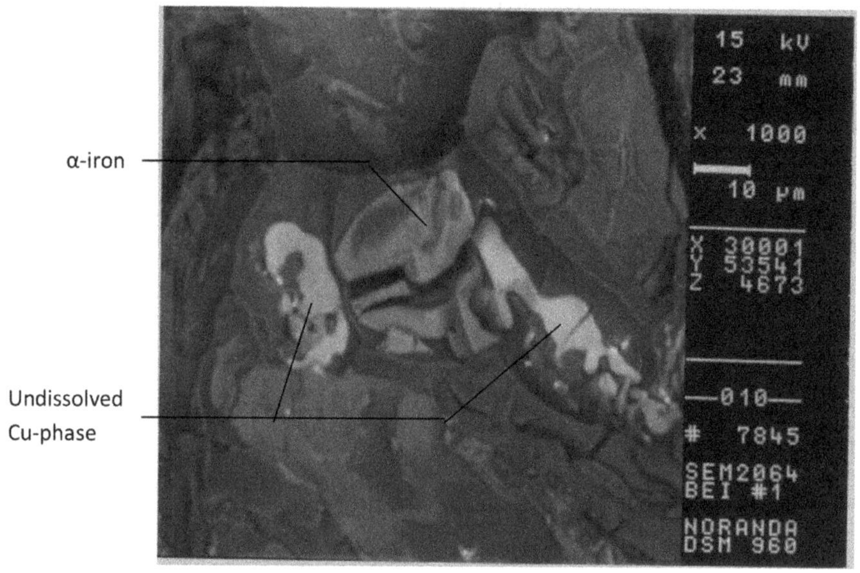

Figure 5. SEM micrograph showing fracture surface of alloy 319.2 (0.4% Fe, DAS 23 μm, unmodified, center).

At the highest iron level (0.8%) and lowest cooling rate, the SEM fractograph of **Figure 6** reveals that crack initiation occurs by massive cleavage of the brittle β-Al_5FeSi phase. A profusion of β-platelets is expected to be observed in the alloy microstructure at such Fe levels and, as can be seen, the crack propagates from one β-platelet across the other, along the direction delineated by the open arrows. In an ordinary two-dimensional optical micrograph, these β-platelets would appear in the form of branched needles. The much larger dimensions of the plate-like β-Al_5FeSi phase would make it more susceptible to crack initiation compared to the Si particles, even if the latter occurs in brittle, acicular form in the unmodified alloy.

In cleavage fracture, the fracture path follows a transgranular plane that is usually a well-defined crystallographic plane, as is judged to be the case by the relatively smooth surfaces of the β-platelets seen in **Figure 6**. The central area of the fracture surface of the same sample, **Figure 7**, shows instances of the Cu-containing (brighter) phase, precipitated on the β-platelets, amplifying the likelihood for crack propagation through these locations. The crack propagates in the direction denoted by the open arrows, along the cleaved intermetallic particles. A number of secondary cracks are also observed in the vicinity. In keeping with the brittle fracture features observed in **Figures 6** and **7**, this sample exhibited a low impact energy of 1.90 J.

In the case of the modified 319.2 alloy containing 0.4% Fe and obtained at the highest cooling rate, the SEM fractographs, **Figures 8** and **9**, revealed considerably different characteristics. Upon application of a high deformation rate during impact loading, **Figure 8**, the β-platelet at the edge of the sample fragmented into several smaller segments, rather than undergoing cleavage as was observed in **Figure 4** for the unmodified alloy. Keeping in mind the effect of Sr (in the modified alloy) on the dissolution and fragmentation of the β-Al_5FeSi phase, the

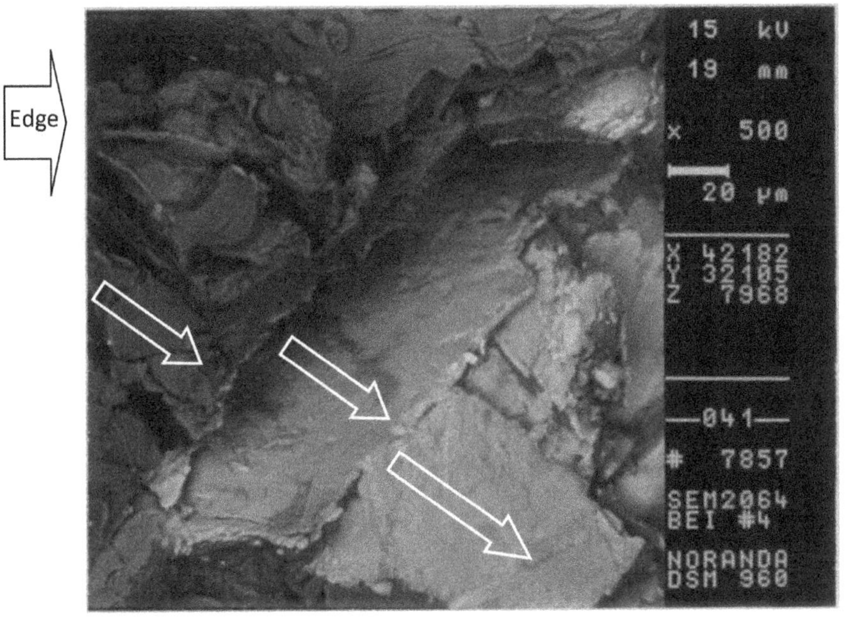

Figure 6. SEM micrograph showing fracture surface of alloy 319.2 (0.8% Fe, DAS 83 μm, unmodified, edge).

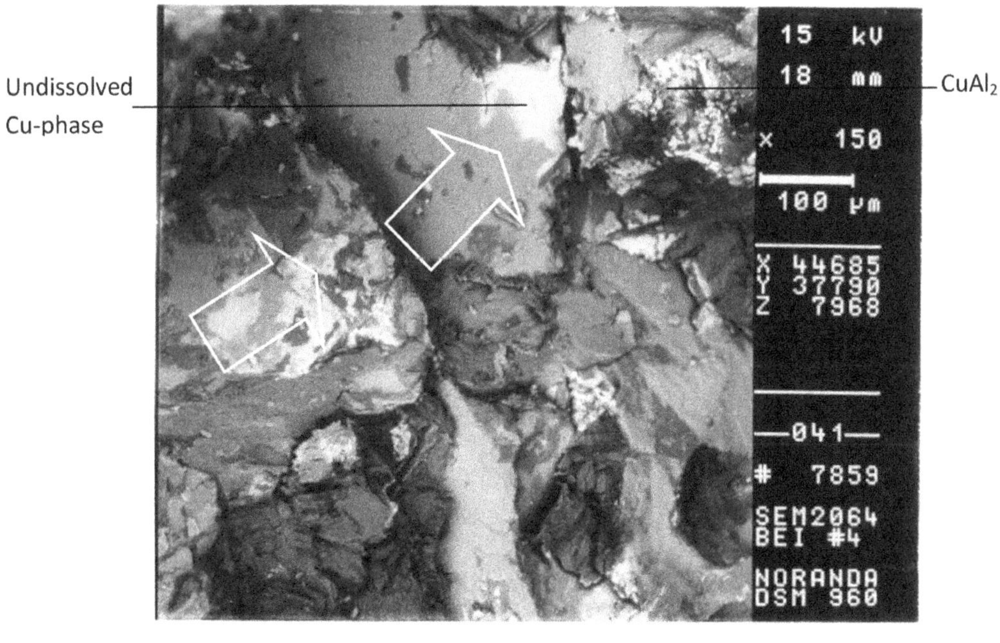

Figure 7. SEM micrograph showing fracture surface of alloy 319.2 (0.8% Fe, DAS 83 μm, unmodified, center).

deterioration of the β-phase caused by Sr addition would more likely result in the fragmentation of the β-platelets rather than their cleavage (as observed in the case of solid β-platelets in the unmodified alloy). Consequently, crack propagation would proceed at a slower rate, leading to higher impact energy, as indicated by the E_I, E_P, and E_T values of 16.39, 14.57, and 30.96 J, respectively. In addition, the transformation of the eutectic Si particles from acicular to fibrous form would also result in a more ductile fracture mode.

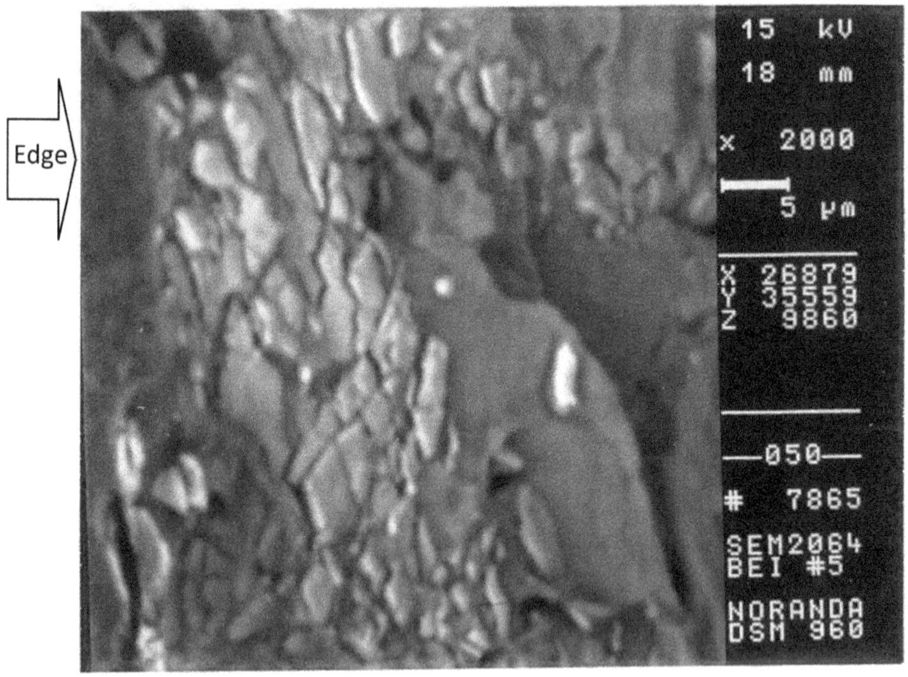

Figure 8. SEM micrograph showing fracture surface of alloy 319.2 (0.4% Fe, DAS 23 μm, modified, edge).

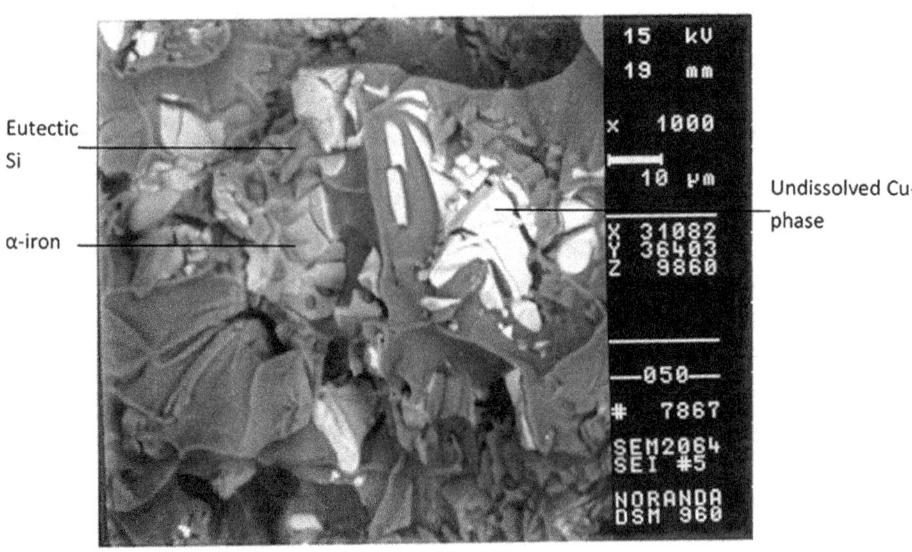

Figure 9. SEM micrograph showing fracture surface of alloy 319.2 (0.4% Fe, DAS 23 μm, modified, center).

Comparing the much higher magnification image of **Figure 8** (2000X) with that of **Figure 4** (500X), one may also estimate the much smaller sizes of the β-platelets obtained in the modified alloy. **Figure 9** shows fractured α-iron script and Cu-containing phase (bright) particles. Due to the surrounding eutectic structure, the fracture surface is more reminiscent of ductile, rather than brittle fracture, in spite of the cracks observed in the $CuAl_2$ particles.

In summary, therefore, Sr-modified Fe-containing 319.2 alloys would be expected to exhibit a mixed fracture mode. This is also reflected by their increased ductility, as was observed from the tensile test results.

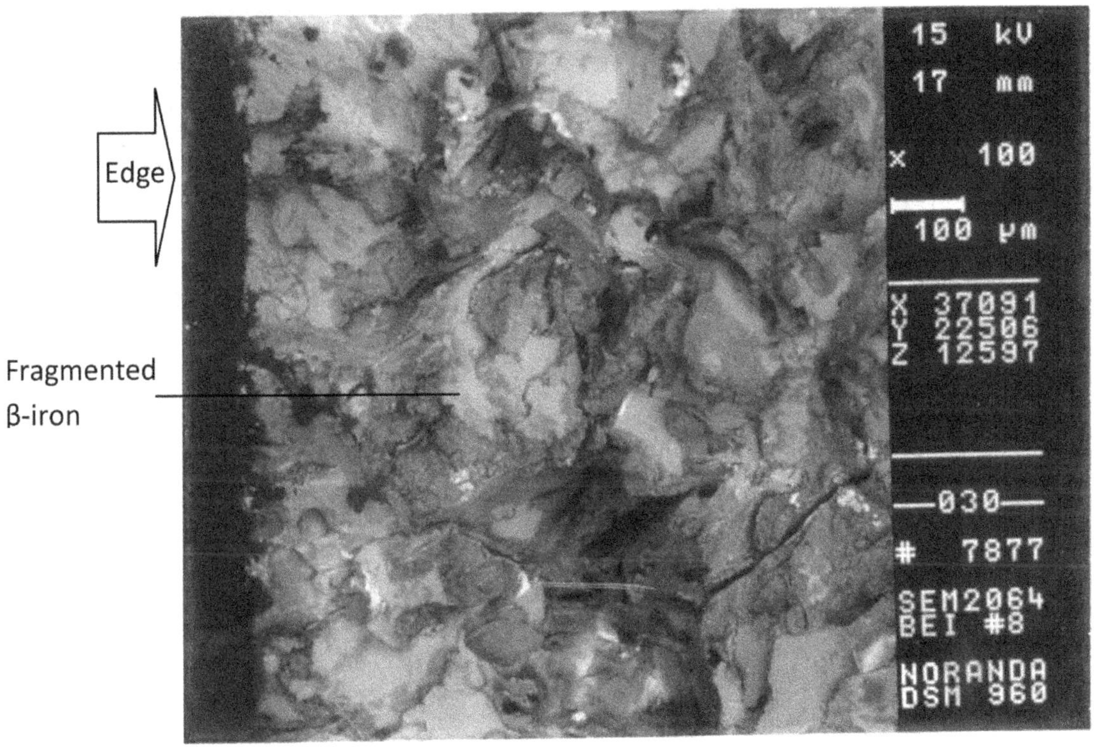

Figure 10. SEM micrograph showing fracture surface of alloy 319.2 (0.8% Fe, DAS 83 μm, modified, edge).

Figures 10 and **11** show SEM fractographs taken from the modified 319.2 alloy sample containing 0.8% Fe and obtained at the lowest cooling rate. Comparing **Figure 10** with **Figure 6** corresponding to the same alloy in the unmodified condition, the effect of Sr on the β-Al_5FeSi phase is evidenced, in the deteriorated appearance of the latter (see medium gray regions all over the fracture surface). The small bright spots in **Figure 10** correspond to particles of undissolved $CuAl_2$. In **Figure 11**, however, the bright regions (marked C) viewed on top of the β-platelets are Cu-containing phases, most likely Al_7Cu_2Fe. As observed by Li et al. [10, 33], this phase is insoluble even after 100 h solution heat treatment at 500°C. Compared to the flat surface of the β-platelet marked B, the perforated nature of those marked A is quite clear. Due to the high Fe content of the alloy, the volume fraction of β-Al_5FeSi phase, although considerable, appears distributed as smaller platelets all over the microstructure. It is interesting to note from the smooth surface of the β-platelet at B (and that at C below) that the β-platelets not modified by Sr cracked by cleavage quite easily, whereas elsewhere, the crack passed through other regions (e.g., eutectic Si) as well, as evidenced by the intergranular cracking.

Both cleavage fracture and intergranular fracture are characteristically low-energy mechanisms, and the simultaneous operation of both can occur when the resolved stresses for transgranular cleavage are approximately equal. Also, if the preferred grain boundary fracture path is not continuous, and if the cleavage stress is relatively low, the regions that do not fracture intergranularly can fracture by cleavage. **Figure 11** shows an example of this type of fracture.

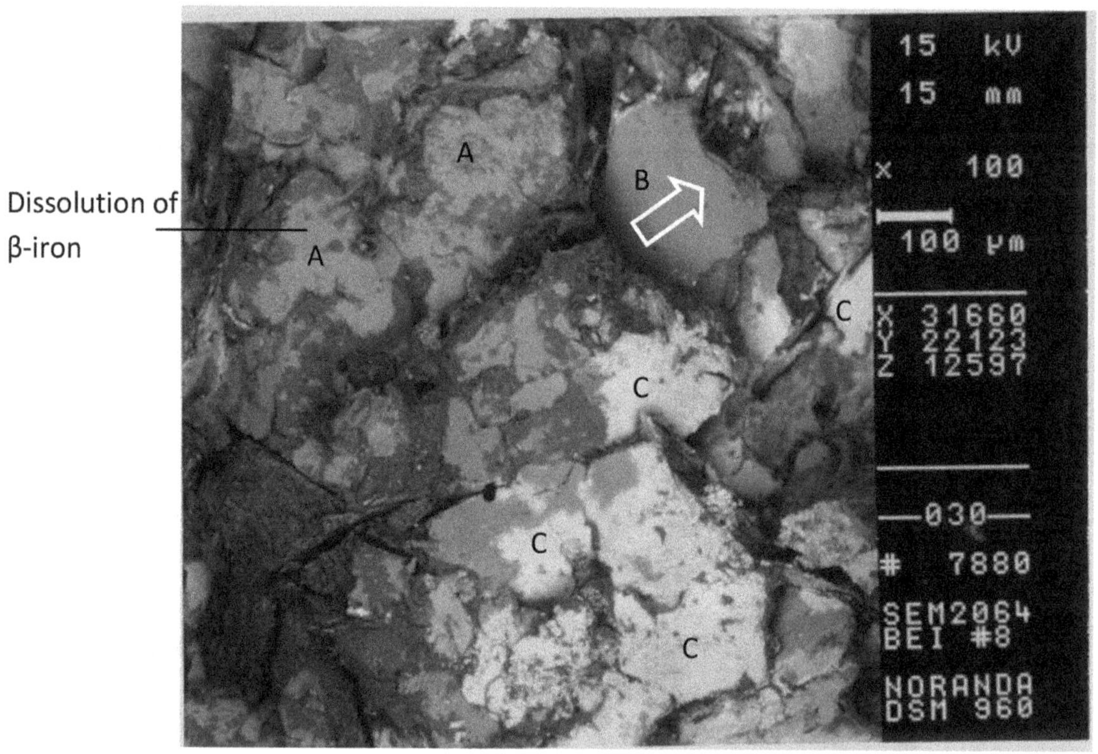

Figure 11. SEM micrograph showing fracture surface of alloy 319.2 (0.8% Fe, DAS 83 μm, modified, center).

From the overall mode of fracture observed in **Figures 10** and **11**, the modified alloy exhibits more resistance to crack propagation than that offered by the unmodified alloy (cf. E_I and E_P of 1.30 and 0.60 J with 1.63 and 1.43 J in the unmodified and modified alloys, respectively) where the brittle fracture/cleave features reflect the ease of crack propagation in that sample.

In what follows, some of the interesting features that were observed in the analysis of the fracture surfaces of various 319.2 alloy samples studied will be shown. These samples were selected to highlight other microstructural aspects such as inclusions or porosity, besides the α- and β-iron intermetallics and Si particles, which could affect the fracture behavior.

The SEM micrograph of **Figure 12** shows the cleavage fracture of an α-AlFeSi intermetallic particle in the modified 319.2 alloy sample containing 0.8% Fe and obtained at the highest cooling rate. The fracture consists of several cleavage steps that provide an indication of the local direction of crack growth. Further above, the cleavage planes followed by the crack show various alignments, as influenced by the orientations of individual grains.

Another feature of cleavage fracture, the "tongue" appears on cleavage facets as very fine slivers (see arrows marked T) that result from cleavage across micro-turns formed by plastic deformation at the tip of the main propagating crack. Around the α-AlFeSi particle, the surrounding eutectic Si regions are identified by the small microvoids and dimpled nature of the fracture surface, indicating the mixed fracture mode of the sample. The E_I, E_P, and E_T in this case are 8.71, 1.31, and 10.02 J, respectively.

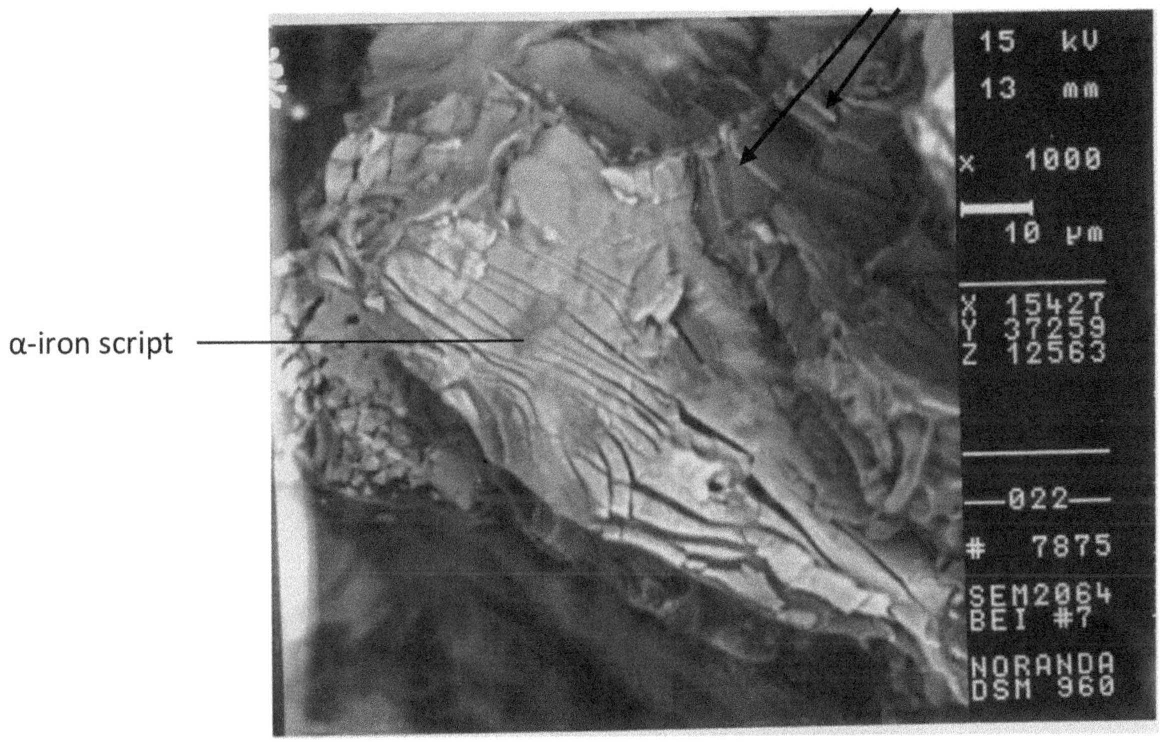

Figure 12. SEM micrograph showing fracture of α-AlFeSi intermetallics in alloy 319.2 (0.8% Fe, DAS 23 μm, modified).

It should be mentioned here that the iron intermetallic phases/particles referred to in the different fractographs presented in this chapter were identified using EDS analysis. The flat cleaved surfaces of these iron intermetallics facilitated identification. For the Cu-containing particles, it was difficult to obtain the exact composition of the $CuAl_2$ particles due to their size and the uneven nature of the surrounding fracture surface. In the case of the Al_7Cu_2Fe phase, as it occurred on top of the β-Al_5FeSi phase (the precipitation of the Al_7Cu_2Fe phase follows that of the β-Al_5FeSi phase [34]), its composition could not be verified correctly, as well.

Figure 13 shows another fractograph of the same sample, where the fracture surface consists of interlinked β-phase regions that would probably correspond to more than one branched β-platelets. Alpha-Al dendrites appear through parts of the β-phase where dissolution of the platelets has occurred. Dimpled rupture of the eutectic Si regions interspersed in between is also observed. The shallow, depressed nature of the dendrite observed near the white arrow, together with the darker void region below (black arrow), indicate the existence of a pore in this area, which would have facilitated crack propagation. The smooth interface between the β-phase and the α-Al dendrite and the pore linked to them affirms the feedability-related characteristics of the β-platelet phase, aiding in the formation of porosity.

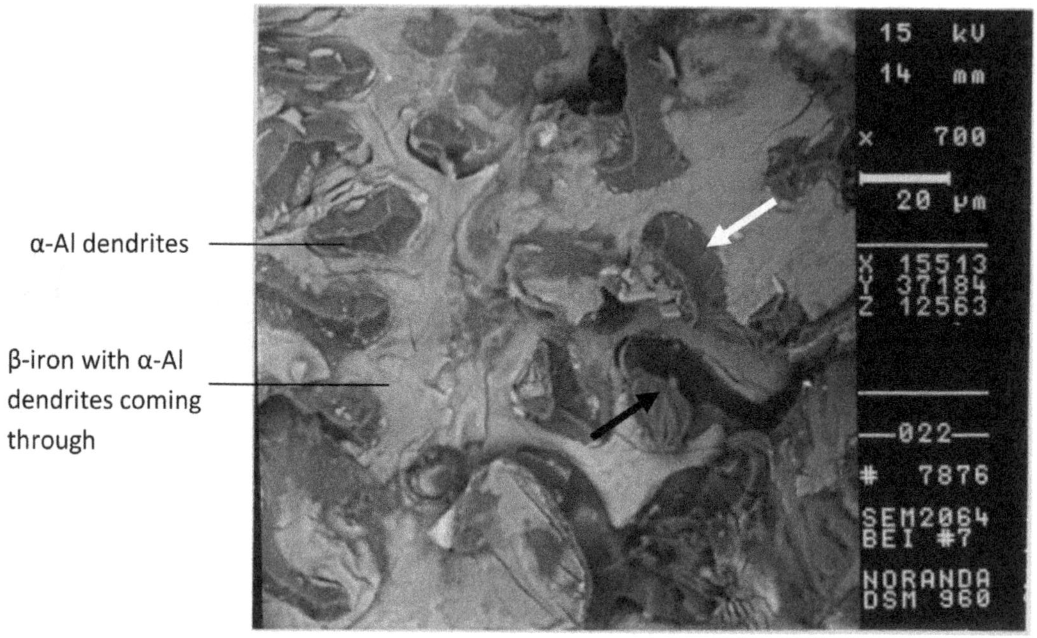

Figure 13. SEM micrograph showing fracture of β-AlFeSi intermetallics in alloy 319.2 (0.8% Fe, DAS 23 μm, modified).

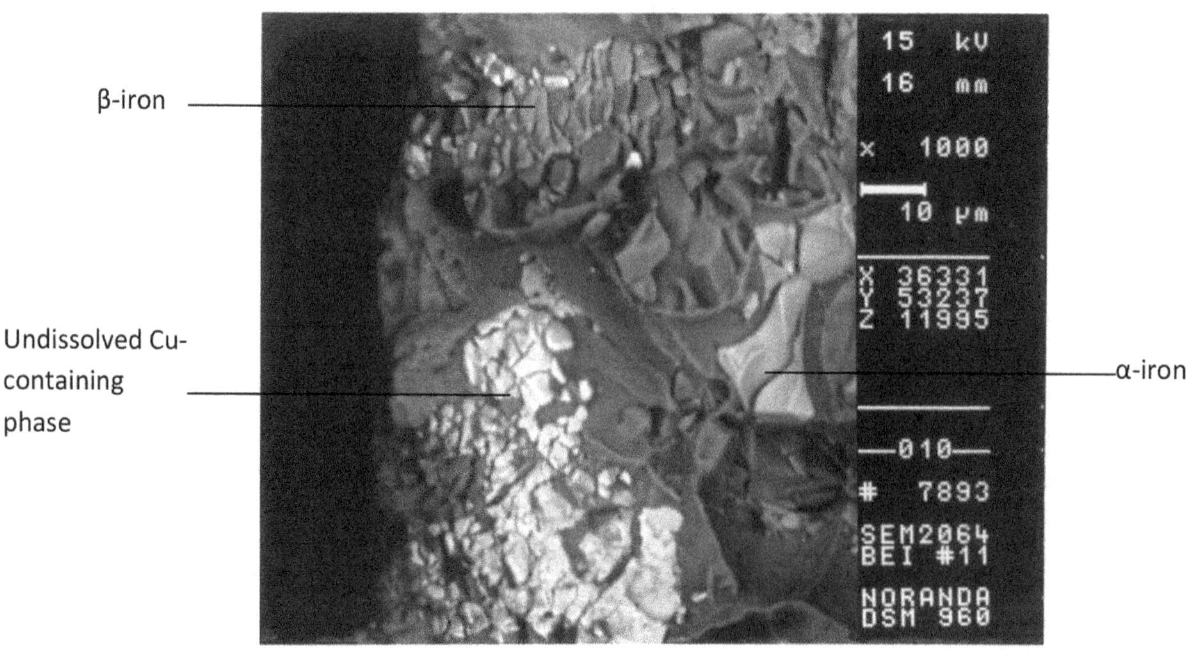

Figure 14. SEM micrograph showing fracture of both iron and copper intermetallics in alloy B319.2 (0.8% Fe, DAS 23 μm, unmodified).

Figure 14 shows the fractograph taken from the unmodified B319.2 alloy sample containing 0.8% Fe, obtained at the highest cooling rate. Transgranular fracture of all intermetallics viz., the α-Fe, β-Fe, and Cu-containing phases is observed. In this alloy, containing a higher level of Mg, the Cu-phase would include $Al_5Mg_8Cu_2Si_6$ and $Al_{15}(Mn,Fe,Cu)_3Si_2$, in addition to the $CuAl_2$ phase. The presence of cracked Si particles and associated microvoids are also seen.

A comparison with the fractograph shown in **Figure 6** (same Fe level, but lowest cooling rate) shows how a high cooling rate is more effective than the Fe content in determining the size of the β-platelets that result and, hence, the nature of the fracture, i.e., fragmentation or cleavage (this point is further brought out in the context of **Figure 16**). The corresponding E_I, E_P, and E_T values of this sample are 6.18, 1.40, and 7.58 J, respectively. However, as **Figure 15** shows, the β-platelet can still fracture by cleavage, particularly in regions where several platelets have precipitated together. The smooth surface of the various β layers in the figure clearly indicate that fracture in this part of the specimen occurred by cleavage. It is not hard to imagine the branched appearance of these several platelets in a two-dimensional optical micrograph.

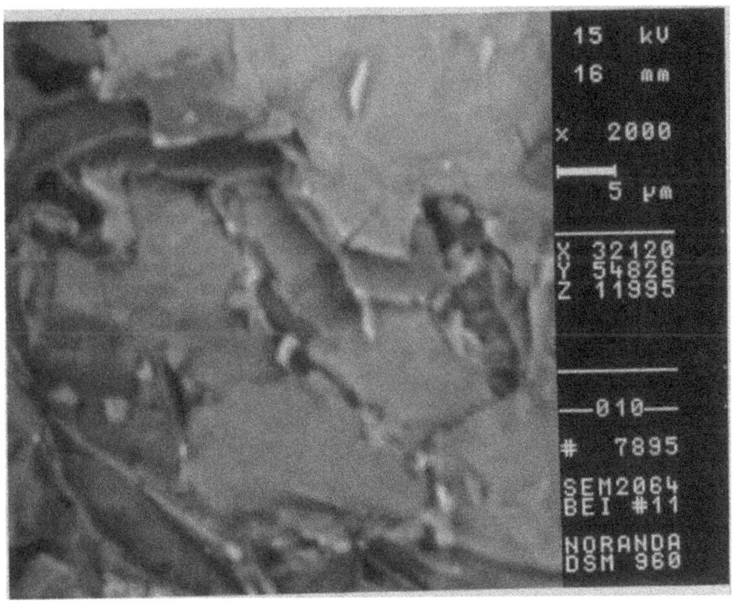

Figure 15. SEM micrograph showing fracture of β-platelets in alloy B319.2 (0.8% Fe, DAS 23 μm, unmodified).

The SEM fractograph, shown in **Figure 16**, of the same 0.8% Fe-containing unmodified B319.2 alloy obtained, however, at the lowest cooling rate shows clearly the massive dimensions that the β-Al_5FeSi platelets can attain at the same 0.8% Fe level, depending upon the solidification conditions. Again, the platelet fractures by cleavage, the open arrow indicating the cleavage direction. Secondary cracks (white arrows) are also observed when the crack propagates through the eutectic Si regions.

Figures 17 and **18** are SEM fractographs obtained from modified 319.2 and B319.2 alloys, respectively, under the same Fe and solidification conditions (0.8% Fe, lowest cooling rate). The rather large dimensions of the pore observed in **Figure 17**, as evidenced by the untouched α-Al dendrites sitting inside, show how the β-phase platelets lodged beside (1) and between

(2) the dendrites restrict the flow of the liquid metal in the surrounding region and hence lead to the formation of porosity. The presence of the pore facilitates fracture. In contrast to the fractured β-platelets (e.g., 3 and 4) around the pore area, the β-platelets within the pore did not participate in the fracture process and remained quite intact.

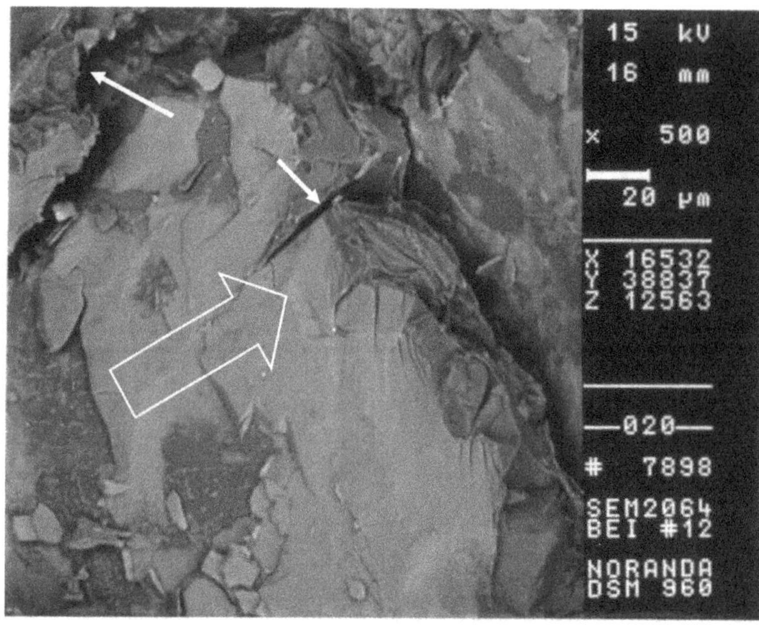

Figure 16. SEM micrograph showing fracture of β-AlFeSi intermetallics in alloy B319.2 (0.8% Fe, DAS 83 μm, unmodified).

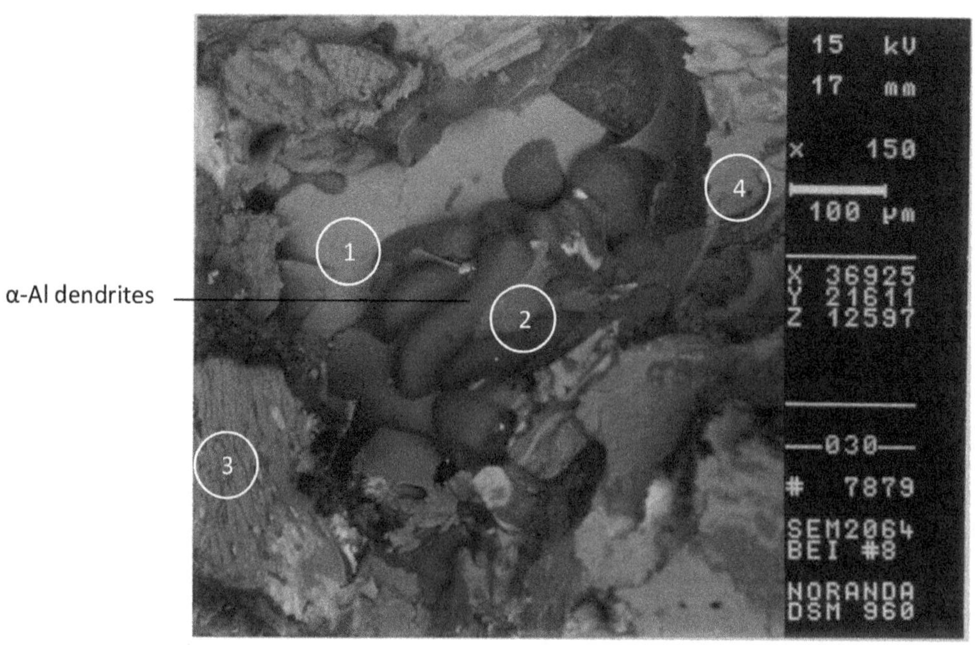

Figure 17. SEM micrograph showing the effect of porosity participating in the fracture behavior of alloy 319.2 (0.8% Fe, DAS 83 μm, modified).

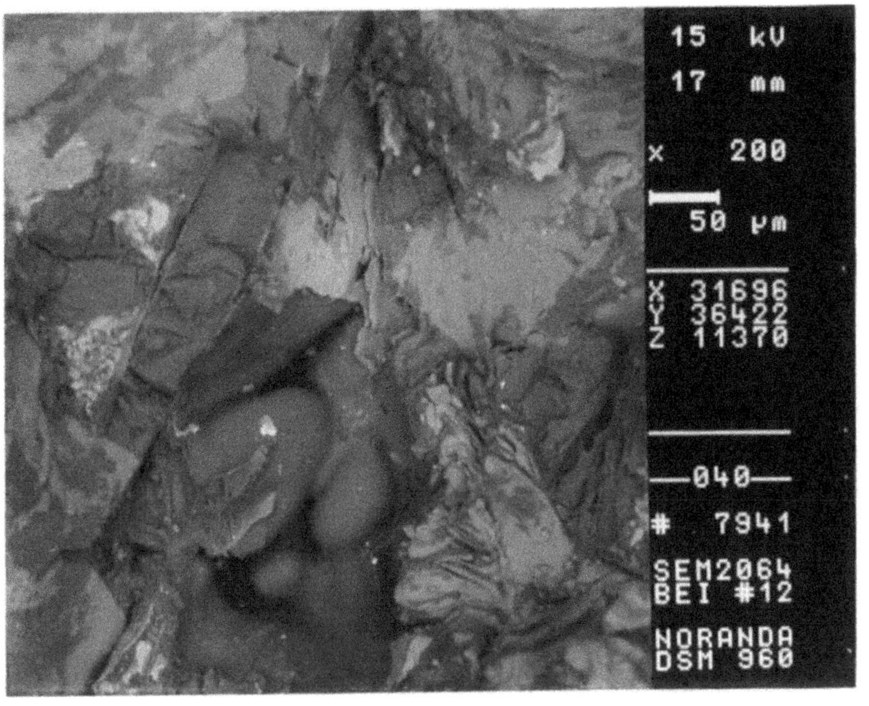

Figure 18. SEM micrograph showing the effect of porosity participating in the fracture of alloy B319.2 (0.8% Fe, DAS 83 µm, unmodified).

Similarly, **Figure 18** shows an example of a pore observed on the fracture surface of the modified B319.2 alloy. In this instance, the pore appears to be much smaller in dimension and more shallow than the one depicted in **Figure 17** with fragments of cleaved β-phase platelets on the surrounding fracture surface. The dissolution of the β-platelets due to the effect of Sr is more evident in this fractograph.

3.3. Alloy A356.2 (alloy C)

As was done for the 319 alloys, the fractographs of A356.2 alloy samples covering the lowest Fe content—highest cooling rate/optimum property and highest Fe content—lowest cooling rate/minimum property conditions have been selected for discussion purposes, to bring out the contrast in their fracture modes and hence comprehend the difference in their impact properties.

As Voigt and Bye [35] have commented, features appearing on the fracture surfaces of Al-Si-Mg casting alloys such as A356 alloys can be very difficult to interpret due to the lack of SEM contrast between the α-Al and the eutectic Si phases appearing on the surface. As Al and Si are close in their atomic numbers, even the use of SEM techniques such as backscattered electron imaging cannot provide additional phase contrast. However, the features observed on the fracture surfaces of A356 alloys can be more clearly interpreted based on a study of the surface crack initiation and propagation process. Usually, by observing the microstructure just below the surface that can be observed inside the crack using SEM techniques, or else by studying longitudinal sections of the fractured samples (perpendicular to the fracture surface) using optical microscopy, additional information can be obtained that is useful in analyzing the fracture behavior.

Figures 19 and **20** are the SEM fractographs taken from the modified A356.2 alloy sample containing 0.1% Fe and obtained at the highest cooling rate. **Figure 19** shows the fracture to be of the transgranular dimpled rupture type. As mentioned before, the BEI image does not provide a good phase contrast.

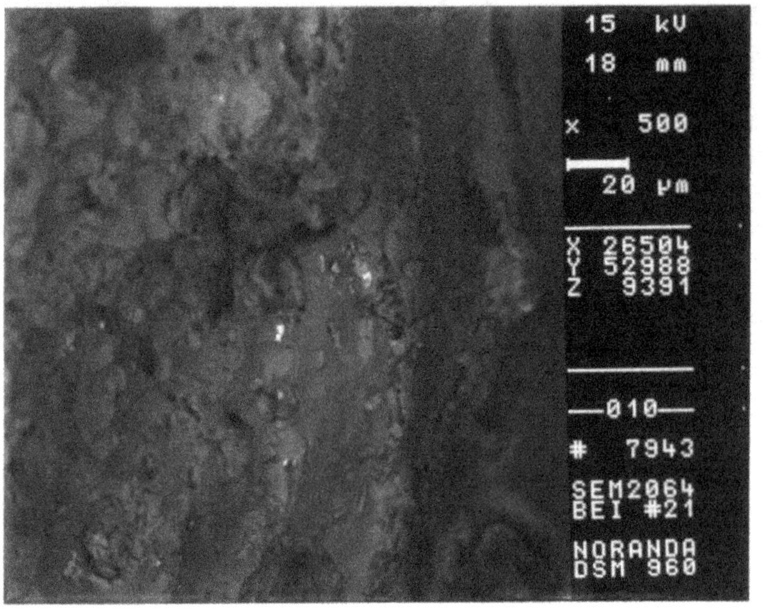

Figure 19. SEM micrograph showing fracture surface of alloy A356.2 (0.1% Fe, DAS 23 μm, modified, edge).

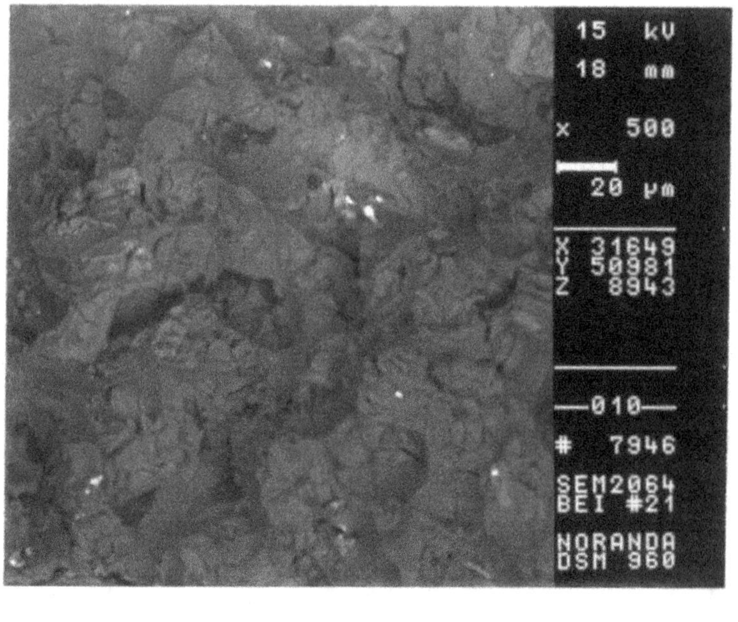

Figure 20. SEM micrograph showing fracture surface of alloy A356.2 (0.1% Fe, DAS 23 μm, modified, center).

The fractograph of **Figure 20** taken from the center of the sample shows these features more clearly. Note the fine Si particles in the structure. The absence of intermetallics, due to the low Fe content, together with the fine Si particles obtained under the optimum alloy conditions of a high cooling rate and Sr modification, results in the sample exhibiting the highest impact energy (80.88 J).

Figures 21 and **22** depict the SEM fractographs of the unmodified A356.2 alloy sample containing a high Fe level (0.6%) and obtained at the lowest cooling rate. The presence of massive β-Al_5FeSi platelets in the alloy microstructure under these conditions results in the brittle fracture of the sample. As **Figure 21** shows, crack initiation takes place by the fragmentation of the β-phase at the edge of the sample and some amount of transgranular fracture, as well, whereas the crack propagates by cleavage fracture, as evidenced by the relatively smooth surfaces of the β-platelets observed in **Figure 22**, as also by intergranular fracture. Accordingly, the impact energy of this sample was 2.45 J.

When the same alloy is modified, crack initiation occurs by fragmentation of the β-Al_5FeSi platelet, followed by cleavage, as seen in **Figure 23**. Modified silicon eutectic regions around the β-platelet can also be distinguished. **Figure 24** shows that the fracture mode is a mixture of cleavage and intergranular fracture.

The high magnification SEM fractograph of an unmodified A356.2 alloy sample containing 0.1% Fe and obtained at the lowest cooling rate, **Figure 25**, reveals the eutectic Si regions much more clearly, where the fracture is seen to occur by the transgranular brittle fracture of the acicular Si particles (see circled areas).

Figure 21. SEM micrograph showing the role of β-Al_5FeSi intermetallics in initiating fracture in alloy A356.2 (0.6% Fe, DAS 75µm, unmodified).

Figure 22. SEM micrograph showing crack propagation in alloy A356.2 (0.6% Fe, DAS 75 μm, unmodified).

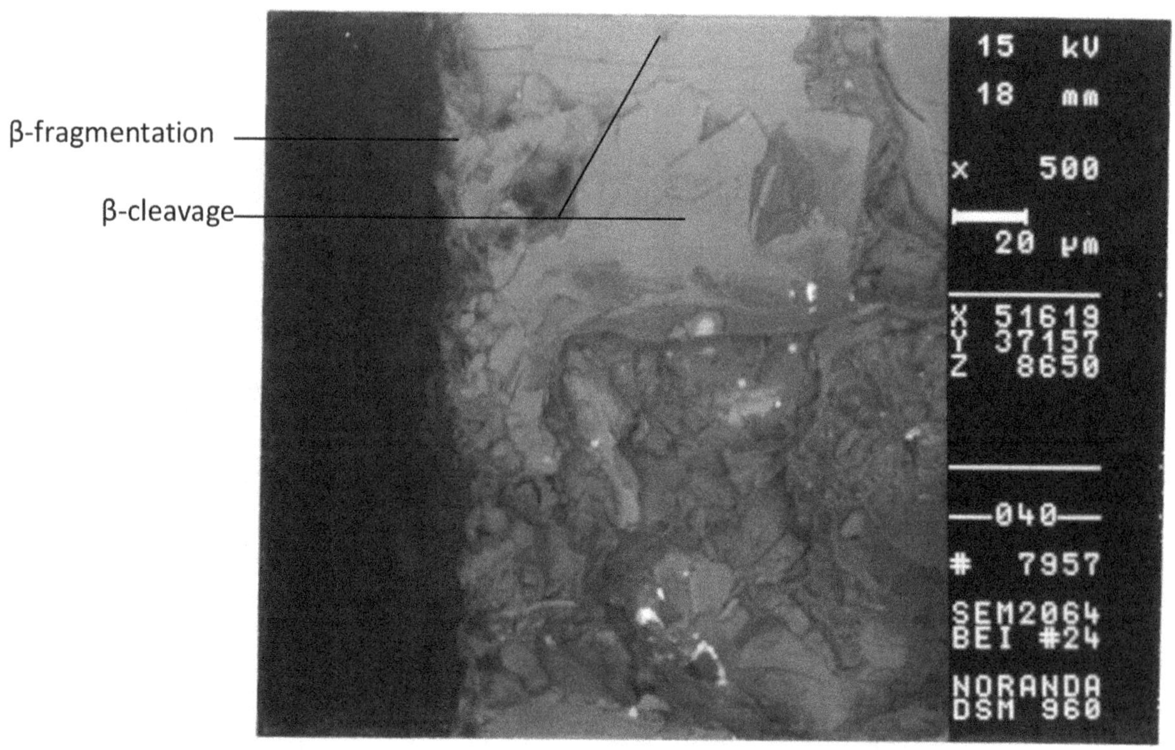

Figure 23. SEM micrograph showing fracture surface of alloy A356.2 (0.6% Fe, DAS 75 μm, modified, edge).

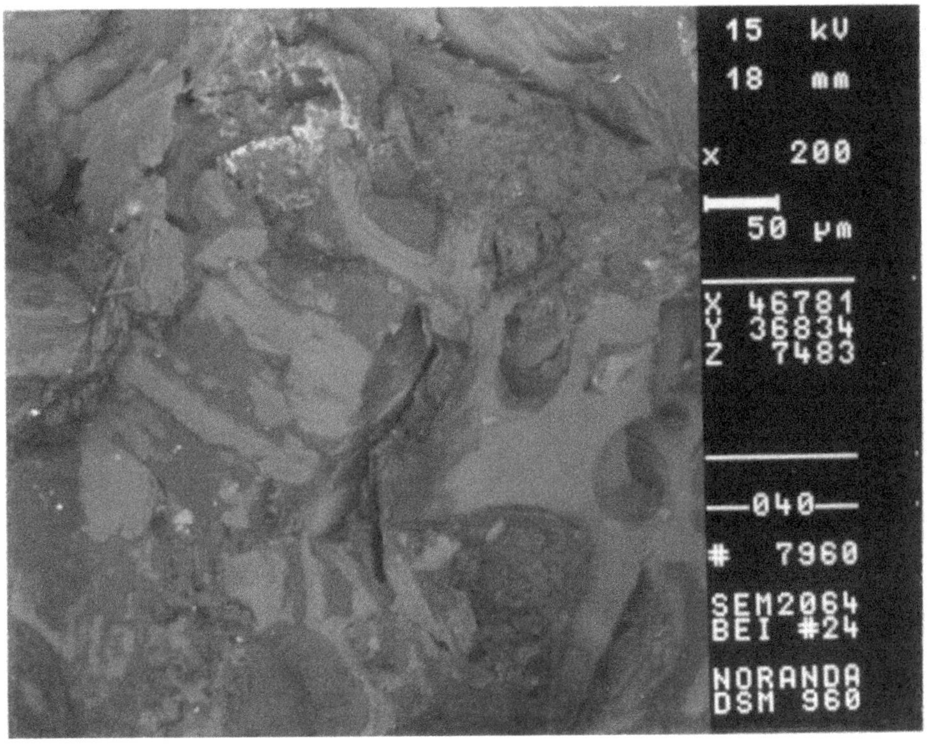

Figure 24. SEM micrograph showing fracture surface of alloy A356.2 (0.6% Fe, DAS 75 μm, modified, center).

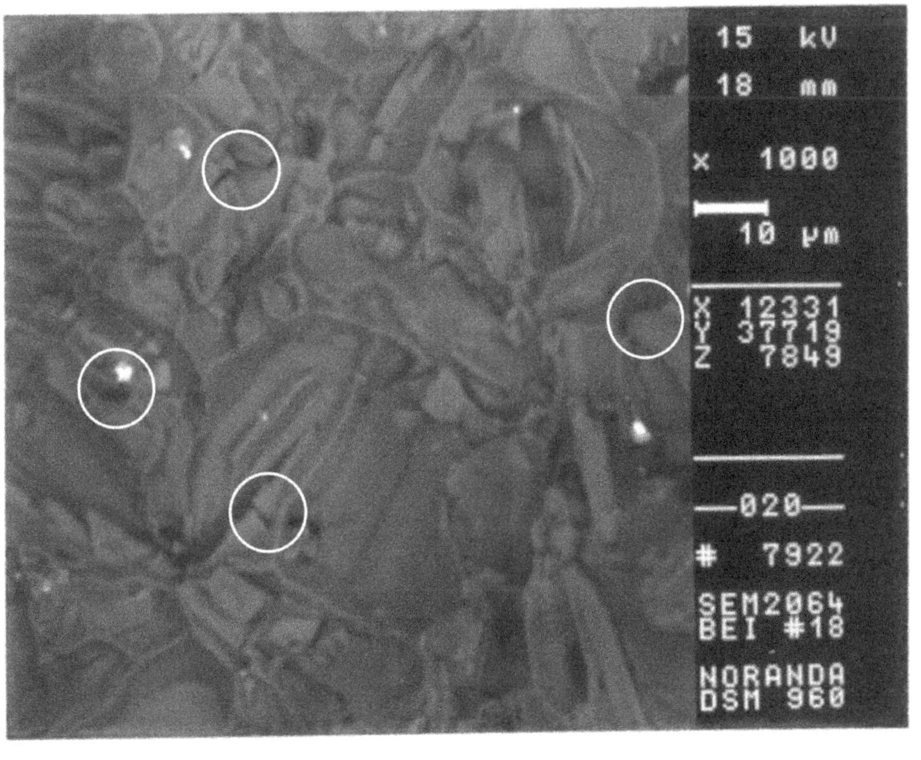

Figure 25. SEM micrograph showing fracture of eutectic Si particles in alloy A356.2 (0.1% Fe, DAS 75 μm, unmodified).

3.4. Optical microscope fractography

Regardless of SEM fractography, it is also quite useful to examine the fracture profile on sections perpendicular to the fracture surface. This can be carried out easily using an optical microscope. In this way, the origin of the fracture can be examined to determine if important microstructural abnormalities are present that either caused or contributed to fracture initiation. It is also possible to determine if the fracture path at the initiation site is transgranular or intergranular and to determine if the fracture path is specific to any phase or constituent present [36]. Some examples of the longitudinal sections of the A356.2 alloy samples are presented in this section with a view to further clarify the fracture process in these samples as their SEM fractographs did not display as good a phase contrast as those shown by the 319 alloy samples.

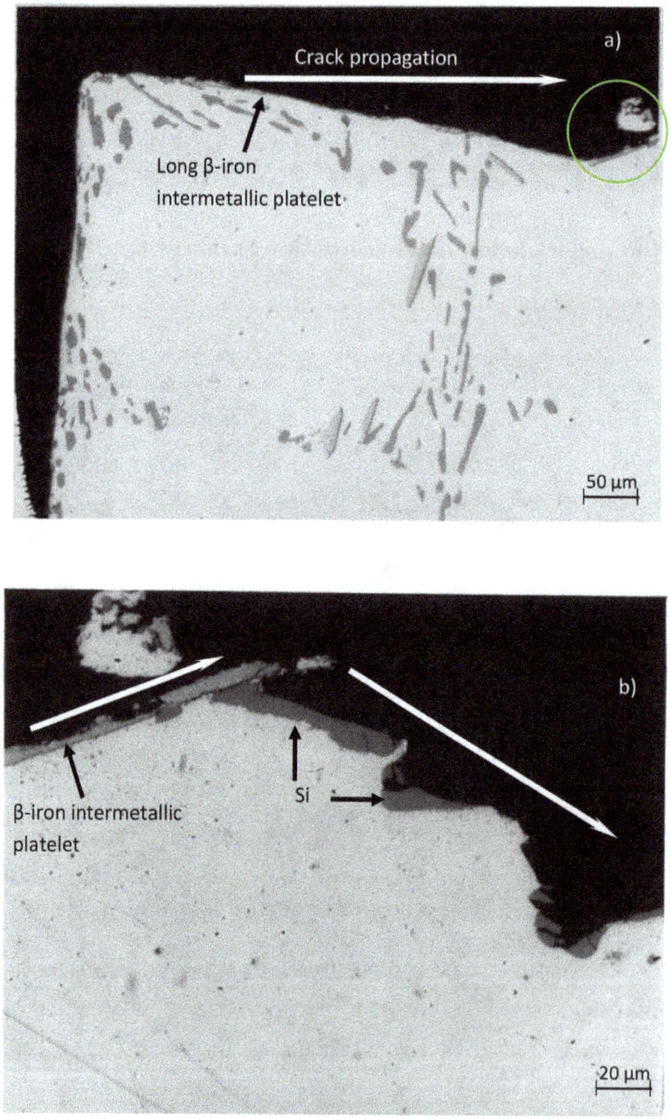

Figure 26. Optical micrographs showing fracture behavior of alloy A356.2 (0.6% Fe, DAS 75µm, unmodified, edge). A high magnification micrograph of the circled area in (a) is shown in (b).

The optical micrographs presented in **Figure 26** show the polished longitudinal section of the fractured A356.2 unmodified alloy sample containing 0.6% Fe, obtained at the lowest cooling rate. In **Figure 26(a)**, the edge of the sample on the left shows how crack initiation occurs by cleavage and/or transgranular fracture of the Si particles and β-Al_5FeSi platelets. The sample edge corner is relatively straight, indicating that cracking occurred almost immediately, i.e., by brittle fracture mode, and the crack propagated along the direction shown.

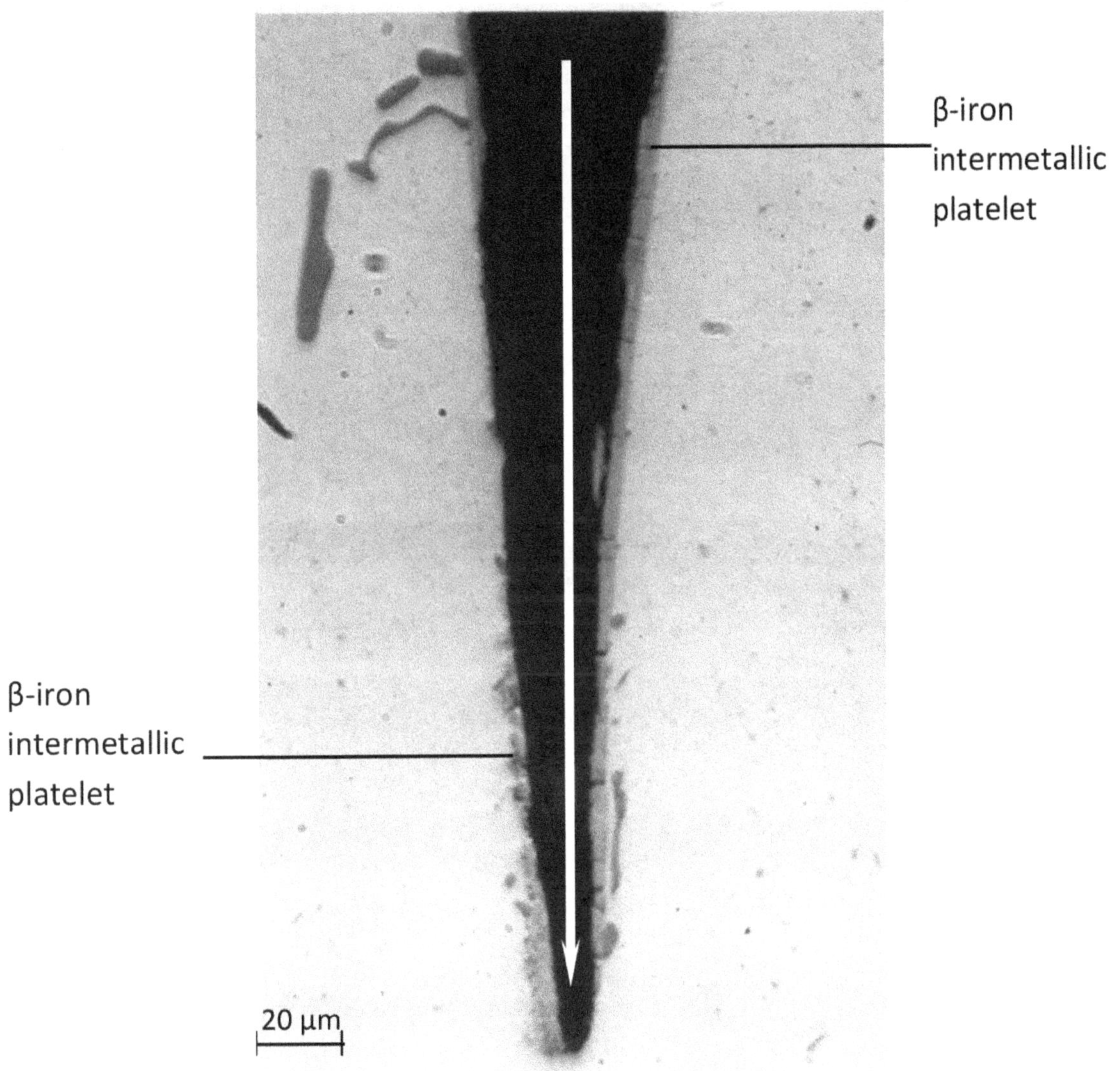

Figure 27. Optical micrograph showing fracture behavior of alloy A356.2 (0.4% Fe, DAS 23 μm, unmodified, edge, 500×).

A higher magnification micrograph, **Figure 26(b)**, of the fracture surface corresponding to the circled area in **Figure 26(a)** shows how the crack propagates further along the fracture surface by cleavage of the β-iron intermetallic and coarse Si particles. The persistence of such coarse Si particles, even after solution heat treatment at 540°C/8 h, aids in inducing the brittle fracture of the sample. Correspondingly, under such alloy conditions, the sample exhibited the lowest impact energy (2.45 J). **Figure 26** provides a good example of how optical microscopy can aid in providing a clearer interpretation of the crack initiation and propagation in A356 type Al-Si-Mg alloys.

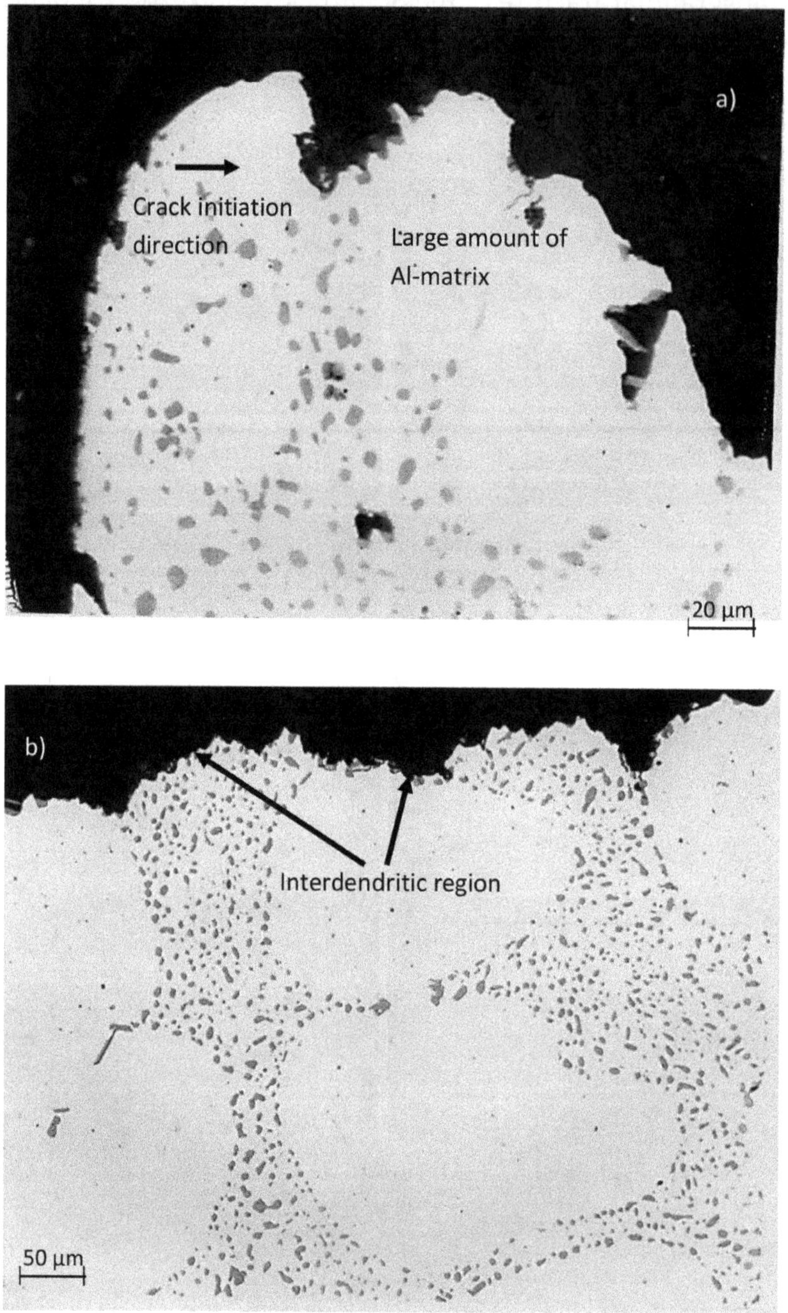

Figure 28. Optical micrographs showing fracture behavior of alloy A356.2 (0.1% Fe, DAS 23 μm, Sr-modified, edge).

The optical micrograph in **Figure 27**, taken from an A356.2 unmodified alloy sample containing 0.4% Fe and obtained at the highest cooling rate, shows the propagation of a secondary crack (see white arrow) that splits a β-Al_5FeSi platelet right through the middle. The crack propagates to the end of the platelet, till it reaches the aluminum matrix which, being ductile, allows for stress relaxation at the crack tip by local plastic flow.

Compared to the micrographs shown in **Figure 26**, those obtained from the A356.2 alloy sample corresponding to optimum energy conditions (viz., 0.1% Fe, lowest cooling rate, Sr-modified alloy), **Figure 28(a)** shows how the sample edge is more curved or rounded , rather than flat as was observed in the case of the unmodified alloy sample of **Figure 26**. The curvature of the fracture surface, also clearly observed in **Figure 28(b)**, indicates a ductile mode of rupture as the crack propagates through the well-modified Si eutectic regions. The E_I, E_P, and E_T values in this case are 60.44, 20.44, and 80.88 J, respectively, compared to 1.41. 1.31, and 2.45 J in the case of the sample shown in **Figure 26**.

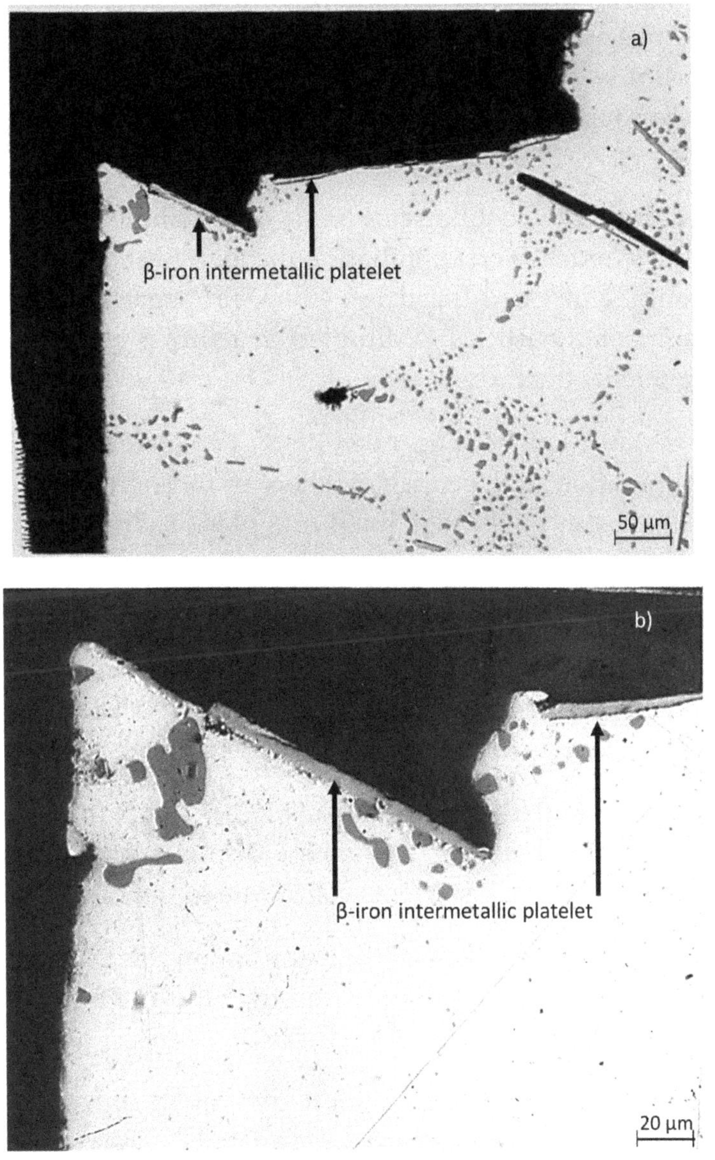

Figure 29. Optical micrographs showing fracture behavior of alloy A356.2 (0.6% Fe, DAS 75 µm, Sr-modified, edge).

However, when the alloy Fe level is increased and the cooling rate decreased to its lowest value, even in the Sr-modified alloy, the fracture still occurs mainly by cleavage of the brittle β-Al_5FeSi platelets, as seen in **Figure 29(a)**. Note the sharp edge of the sample fracture surface in this case, compared to **Figure 28**. **Figure 29(b)** shows these features much more clearly at a high magnification. In instances when the crack propagates through the more ductile Al-Si regions, the fracture surface is more rounded than sharp. The corresponding E_I, E_P, and E_T values are 3.41, 1.46, and 4.87 J, respectively.

4. Conclusions

The present research was carried out on two of the main alloys used in automotive applications, namely 319 and 356 alloys, where the effects of iron content, Sr modification, and cooling rate on the impact properties were studied. The main observations on the fracture behavior in relation to the impact properties are presented in this chapter and may be summarized as follows.

1. At low iron levels and high cooling rates (0.4% Fe, 23 μm DAS), and in the unmodified condition, 319 alloys undergo crack initiation through the fragmentation of Si particles, β-iron intermetallics, and $CuAl_2$ particles. Crack propagation takes place through the linking of the fragmented particles. A dimpled structure is observed for the Sr-modified alloys, indicating ductile fracture.

2. At high iron levels and low cooling rates (0.8% Fe, 83 μm DAS), in the unmodified condition, crack initiation and propagation occur through the cleavage of β-Al_5FeSi platelets (rather than by decohesion of the β-iron platelets from the matrix).

3. The morphology of the β-iron intermetallic platelets (individual or branched) is important in determining the direction of crack propagation. Cracks also propagate through the fracture of undissolved $CuAl_2$ or other Cu-intermetallics such as Al_7Cu_2Fe, as well as through fragmented Si particles. In the Sr-modified alloys, cracks are initiated mostly with the fragmentation or cleavage of perforated β-phase platelets.

4. A small amount of crack initiation can also take place through the fragmentation of coarse Si particles that have not fully undergone spheroidization or undissolved Cu-intermetallics. Thus, all microstructural constituents contribute to the fracture process.

5. In the A356.2 alloys, in samples obtained under optimum conditions (0.1% Fe, 23 μm DAS, Sr-modified alloy), cracks initiate mainly through the fracture of Si particles or the debonding of the Si particles from the Al matrix.

6. In the unmodified condition, cracks mainly propagate through the coalescence of fractured Si particles, except where β-iron intermetallics are present, in which case the latter takes priority in fracture propagation.

7. In the Sr-modified condition, more of the Al matrix is involved when cracks propagate through the linkage of fractured/debonded Si particles, as well as fragmented β-iron intermetallics. For samples characterized by low-impact energies, crack initiation and propagation occurs mainly through the cleavage of β-iron intermetallics.

Acknowledgements

The authors would like to thank Ms Amal Samuel for enhancing the artwork presented in this manuscript.

Author details

Zheyuan Ma[1], Agnes M. Samuel[1], Herbert W. Doty[2] and Fawzy H. Samuel[1*]

*Address all correspondence to: fhsamuel@uqac.ca

1 University of Québec at Chicoutimi, Québec, Canada

2 Materials Technology, Pontiac, MI, USA

References

[1] Gabriel BL. SEM: A User's Manual for Materials Science. Metals Park, OH: American Society for Metals; 1985, p. 97.

[2] Gerberich WW. Microstructure and Fracture. In: Metals Handbook, Vol.8: Mechanical Testing. 9th ed. Metals Park, OH: American Society for Metals; 1985, p. 476–491. ASM Handbook Committee. John R. Newby, Coordinator.

[3] Ma Z, Samuel AM, Doty HW, Valtierra S, Samuel FH. Effect of Fe content on the fracture behaviour of Al–Si–Cu cast alloys. Materials & Design. 2014;57:366–373.

[4] Kerlins V, Phillips A. Modes of Fracture. In: Metals Handbook, Vol. 12: Fractography. 9th ed. Metals Park, OH: American Society for Metals; 1987, p. 12–71. ASM Handbook Committee. The Editors.

[5] Wang QG, Cáceres CH. The fracture mode in Al–Si–Mg casting alloys. Materials Science and Engineering A. 1998;241:72–82.

[6] Ammar HR, Samuel AM, Samuel FH, Simielli E, Sigworth GK, Lin JC. Influence of aging parameters on the tensile properties and quality index of Al-9 Pct Si-1.8 Pct Cu-0.5 Pct Mg 354-type casting alloys. Metallurgical and Materials Transactions A. 2012;43(1):61–73.

[7] Gall K, Yang N, Horstemeyer M, McDowell DL, Fan J. The debonding and fracture of Si particles during the fatigue of a cast Al-Si alloy. Metallurgical and Materials Transactions A. 1990;30A(12):3079–3088.

10

Fractography on Rigid Ceramics with Ultra-High-Molecular-Weight Polyethylene Fabric after Ballistic Impacts

Giovanni Barrera Torres, Juan Manuel Vélez and Carlos Mario Gutierrez

Abstract

The impact protection systems have traditionally been developed with metallic materials and structural protection applied in the automotive sector; while personal protection systems and composite systems evolve very fast, these systems have a large increase in applications due mainly to the amplitude of the manufacturing process. In the composite system for impact protection, it exposes both functionality and the rigid systems, aimed at structural protection, and flexible systems for personal protection. Nowadays, the development of materials with ballistic applications has emphasized protection of lightweight materials, for protection against projectiles of high and low speeds, among which are the bullets from weapons, fragments of tempered steel grenade hand or aircraft fragments, or a vehicle at high speed directly against housing and human integrity. It is necessary to investigate the mechanisms of fracture of the materials usually used for protection against impact, thus, it is possible to obtain important design systems that reduce the probability of failure and protect human lives and reduce damage to infrastructure information. In this chapter, the behavior of laminate sandwich-type systems, made from handcrafted ceramic plates with sheets of polyethylene (**ultra-high-molecular-weight polyethylene**, UHWMPE) against the impact of a metallic projectile, has been explored. The experimental work was made of two groups with different arrangements: the first group with the side that receives the impact of ceramic material (silicon carbide, SiC)-backed polymeric material and the second group with the side that receives the impact on polymer-backed ceramic material, the plates had dimensions 200 mm^2 of thickness to 5 mm for single plates and 20 mm for double plates. The experimental test was performed following the parameters of impact of NIJ III A standard. Some mechanisms (morphologies) of dissipation of kinetic energy received in the components were identified, as cited below, first in the ceramic material the formation of a crater, fracture, and delamination was observed, formed in the double and simple plates. Later, in the fabric polymeric material deformation mechanisms, such as the origination of defibrillation, conical geometry formation, delamination, twisting, and melting fibers due to the tribological contact of the metal shell impact, were observed. The exploration was culminated making a comparison between the arrangement that had higher energy absorption compared to an additional system designed with a ceramic (SiC) with less porosity, also aside to this chapter, it shows some of the deformed projectiles with a basic description of the fracture obtained after the impact, complementing the overall analysis of the systems used.

Keywords: ballistic impact, UHWMPE, SiC, fractography, delamination

1. Introduction

Currently, the need to use protection systems has increased sudden impacts that may cause damage to human or structural integrity, for example, the fall of a hammer, splashing chip result of the fall or 1indirect impact of a solid on a tornado, and the possibility of metallic projectile impact, in military use. In industry, some of the energy dissipation mechanisms and phenomena that accompany projectile impact environment-specific materials are still unknown. In order to advance the development of materials or focus on the protection of these eventual systems, it is important to take steps to identify environmental problems, to advance the possibility of effective impact protection solutions using known materials, using research like this, but with effective collaboration to develop solutions and further reducing noise by lack of resources. It is necessary to obtain collaboration of several entities, such as state and university-industry scientific and further technological development. The research on ballistic protection has restricted nature; due to the qualification of the subject as strategic, it is necessary to continue in these strategic research for the preservation of life, and the use of high-tech materials, call the research the development of materials that have a minimal negative impact on the environment, which would be a second function of protecting life.

2. Protection systems

A protection system emerges as an assembly of components where each has its own functions to protect an item or human being through the absorption of kinetic energy derived from direct or indirect collision with a projectile. Systems can be flexible, formed by rigid or polymer fabrics or metals, or ceramic plates and rigid polymers. Systems can be formed by various materials as sandwich panels; an example applied is usually illustrated in **Figure 1**. The penetration mechanism is shown in a system designed by a combination of plastic, center of ceramic material, with a backrest-purpose flexible application–specific protection material.

The polymers have been occupying spaces where metal and ceramic materials were predominant, now in response to the need for mass reduction, polymeric materials have applications in light protection such as suits, military [1] hardware, and rigid applications in specific parts of helicopters and land vehicles. Compounds systems aim at combining the properties of individual resistance of each material to obtain a higher property required. In recent years, with the ultrahigh molecular weight polyethylene (**UHWMPE**) sheet and addressing, different fibers get increased kinetic energy absorption [2]. The composite systems may be obtained in a variety of ways with advanced manufacturing processes [3]; with the fibers such as aramid and ultrahigh molecular weight polyethylene, it is possible to develop products for protection against projectiles, whereas ceramic materials without compromising the volume thereof do not represent effective protection and need to be combined with metal and/or polymers for designs in areas such as aerospace or personal protective competing with kinetic energy absorption of steel [4].

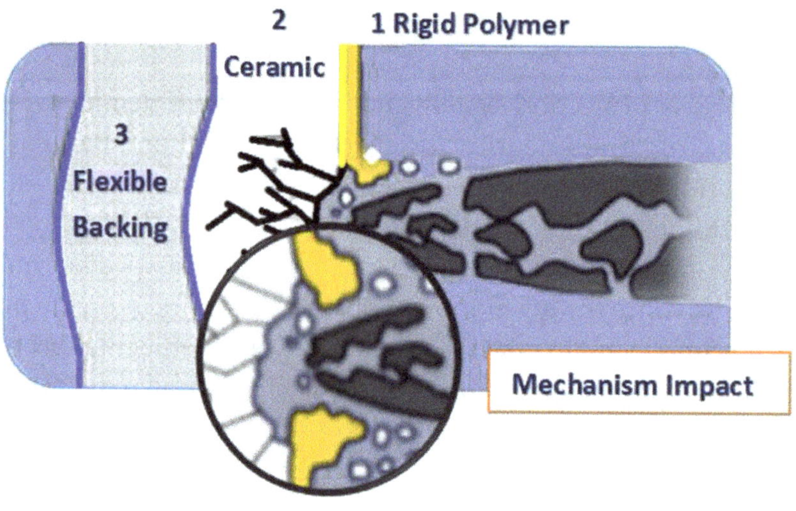

Figure 1. Composite panel (1, polymer; 2, ceramic flexible; 3, backing).

3. Materials and methods

The selection of materials was made from knowledge of the properties of polymers and ceramics used in ballistic applications [5–8]. Silicon carbide (SiC) was selected as the ceramic material because of the high hardness, high abrasion resistance, and its history in ballistic applications [9], besides being a commercial ceramic. The properties of SiC are due to the type of atomic bonding and crystalline structure hexagonal designed that cubic form α– SiC, the structure obtained depend on the method used for their production; in this case, the powder of silicon carbide used is produced by the Acheson process, which involves the reduction of silica sand in contact with petroleum coke or anthracite to a temperature close to 2400°C for 36 h, to form the commercial grade SiC α [10]. The properties of the ceramic used are as follows:

- Grain size: 10–50 μm
- Purity of material: silicon 80%, 20% silica binder based on cellulose apparent porosity of 40%

- Average true density: 1.60 g/cm³, obtained by making direct mass and volume of samples
- Temperature: 1250°C
- Plate dimensions: 197 ± 197 ± 3 mm × 3 mm × 5 ± 1 mm
- Average mass of plate: 293.4–312.9 g
- Theoretical density of SiC: 3.1–3.2 g/cm³ [11]
- Color: green
- Composed of compaction at atmospheric pressure, subsequently sintered in an electric furnace at 850°C

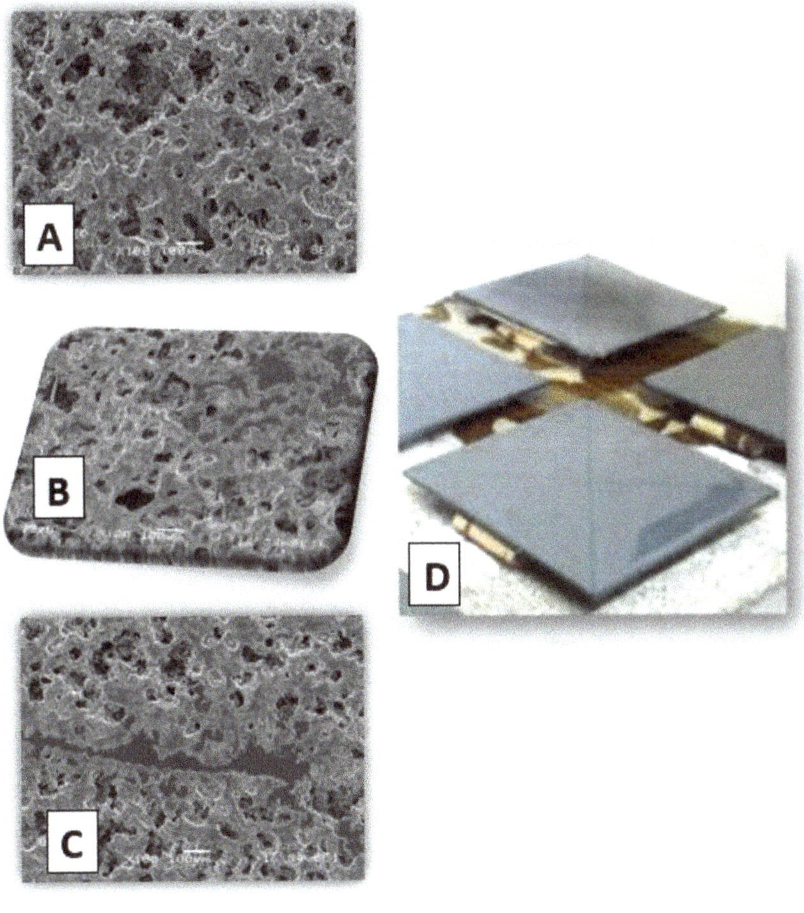

Figure 2. Morphology of the ceramic plates.

To determine the porosity of the samples, initially the specimens were dried and sintered in an oven at 850°C, then these were tested according to the ASTM C1039 standard for the apparent porosity of 40%; the surface porosity of the plates was 46% obtained using image analysis software. Image A in **Figure 2**, the plate obtained is shown unmodified, exhibits the surface porosity resulting from compaction process. Image B is observed at the surface of the plate with epoxy resin coating after dipping process, the isometric view. In image C, a portion of the plate with magnification 100× microscopy, where the resin has achieved infiltration into the porous body thereof, image D, as can be seen to the shaped plate bodies, is observed to be adhered to the polyethylene sheet (UHWMPE).

The flexible material selected for the system embodiment was laminated fabric of ultrahigh molecular weight polyethylene (UHMWPE) given the properties of high kinetic energy absorption in ballistic protection systems published in the current literature. In **Figure 3**, the morphological composition of the panel, in which each of the layers consisting of parallel threads, in the configuration of two successive layers in the same orientation panel is observed; the next two layers are oriented perpendicularly to the direction of previous threads. In **Figure 3**, image A, the panel is illustrated with an impact recording, seen isometrically. In images B and C, the panel approach is illustrated; it is evident that the diameter of each wire was about 18 μm and the size of the fabric to the development of the system was 200 mm × 200 mm ±3.

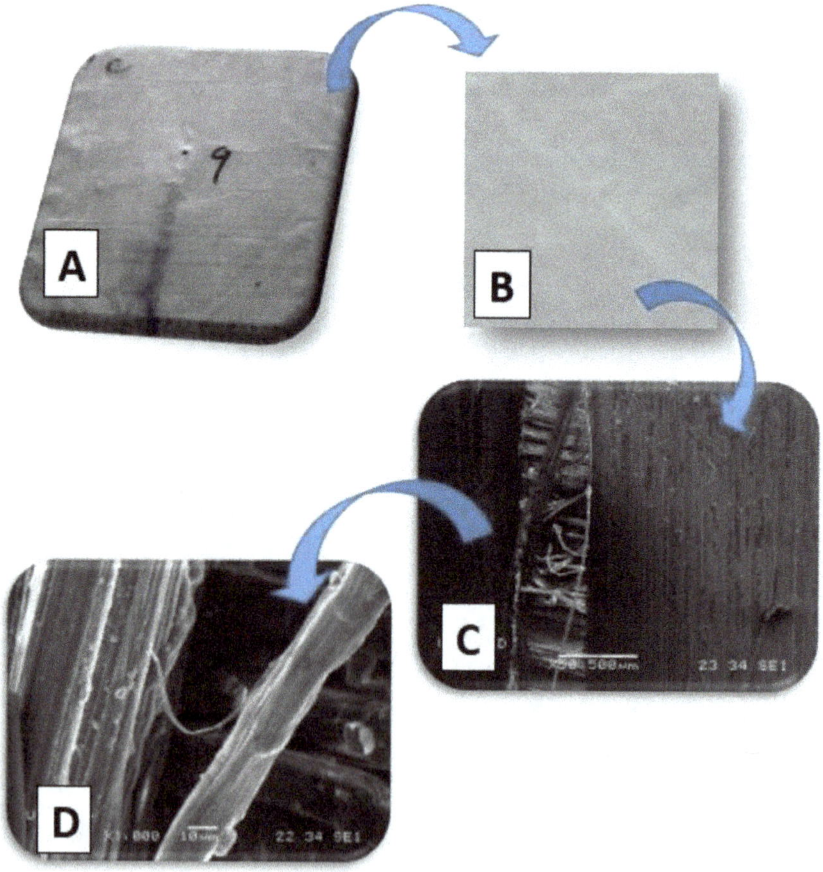

Figure 3. Images polyethylene ultrahigh molecular weight used.

The adhesive used in exploration is epoxy resin due to its good response, good heat sink residual stress, and compression stress [12], with better properties than polyester or phenolic resins have high hardness and very good adhesion in inhomogeneous materials [13]. Some of the properties of the epoxy resin used in this work are as follows:

Density: 1.1–1.4 g/cm^3.

Glass transition temperature: 120–190°C.

Viscosity: 113 centipoise at 22°C, viscosity taken in laboratory viscometer Brookfield of the Instituto Tecnológico Metropolitano, Medellin, Colombia. Main features are as follows: low shrinkage, good mechanical performance up to 180°C, good chemical resistance, high wear resistance, and high crack resistance.

3.1. Ballistic essays

As a destructive testing protocol, Ballistic Resistance of Body Armor NIJ Standard-0101.06, Section 4.2.1.2 [14], where the configuration of ballistics test is conditioned, is considered. In **Figure 4**, the graphical representation of the assembly for performing the impact test is observed. In **Figure 5**, the projectile used is shown, in side view, according to the standard for conducting the tests.

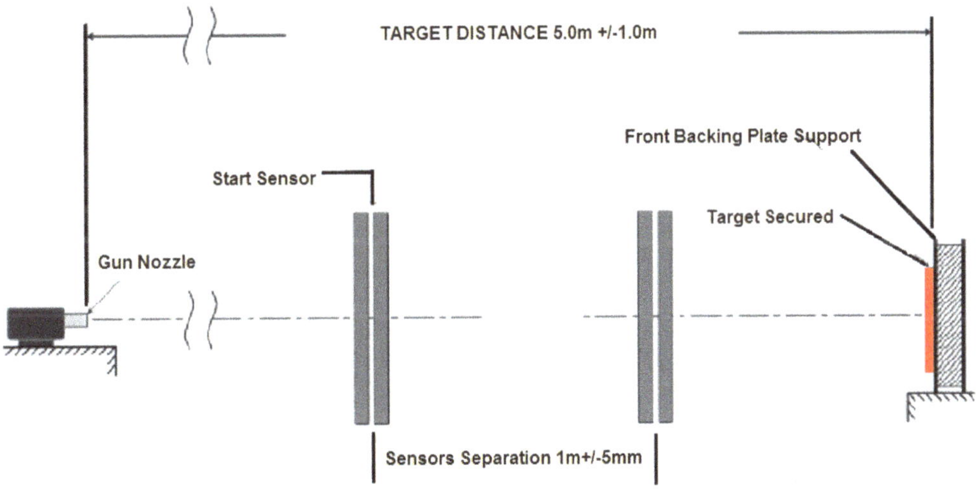

Figure 4. Assembly required for testing ballistic impact, NIJ standard.

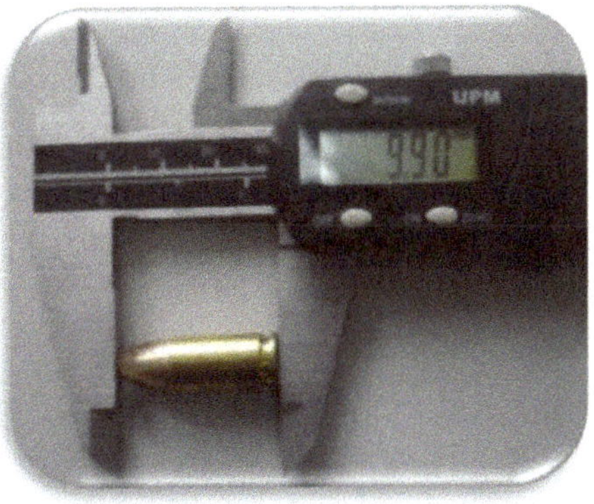

Figure 5. Longitudinal dimensioning of the projectile used.

3.2. Fractography

For the characterization of the mechanism fault at macrometric scale, a calliper gauge to measure the diameters of the input crater and the output crater cone formed in the flexible phase of the system was used. To observe the cross section of a system, hydrocutting was used and the impact zone was analyzed, and to observe the details in the deformations through optical microscopy, the results are presented in the following sequence.

To manufacture the laminated composite systems the following process was carried out: they were placed, plates and sheets organized by polyethylene, in a cubic container that contains enough volume to be immersed in the epoxy resin, which was poured manually. The curing process of the resin is carried out at a room temperature of 19–25°C with a time period of 24 h because there it was recommended to use furnace having ceramic covered surface plates with epoxy resin; and the process contraction of the resin may crack the plates.

As a response variables to the impact factors such as diameters generated (input, output, and plastic cone on flexible material) are considered. In **Figure 6**, designs for impact analysis are shown. Group A (see from right to left) corresponds to the arrangement of the system with main layer receiving the projectile in woven polyethylene with ceramic backing and group B (see from left to right) corresponds to the arrangement of the system with main layer receiving the projectile in ceramic-backed polymer fabric. To perform the ballistic test for each impact system and perform characterization using optical microscopy, the samples were selected randomly to observe the predominant interaction in each of the systems.

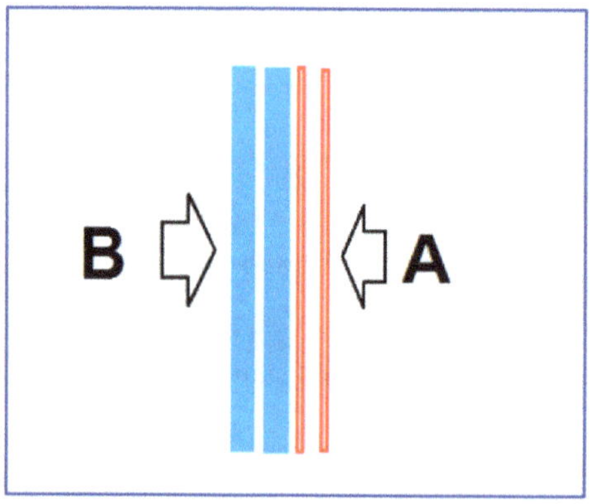

Figure 6. Impact sides. Direction side A: impact through UHWMPE; direction side B: impact through ceramics.

4. Results and discussions

4.1. Case I. Ceramic plate in front and polyethylene flexible as backing

In this case, the experiment was carried out placing the ceramic plate and receiving the metal projectiles, which were obtained with different morphology of some fractures as cited below. The bullet entered forming a crater that eroded an area of ≈3 mm around the crater and it is recalled that the initial diameter of the projectile was 9 mm. In **Figure 7**, drilling obtained as a result of the impact is illustrated in the left panel, the image formation of five cracks with perpendicular impact is illustrated in the right image; such fractures that move only by the surface plate without presenting complete rupture were observed. Possibly the presented erosion was due to the deformation process of the projectile, whereas the drilling system was due to their characteristics that greater drag is not imposed as a result if a large number of traces are drilled. The highlighted panel is uniform along the axial symmetrical quasierosion drilling.

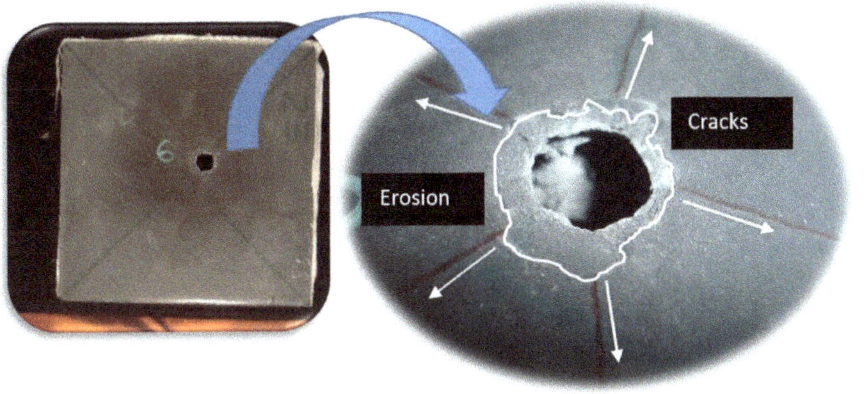

Figure 7. Footprint of the projectile on the ceramic plate.

4.1.1. Fractography of Case I

The footprint of the projectile after passing through the system, the flexible fabric material ultrahigh molecular weight polyethylene, important evidence deformations caused by the moving projectile, and fractures shown are reflected in these images.

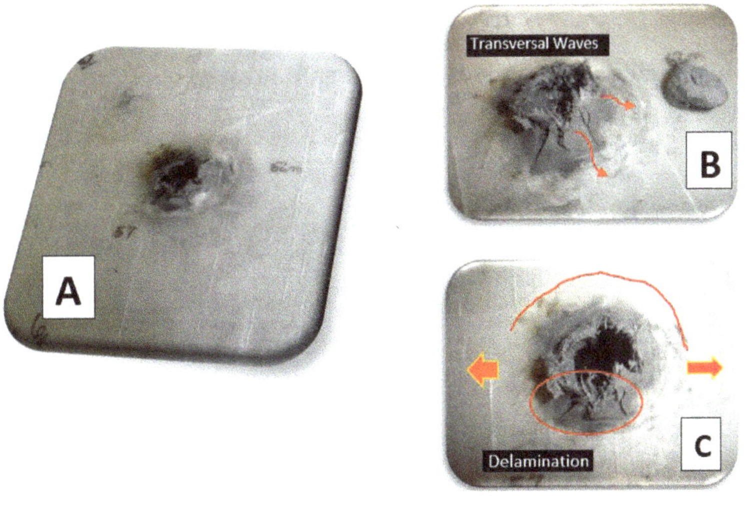

Figure 8. Images to backing fractography.

The isometric view of the composite system is observed in **Figure 8**. The overall damage done by projectile in the material can be seen in image A; an approach breakage of the fibers is shown in image B, along with the tag marked with red arrows in the footprint direction taken by transverse waves, which end with the formation of a neck in the laminate plane (image D), total rupture of the primary fibers that come into direct contact with the projectile, fracturing the flexibility scheme delamination near the area impact. The deformation of the polyethylene layers results in a 12 mm high cone, with almost 40 mm base diameter and 18 mm in the outlet of the projectile; the fiber tear around the hole and image B together with the cone can be seen; deformation caused to the projectile as a result of the penetration process.

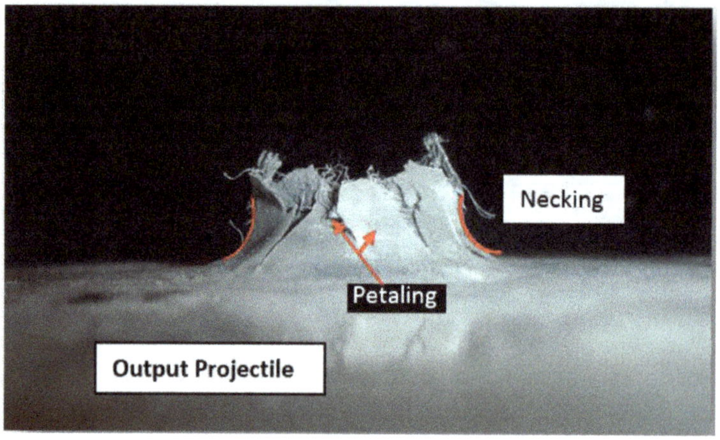

Figure 9. Output of the projectile.

Figure 9 shows the formation of a neck (necking) in the laminate plane, the total disruption of the primary fibers that come into direct contact with the projectile; the projectile completely passes through detailed, they deform the polyethylene layers forming a 12 mm high cone with resulting delamination process with cross section of the projectile on the system; the deformation petals on the flexible layer is called petaling.

4.2. Case II. Flexible fabric in front and ceramic as backing

4.2.1. Fractography of Case II

Figure 10 shows the location of the plates and the direction of impact test that was obtained after subsequent analysis of fractography.

In **Figure 11**, as shown from top to down, the plate is in isometric view, where the penetration hole of the projectile, in image A, shows the detailed view with increased detail in which the crater is formed with broken fibers, the shearing caused tissue and adjacent tissues delamination of the crater. Image B is observed, twisting and fusing the fibers as the result of mechanisms present after impact stress, image C is detailed in the image cutting and twisting the fibers of polyethylene during ductile voltage departure.

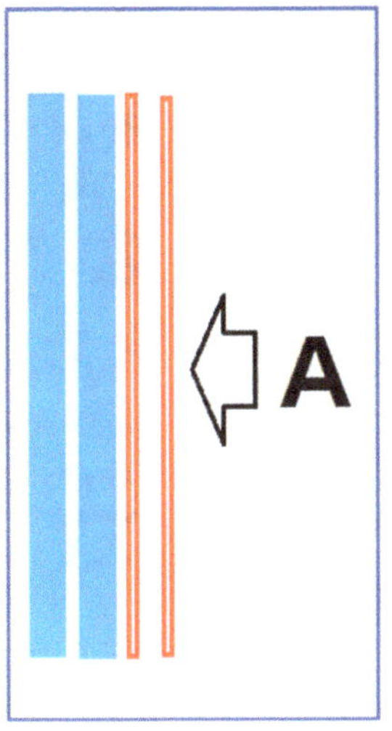

Figure 10. Projectile input direction side A, impact thought UHWMPE.

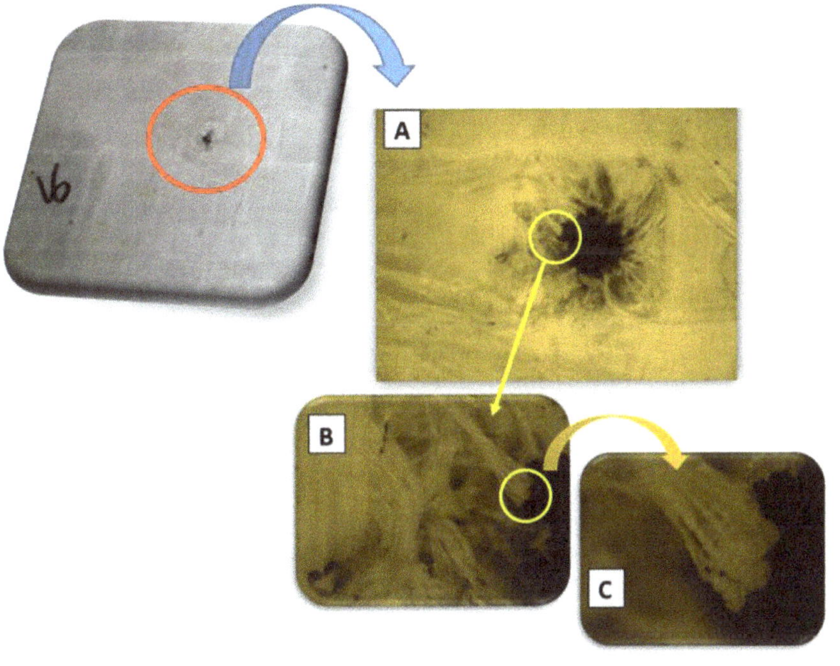

Figure 11. Fractography in the projectile output zone.

In **Figure 12**, the fractography of a system receiving the projectile double ceramic plate that is 10 mm thick and double back sheet flexible polyethylene is shown; image A is highlighted by the mark left by the projectile detailed, which enters flexible face with a hole diameter of the projectile (9 mm), greater penalties, and the rear face forming a crater that is eroded in an inhomogeneous way around the inlet port, cracks occur, which travel only by presenting external ceramic plate breakage. Rupture plates with crater formation in stages with the passage of the projectile are observed. The geometry of the crater [6, 15, 16] has formed in the radial symmetry breaking evenly.

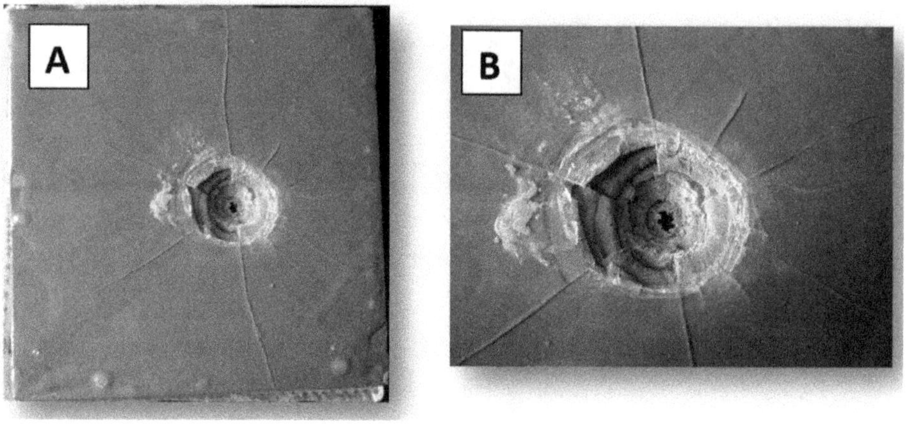

Figure 12. Fractures and cone on the plate "Backing".

In **Figure 13**, a frontal approach presents originated crater, streaking in the resin layer with a radial direction and erosion on the area near the impact surface is highlighted. The strain energy breaks the crater forming cracks from the edge to the back of the panel, which travel through the ceramic material, the breakage of fibers near the orifice, and close to delamination contact area.

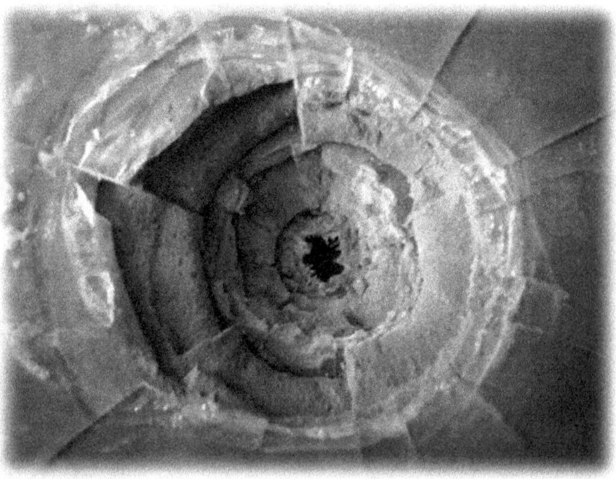

Figure 13. Crater originated in the double ceramic plate.

In **Figure 14**, the morphology of the system in cross section cab be observed, which was created using water jet cutting. The image in the input direction of the projectile, which originated a large area delamination and spalling inside the plate, and in the radial direction of the projectile is observed. It is observed that vacuum is created conically. Interlayer resin deformation only in the input diameter of the projectile in the plate target can be seen; it can be ensured that achieving attenuate kinetic energy, which is distributed in the ceramic plates, meaning through spalling, but without stopping the projectile, shows that the resin layer remains intact and shows good adhesion with the ceramic. It can be stated that the ceramic dissipates energy with the formation of the crater, along with the flexible system, which is seriously affected.

Figure 14. Fractography cross-sectional view.

In **Figure 15**, some failure mechanisms as spalling between the ceramic layers and the resin layer can be seen in cross-section, which can be realized as the intraseparation and interlaminar ceramic where the fracture material arrangement similar to leaves is superimposed without total detachment. Also, the crack initiation deforming plate from contact with the projectile is observed traveling environment symmetrically radial through the complete system. Some fractures travel lengthwise through porosity ceramic into contact with the gaps or vacancies in the resin layer, which ends up slowing the advance of the fracture plate in the bottom plate continuous plastic deformation with continuity conical-shaped crater to the total output of the projectile, with the red arrow penetrating the sense indicated.

A detail of the cross section of image A spalling is illustrated in **Figure 16**, which is mentioned in the previous paragraph describing environment puncture by the displacement of the plates; this dislocation of the solid layers is possibly the result of the trip transverse waves of energy during the compression exerted by the projectile. The eroded layer is the result of pressure and high temperature as a mark left by the passage of the projectile during contact with the ceramic plates. In image B, the detailed projectile outlet and a portion of tissue shearing during impact stress as observed drag is shown.

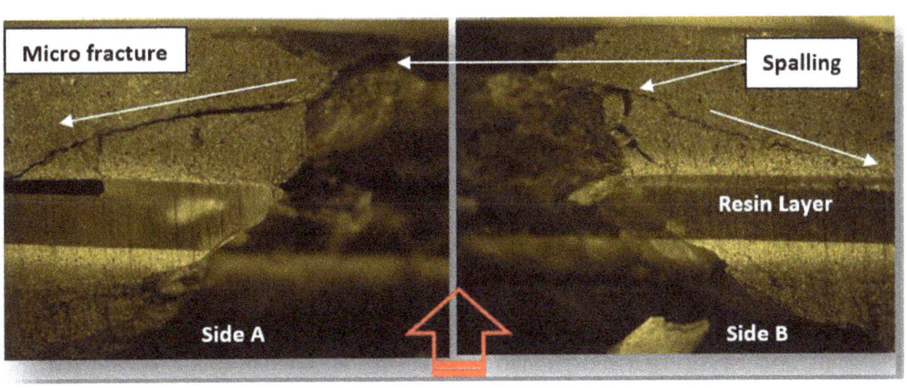

Figure 15. Fractography double cross view on the ceramic plate.

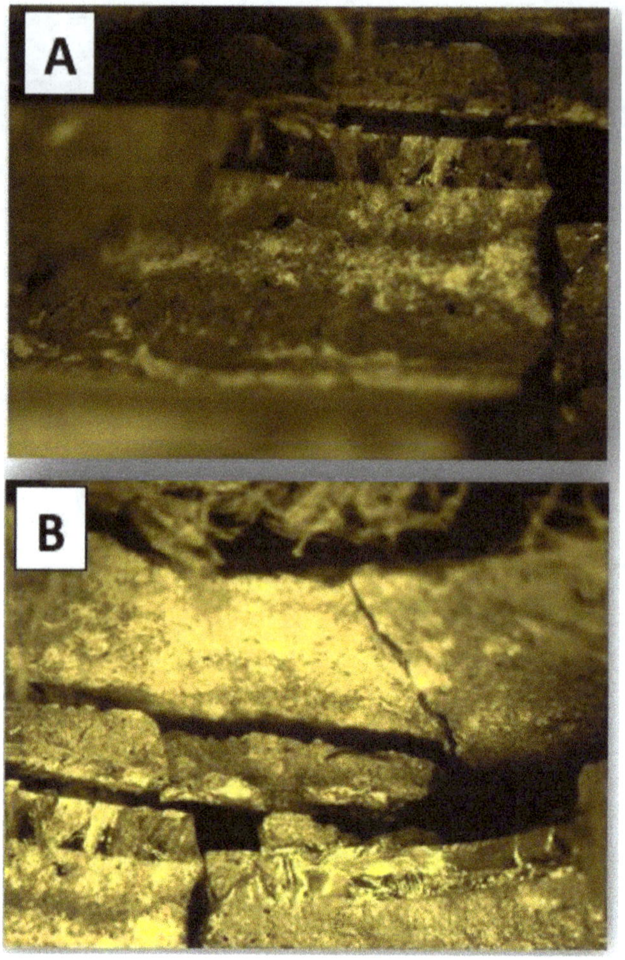

Figure 16. Detail of the passage of the projectile by the ceramic body.

4.3. Fracture mechanisms in the epoxy resin layer

In **Figure 17**, a trace of the passage of the projectile, which crosses the plate, causing a crater with a diameter greater than the projectile penalties, erosion causes environment entry system, leaves a marked trace of the shear wave, is observed as deformation caused by high contact temperature and the velocity of the projectile, which dissipates some of the kinetic energy in the radial cracks from the impact point.

In **Figure 18**, the footprint on the plate epoxy resin is shown in detail in image A, the projectile enters the plate melting the resin perimeter of hole, and waves are formed due to the rapid solidification of the material before the passage of the projectile with high temperature. In image B, the morphology of details of one of the waves and the rough texture caused by the process of rapid cooling of the material is observed. In image B, the crater is observed and chipping caused in the layer of flexible polyethylene is shown, which solidified with resin around the crater erosion caused in response to the opposition of the material passage of the projectile.

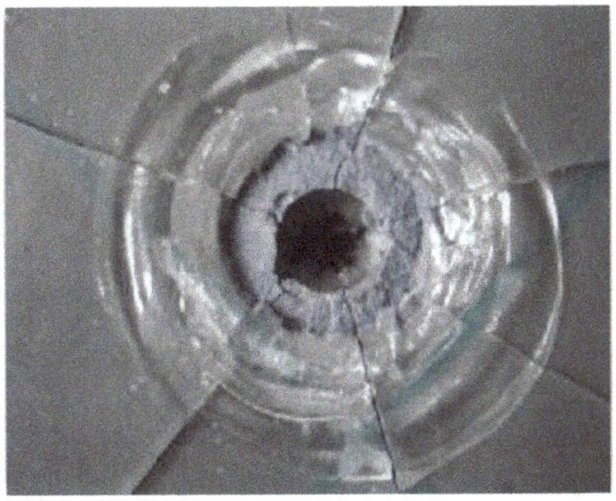

Figure 17. Footprint of the projectile on the layer epoxy resin.

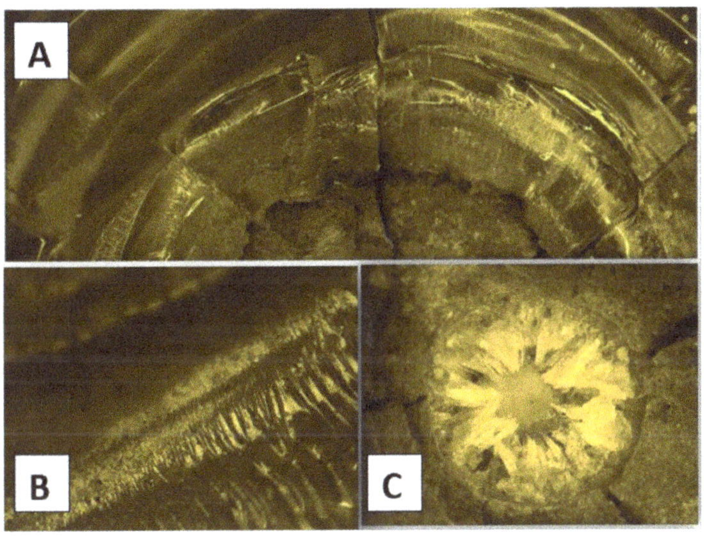

Figure 18. Footprint waves on the layer of the epoxy resin.

4.4. Fracture mechanism in ceramic low-porosity plates in front and UHWMPE as backing

In **Figure 19**, the result of the impact on a plate of silicon carbide with 5% porosity is closely illustrated; it was manufactured by means of high pressure and temperature. In image A, a view of the isometric plate is generally observed, in which in the center the product of impact deformation can be seen, with the distribution transverse cracks the crater. In image B, the front view of the crater on the ceramic plate is observed in the joint caused by shear waves that completely fractured ceramic cracks. In image C, the center of the impact is displayed, which completely distorts the projectile, completely dissipating the kinetic energy, and the fracture was complete. The silicon carbide plate was very efficient, it was observed that totally absorbed kinetic energy of the projectile, the plate with geometric characteristics and low porosity compared to original plates, was used in the exploration, whereas the rear layer of polymeric fabric did not present deformation and remained intact against impact, worked to contain remaining fragments of ceramic originating from direct contact with the projectile.

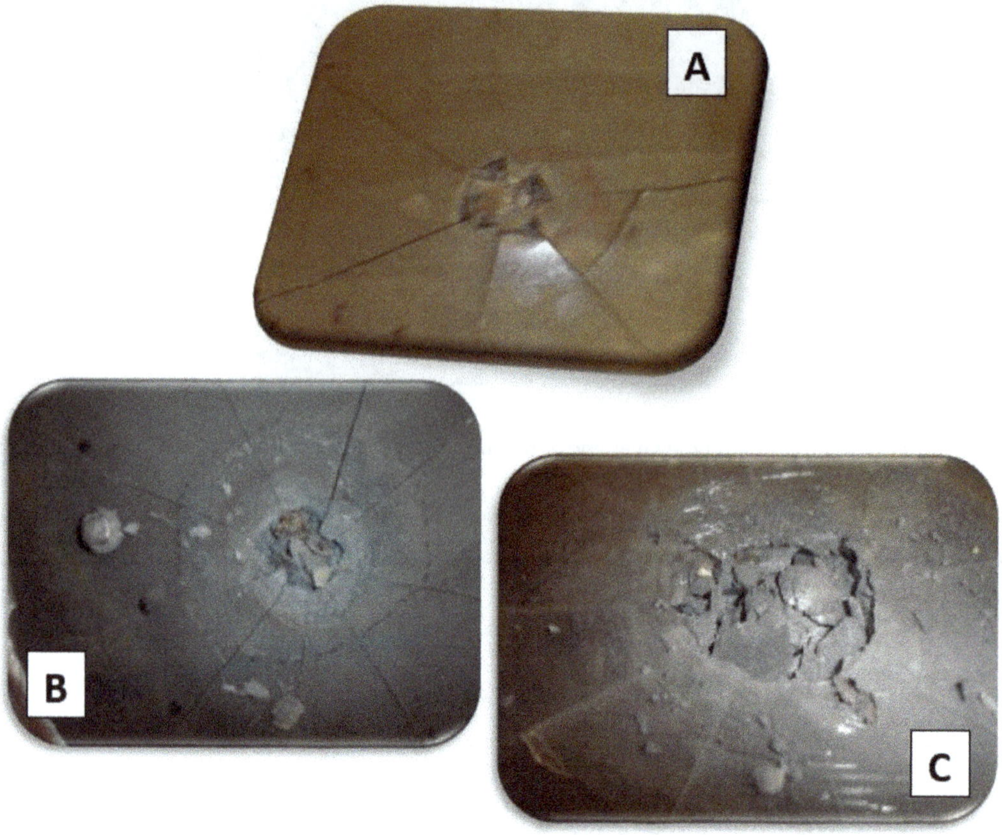

Figure 19. Fractography in the ceramic plate with reduced porosity.

The failure mechanisms generally observed in this exploration have been reported, a brief description of different types of failure is presented below. The formation of a crater followed by crack propagation is reported in the rigid ceramic systems; Horsfalla et al. [17] and Medvedovski et al. [18] found that it was also possible to identify the individual component in a compound system with different functions during the deformation of the panel in these publications and that the form of energy dissipation also depends on the materials and the manufacturing processes. The compaction process and pressing of ceramic powders determine the porosity in the final product, with these processes it seeks to maximize the mechanical protection systems with high-capacity energy dissipation properties, characteristic involves reducing the porosity at least; the pores are small defects that may act as stress concentrators and initiators for material failure [19].

It has been found that the composite systems surpass the dissipation of energy capacity; at the conventional systems used without any combination, Sherman and Brandon [20] detected a sequence in the mechanisms of energy dissipation in a ceramic plate, such as the formation of radial cracks during traction associated with this downward tensile strength of ceramics. Cunniff et al. [21] presented the performance of ballistic polymer fiber; they also identified the mechanisms of fracture in flexible fibers and its potential energy absorption. In the flexible system, another predominant mechanism in tissue polymers during the dissipation of kinetic energy is the formation of a conical pyramid backing material; it was observed that the strain and deformation are function of the distance of the impact; therefore, the wires are closer, experience stress failure, whereas the most distant point of impact wires have no tension [22].

4.5. Kinetic energy dissipation

The basis for the model applied to the evaluation of these analyses was developed by Morye et al., being the most similar characteristics presented for the study [23]. This model analyzes the tensile failure of the primary fibers, elastic deformation of the secondary fibers, delamination, breakage of the system matrix, and the formation and movement, which are five deformation mechanisms that contribute to the dissipation of energy from the projectile and are considered a cone on the rear side of the plate of the composite. The five energy absorption mechanisms that were applied are as follows:

Equation	Application	
$ET = E_{FP} + E_{FS} + E_{KC} + E_{DL} + E_{RM}$	Explained in the previous paragraph	Eq. (1)
$ET = E + W$	E is the difference between the initial and residual kinetic energies, this corresponds to the energy lost by heat and deformation during impact is calculated as follows (Eq. (3))	Eq. (2)
$E = \frac{1}{2}mp(v_s)^2$ $E = \frac{1}{2}mp(v_s)^2 - \left[\frac{mp}{mp + ms}\right]$	V_s is the initial velocity, M_p is the mass of the projectile, and M_s is the plug mass	Eq. (3)
$\frac{1}{2}mp(v_s)^2 + E + W + (mp + ms)Vr^2$	The additional kinetic energy (*W*) can be lost through deformation during drilling due to the presence of the circumferential cutting zone	Eq. (4)
$E_{kci} = \frac{1}{2}m_{ci}Vi^2$	m_{ci} is the mass of the cone formed on the rear face of the composite plate and is determined by Eq. (5)	Eq. (5)
$m_{ci}Vi^2 = \pi(R_{ci}^2)(e)(\rho)$	ρ is the density of composite laminate, *e* is the flexible laminate (UHWMPE) thickness	Eq. (6)
$E_{kci} = \frac{1}{2}\pi(R_{ci}^2)(e)(\rho)(v_i^2)$		Eq. (7)
ET = E + W + Ekci	Empiric equation used for this exploration	Eq. (8)

Table 1. Equations used in the analysis of kinetic energy dissipation.

The total energy absorbed by the laminate composite in Eq. (1): where EFP is the energy absorbed by the failure of the primary fibers; SAI, energy absorbed by the elastic deformation of the secondary fibers; EKC, energy absorbed in the formation and movement of the cone at the rear face of the panel; EDL, energy absorbed due to delamination of the material; ERM, energy absorbed due to breaking of the matrix.

Another model studied was presented by Retch and Ipson [23, 24]; they presented a semiempirical model where a balance of kinetic energy is analyzed and the energy is dissipated by the generation of the crater during the impact (Eqs. (6) and (7)). For the study of energy, balance takes into account the following characteristics of the composite system: mass, density, and dimensions of the system; mass, diameter, and velocity of the projectile; and residual impact velocity and mass of the plug (product ejected by impact); the energy balance formula is presented in **Table 1**.

FMJ (full metal jacket) round tip 9 mm diameter × 19 mm in length with a nominal mass of 8.0 g and a speed of 436 m/s					**Projectile**
Case study	**Input projectile side**	**Mechanism observed**	**Kinetic energy projectile (J)**	**Kinetic energy dissipated (J)**	**Kinetic energy to retain (J)**
Case A	Double flexible fabric in front and ceramic double plate as backing	Crater formation on ceramic backing, ductile fracture in the fibers, defibrillation	577.6	342	237
Case B	Ceramic double plate in front and double Flexible fabric as backing	Low crater formation on ceramic backing, fracture in he fibers nearly to crater and conic formation on the UHWMPE backing		437.382	140
Case C	Ceramic plate with less porosity Single in front and double Flexible fabric as backing	Crater formation on the ceramic plate with multiple fracture, neither deformation in the backing face.		577.6	0

Table 2. Results of kinetic energy dissipated.

The analysis complemented with the study of flexible and rigid systems, abstracted to empirical analysis, modeling the energy absorption by the mechanism of formation of the cone on the rear side of the laminate, whose vertices move at the same speed of the projectile (V_i) and the depth thereof is equal to the total displacement of the projectile (D_i), during contact, therefore the kinetic energy of the cone E_{kci} can be determined using Eq. (6). In this model, a general energy balance where we can analyze the absorption due to the formation of the crater and the mass displaced by the projectile, valuable information for analysis systems in predominantly rigid exploration; this model can be complemented performing the model presented by Morye et al. with Eqs. (6) and (7), assuming that the flexible polyethylene backing (UHWMPE) remains bonded to the ceramic and can be analyzed as a circular membrane with an initial mass that is produced in the system by reducing the velocity of the projectile. From Eqs. (5) to (7), the kinetic energy is obtained due to the formation and movement of the cone on the rear side of the composite plate. The results are shown in **Table 2**.

5. Conclusions

The fractography found in the ceramic component is defined in the formation of a crater, erosion, and fracture, while the polymer components were observed as mechanisms for energy dissipation, delamination, cone formation, petals, breakage, and melting fibers.

A large number of failure mechanisms were observed in the design sandwich double ceramic plate in the front and double sheet of UHWMPE in the back; they were observed as dissipation mechanism energy, fragmentation and erosion on ceramic plates, forming petals, delamination and cuts in polymer sheets; these evidences in the photographs were indispensable data for the analysis of kinetic energy dissipation using the semiempirical models of Reich and Ipson and Morye et al.

In this exploration, it was possible to observe the influence of the manufacturing process in a system of high requirements in terms of dissipation of kinetic energy, which is why the analyzed total dissipation of energy system was obtained by high pressure and high temperature of case C as compared to cases B and C, which were manufactured by traditional techniques.

Acknowledgements

The authors specially thank to Knight Group Owners Caballero SAS International, headed by the CEO engineer Miguel Caballero, for their decisive collaboration, disposition, and financing of ballistics tests, laboratories and polymeric material. They also thank the company Quimiresinas for its contribution in the epoxy resin and to the Universidad Nacional de Colombia, Medellin, at the laboratory metallography and materials testing by providing resources for the acquisition of most of ceramic plates used in this exploration. Acknowledgments in memory to Military Engineer Edgar Caicedo for was showing me the way of success in the Engineering Design.

Author details

Giovanni Barrera Torres[1*], Juan Manuel Vélez[2] and Carlos Mario Gutierrez[1]

*Address all correspondence to: gbarrerat01@gmail.com; giovannibarrera@itm.edu.co

1 Metropolitan Institute of Technology (ITM), Medellin, Colombia

2 National University of Colombia, Medellin, Colombia

References

[1] H. Harel, G. Marom. Delamination Controlled Ballistic Resistance of Polyethylene/ Polyethylene Composite Materials. Jerusalem, Israel/52526 Ramat Gan, Israel: Casali Institute of Applied Chemistry, The Hebrew University of Jerusalem/S. Kenig Department of Plastics Engineering, Shenkar College. 2000.

[2] A. Bhatnagar. Lightweight Ballistic Composites, Military and Law-Enforcement Applications. Cambridge, England: Woodhead (Woodhead Publishing and Maney Publishing on behalf of The Institute of Materials, Minerals & Mining). pp. 4–7. 2006.

[3] George Murray, Charles V. White, Wolfgang Weise. Introduction to Engineering Materials. Chapter 1, Classification of materials. 2nd ed. Taylor and Francis Group. pp 13–15. 2008.

[4] Donald R. Askeland, Pradeep P. Phulé, "Wendelin J. Wright. Composites:TeamWork and Synergy in Materials". *The Science and Engineering of Materials* , Stamford, Cengage Learning, pp. 653–687. 2006.

[5] S.K. McGuire, M.G. Jenkins. Ceramics Testing. Stanford, California/Seattle, Washington: Stanford University/University of Washington, Chapter 23. 2007.

[6] F.G. Díaz–Rubio. Mechanical Characterization of Advanced Ceramic Materials at High Strain Rates; Doctoral Thesis, Aeronautical Engineering. pp. 13,15. 1999.

[7] James K. Wessel. Advanced Ceramics materials. Handbook of Advanced Materials: Enabling New Designs. John Wiley & Sons, Inc. pp 65–88. 2004.

[8] Eugene Medvedovski. Ballistic performance of armour ceramics: influence of design and structure. Part 2. Ceramics International 36 (7).pp 2117–2127. 2010.

[9] P. Colombo, F. Zordan, E. Medvedovski. Ceramic–Polymer Composites for Ballistic protection. Padova, Italy: Dipartimento Di Ingegneria Meccanica, Settore Materiali Università Di Padova Via Marzolo. 2005.

[10] F. Vargas, E. López, D. López, J. Areiza, M. Monsalve. Manufacture of Silicon Carbide Tubes for use as Heating in Ovens and Heating Systems Body. Group Pyrometallurgical and Materials research, Universidad de Antioquia. 2008.

[11] R. Obeso Cobo and F.G. Díaz–Rubio. Dynamic properties study of ceramics materials. Application Al– SiC. Canales y Puertos. Madrid: Department of Materials Science. pp 12, 13, 45–,46. 2004.

[12] S.K. Garcia Castillo. Analysis of Laminated Composite Materials in its Plane and Preload Subjected to Impact University Carlos III of Madrid. Department of Continuum Mechanics and Structural Analysis Doctoral Thesis. pp. 12,14. 2007.

[13] F.L. Matthews and R.D. Rawlings. Composite Materials, Engineering and Science. Woodhead Publishing. pp 173, 372, 373. 1994.

[14] Ballistic Resistance of Body Armor, NIJ Standard–0101.06. National Institute of Standards and Technology Prepared For: National Institute of Justice. Washington, DC. July 2008. http://nist.gov/oles/upload/ballistic.pdf.

[15] D. Sherman, D.G. Brandon. The ballistic failure mechanisms and sequence in semi-infinite supported alumina tiles. Journal of Materials Research. 12:1335–1343. 1997.

[16] Zukas, Jonas A., Nicholas, Theodore, Swift, Hallock F., Greszczuk, Longin B., Curran, Donald R. Impact Dynamics. Wiley. 1982.

[17] I. Horsfall, S.J. Austin, W. Bishop. Structural ballistic armour for transport aircraft. Materials & Design, 21(1):19–25. 1999.

[18] H.L. Gower, D.S. Cronin, A. Plumtree. Ballistic Impact Response of Laminated Composite Panels. International Journal of Impact Engineering pp.1000–1008. 2007.

[19] V.A. Greenhut. Characterization of ceramics and glasses: an overview. Engineered Materials Handbook, Ceramic and Glasses. Vol. 4, ASM International. 1991.

[20] Dov Sherman and D.G. Brandon. Supported Alumina Tiles. Journal of Materials Research. V.12. pp 1335–1343. 1997.

[21] P.M. Cunniff, M.A. Auerbach. High performance "m5" ballistic fiber.http://web.mit.edu/course/3/3.91/OldFiles/www/slides/cunniff.pdf. April 04.2016.

[22] A. Bhatnagar. Lightweight Ballistic Composites, Military and Law-Enforcement Applications. Cambridge, England: Woodhead (Woodhead Publishing and Maney Publishing on behalf of the Institute of Materials, Minerals & Mining). Chapter 28, pp 222–223. 2006.

[23] S.K. Garcia Castillo. Análisis de laminados de materiales compuestos con precarga en su plano y Sometidos a Impacto Universidad: Carlos III de Madrid. Tesis Doctoral Año. pp 124–139. 2007.

[24] Ashok Bhatnagar. Lightweight Ballistic Composites: Military and Law-Enforcement Applications. CRC Press (Woodhead Publishing and Maney Publishing on behalf of the Institute of Materials, Minerals & Mining). Chapter 28. pp. 229,230. 2004.

Permissions

All chapters in this book were first published in FMPPB, by InTech Open; hereby published with permission under the Creative Commons Attribution License or equivalent. Every chapter published in this book has been scrutinized by our experts. Their significance has been extensively debated. The topics covered herein carry significant findings which will fuel the growth of the discipline. They may even be implemented as practical applications or may be referred to as a beginning point for another development.

The contributors of this book come from diverse backgrounds, making this book a truly international effort. This book will bring forth new frontiers with its revolutionizing research information and detailed analysis of the nascent developments around the world.

We would like to thank all the contributing authors for lending their expertise to make the book truly unique. They have played a crucial role in the development of this book. Without their invaluable contributions this book wouldn't have been possible. They have made vital efforts to compile up to date information on the varied aspects of this subject to make this book a valuable addition to the collection of many professionals and students.

This book was conceptualized with the vision of imparting up-to-date information and advanced data in this field. To ensure the same, a matchless editorial board was set up. Every individual on the board went through rigorous rounds of assessment to prove their worth. After which they invested a large part of their time researching and compiling the most relevant data for our readers.

The editorial board has been involved in producing this book since its inception. They have spent rigorous hours researching and exploring the diverse topics which have resulted in the successful publishing of this book. They have passed on their knowledge of decades through this book. To expedite this challenging task, the publisher supported the team at every step. A small team of assistant editors was also appointed to further simplify the editing procedure and attain best results for the readers.

Apart from the editorial board, the designing team has also invested a significant amount of their time in understanding the subject and creating the most relevant covers. They scrutinized every image to scout for the most suitable representation of the subject and create an appropriate cover for the book.

The publishing team has been an ardent support to the editorial, designing and production team. Their endless efforts to recruit the best for this project, has resulted in the accomplishment of this book. They are a veteran in the field of academics and their pool of knowledge is as vast as their experience in printing. Their expertise and guidance has proved useful at every step. Their uncompromising quality standards have made this book an exceptional effort. Their encouragement from time to time has been an inspiration for everyone.

The publisher and the editorial board hope that this book will prove to be a valuable piece of knowledge for researchers, students, practitioners and scholars across the globe.

List of Contributors

Guangqing Zhang and Tiegang Fan
Department of Engineering Mechanics, China University of Petroleum in Beijing, Beijing, China

Mügem Aslı Gürel, Bağdagül Helvacıoğlu Kıvanç
Department of Endodontics, Faculty of Dentistry, Gazi University, Ankara, Turkey

Hacer Deniz Arısu
Department of Operative Dentistry, Faculty of Dentistry, Gazi University, Ankara, Turkey

Kunkun Fu and Li Chang
School of Aerospace, Mechanical and Mechatronic Engineering, University of Sydney, NSW, Australia

Youhong Tang
School of Computer Science, Engineering and Mathematics, Flinders University, SA, Australia

Carlos Berejnoi
Faculty of Engineering, National University of Salta/CIUNSa, Avda. Bolivia, Salta, Argentina

Juan E. Perez Ipiña
Fracture Mechanics Group, National University of Comahue/CONICET, calle Buenos Aires, Neuquén, Argentina

Meng Zhao, Huadong Yong and Youhe Zhou
Key Laboratory of Mechanics on Environment and Disaster in Western China, The Ministry of Education of China, Department of Mechanics and Engineering Sciences, College of Civil Engineering and Mechanics, Lanzhou University, Lanzhou, Gansu, PR China

Achouak Elghazel, Rym Taktak and Jamel Bouaziz
National Engineering School of Sfax, Sfax, Tunisia

José António Fonseca De Oliveira Correia, Abílio M.P. De Jesus, Pedro M.G.P. Moreira and Rui A.B. Calçada
Faculty of Engineering, University of Porto, Porto, Portugal

Alfonso Fernández-Canteli
Department of Construction and Manufacturing Engineering, University of Oviedo, Gijón, Spain

Tadanobu Inoue
National Institute for Materials Science, Tsukuba, Japan

Zheyuan Ma, Agnes M. Samuel and Fawzy H. Samuel
University of Québec at Chicoutimi, Québec, Canada

Herbert W. Doty
Materials Technology, Pontiac, MI, USA

Giovanni Barrera Torres and Carlos Mario Gutierrez
Metropolitan Institute of Technology (ITM), Medellin, Colombia

Juan Manuel Vélez
National University of Colombia, Medellin, Colombia

Index

A

Adhesion, 29-34, 36, 40, 44-45, 47-51 74-75, 218, 224

Adhesive Dentistry, 29, 35, 43, 46-47 49-51

Artificial Rocks, 24

B

Bioceramic, 115, 129-130

Bond Strength, 29, 31, 34-40, 43, 47-51

Brittle Thin Film, 53, 60, 65, 74

C

Caliber Rolling, 166, 168-170, 179 181-182

Ceramic Plate, 220-221, 223-225 227-228, 230-231

Cleavage Initiator, 79, 92

Crack Front, 79-80, 92, 126-127, 131, 187

Crack Pattern, 54-55, 65-67, 69-72

Crack Propagation, 41, 54, 59-60, 82 114, 125, 127, 130, 132, 135, 142 147, 152, 156, 166, 169 174, 179, 184, 187, 192, 194 196, 198, 200, 206, 212, 228

Crystal Grain, 167, 178

Cyclic Plasticity, 143-144

D

Dentin, 29-34, 36-39, 41-45, 47-51

E

Elastic Modulus, 30, 33-37, 43, 116-119 124, 128, 132-133

Elastoplastic Analysis, 139-140, 142 149, 152, 159

Elementary Material Block, 152, 155 157-159

Elongated Grains, 171, 179-180

Enamel, 29-32, 34, 36-38, 43-45, 47-49

Energy Dissipation, 215, 228-229, 231

Energy Transition Curve, 174, 180

Epoxy Resin, 32-33, 217-218, 220 226-227, 231

F

Failure Mode, 77-78, 130

Fatigue Behaviour, 50, 136, 149, 161

Ferrite-pearlite Structure, 166, 169

Ferritic Steels, 77-79, 84-86, 92-94, 181

Finite-element Analysis, 42-43, 118 138, 142, 155-156, 159

Flattened Brazilian Specimen, 116, 120 124-125

Flexural Strength, 35-36, 43, 45

Fluorescent Dye, 4-6

Fractography, 184-185, 192, 208 213-215, 220-223, 225, 228, 231

Fracture Behavior, 53, 78, 96, 98-99 112, 166, 169, 172, 174, 177, 179 185-190, 198, 202, 204, 208-212

Fracture Extension, 16, 21, 23

Fracture Initiation, 1, 4-5, 9, 12-13, 24 27, 208

Fracture Mechanics, 29, 38, 41, 50-51 53, 61-62, 64, 69, 85, 93-95, 103 113-114, 135-136, 139, 157, 160-161 164, 170

Fracture Network, 1, 10, 19, 24

Fracture Resistance, 36, 46-47, 77-78 92, 190

Fracture Strength, 29, 34, 39

Fracture Toughness, 30, 32, 35, 41-42 50-51, 53-57, 59, 64, 66, 69, 72-80 82, 84-85, 90, 92, 94-95, 115-117, 125, 129, 167, 169 172, 178, 181-183

Fracturing Fluid, 4-6, 13, 22

G

Grain Refinement, 166-167, 172, 174 176, 180

Grain Structure, 171, 178, 182

H

Hydraulic Fracture, 1-2, 6, 8-10, 12-13 18-19, 21-22, 24-28, 164

I

In Vitro, 36, 41, 43, 46-47

Injection Rate, 1-2, 4-6, 17, 22-23

Interface Crack, 96-100, 103-105 107-108, 113-114

K

Kinetic Energy, 214-216, 218, 224 226-231

L

Local Fatigue Approaches, 135, 139

Low-alloy Steel, 166, 168-169, 179-180 182

M

Master Curve, 77, 79, 84, 91, 94-95

Metallography, 184, 231

N

Nano-indentation, 53-54, 65, 72

Natural Fracture, 1-2, 6, 17, 19, 21-22 24, 27

Naturally Fractured Formation, 1
Numerical Modeling, 164

O
Oblique Fracture, 17
Optical Microscopy, 185, 188, 190, 192 204, 209, 220

P
Plastic Deformation, 103, 143, 167, 176 187-188, 190-191, 199, 225
Polyethylene, 214, 216, 218, 220-223 226, 230, 232
Probabilistic Model, 135, 139

R
Residual Stress Distribution, 141, 155 157, 159
Restorative Materials, 32, 36-38, 43 45-46
Rupture Strength, 118, 129-130

S
Semi-circular Bend, 115-117, 121, 123 127, 131
Shape Parameter, 80, 82, 85, 87, 145
Size Effect, 79, 92-93
Stacked Conductors, 96, 101, 103-105 107
Stimulated Reservoir Volume, 16, 24
Strain Energy Release Rate, 96, 98-100 104-108, 122
Strength-toughness Balance, 166, 169 174, 177-180
Stress Distribution, 37, 39, 41-42, 51, 54 56, 59-62, 64, 67-69, 72, 111, 114, 125 141, 154-157, 159
Stress Intensity, 41, 98, 100, 113-116 119, 122-127, 129-132, 136, 138 141-142, 154-159, 162
Stress-based Model, 53-54, 62, 65, 67, 72
Substrate Effect, 53-54, 56, 65, 70-72

T
Tensile Strength, 12-13, 30, 32, 70 115-118, 128-129, 132-133, 149, 168 172, 228
Tooth Structures, 43

U
Ultrafine Elongated Grain, 166, 168